Phylogeny
and
Classification
of the Orchid Family

Phylogeny
and
Classification
of the Orchid Family

Robert L. Dressler

CAMBRIDGE
UNIVERSITY PRESS

Published by the Press Syndicate of the University of Cambridge
The Pitt Building, Trumpington Street, Cambridge CB2 1RP
10 Stamford Road, Oakleigh, Melbourne 3166, Australia

© Dioscorides Press 1993

First published outside North America by Cambridge University Press 1993

First published in North America by Dioscorides Press
(an imprint of Timber Press, Inc.) 1993

A catalogue record for this book is available from the British Library

ISBN 0 521 45058 6

Contents

Color plates 1–16 follow page 176.

1

Introduction

When I agreed to publish *The Orchids: Natural History and Classification* (1981), I found myself with a dilemma and only limited time for a resolution. The material on natural history was straightforward and relatively easy to handle, but I had to say something about orchid classification. The traditional classification was artificial, but how best to improve the system was not clear. I associated the Neottieae with the Orchideae, convinced myself that the vandoid orchids were a natural group, and treated number of pollinia as an important feature to distinguish the vandoid groups. Now it is clear that none of these was a wise decision.

At the time, I noted that many professional botanists tend to avoid the Orchidaceae. One professional botanist who scoffed at the idea was later heard to say "I never look at an orchid, I just give them to Garay." Those who study the Asteraceae (or Compositae) still greatly outnumber us, but the last decade has seen several good doctoral theses on orchids, and one hopes that the trend will continue. Some new data appeared while the 1981 book was in press, and very useful information has been published since then. One can now attempt a more complete revision of orchid classification, using information that was unavailable in 1979. There are still many questions in need of answers, and we cannot produce a long-lasting, definitive classification of the Orchidaceae with the available information. Still, it is useful to revise our old ideas (especially when they are clearly wrong) and offer new ones. The revised classification can be tested against new information as it appears, and can serve as a guide in the search for new information. Any classification is only as good as (or, at best, no better than) the information available to its author. We need more information on many aspects of the orchids, and I hope that the ideas and hypotheses offered here may stimulate such work.

F. N. Rasmussen (1982) suggests that we need some sort of practical compromise, a family classification (preferably not too unnatural) to use as a base and a filing system. We can then make hypotheses about relationships without changing our baseline every few months. I quite agree with Dr. Rasmussen, but I would like to start with a fairly good classification. The traditional Schlechterian classification, even with the names modernized (see Table 3-6), is still an arrangement of polyphyletic and paraphyletic grades, or a "horizontal" classification. My 1981 classification was intended as a phylogenetic system, but it fell far short of the goal. The classification presented here differs from that used in 1981 especially in breaking up the polyphyletic Vandoideae and removing the Neottieae from the Orchidoideae. By stressing uniquely derived features, one may make out at least the beginnings of a phylogenetic classification of the advanced Epidendroideae.

There are now several groups that appear to be clearly delimited and quite natural, including the Apostasioideae and the Cypripedioideae. The Cranichideae have become well delimited, though it is by no means clear that *Diceratostele* and the Tropidiae are their close relatives. The Orchideae plus the Diseae are quite clear, but we may be unable to separate the Diseae from the Orchideae. Among the Epidendroideae, the cymbidioid phylad and the Epidendreae with the dendrobioid subclade each appear to be quite natural, though there remain problems within each group. In the relatively primitive more or less epidendroid groups treated in Chapter 5, there are major problems, and future workers may well resurrect the Neottioideae or name new subfamilies for some of these groups. Embarrassingly enough, the classification within the familiar tribe Orchideae is especially poor. Kurzweil and Weber (1991, 1992) show that none of the traditional subgroups of the Orchideae is a clearly delimited phyletic group.

THE ORCHIDS

The Orchidaceae is probably the largest family of flowering plants. Estimates range from 17,000 to 35,000 species. Recent efforts to estimate the number of distinct, named species fall very close to 19,000 (18,730, 19,128, and 19,172; Atwood 1986). A new estimate made for Appendix 2 is about 19,500 species. Relatively few orchid groups have been the subjects of taxonomic revisions, and orchids have not been as well sampled as some other plant groups. The publication of new species has accelerated in recent years, with more students working on orchids and with increased collecting in the tropics. What proportion of the family remains unnamed is difficult to guess. Orchids are a major group of epiphytes, and Atwood (1986) finds that about 73 percent of the species are epiphytic. Orchids are very widely distributed, but the greatest diversity, including nearly all of the epiphytes, occurs in the tropics, and especially in tropical mountains.

It is customary to say that the orchids show great diversity in flower structure but their vegetative features are monotonously uniform. Benzing (1986a,b) maintains, on the contrary, that flower structure is relatively stereotyped, while plant structure is remarkably diverse. Few other plant families match the vegetative diversity of the orchids. Flower structure is rather uniform in number and arrangement of parts, but there is great diversity in size and structural details.

Conventional wisdom also holds that the orchids are highly specific in their pollination relationships, implying that most orchid species are limited to single pollinator species. Such highly specific relationships do occur, but they are probably the exception rather than the rule. Many orchids are normally pollinated by a single class of pollinator, and some are quite "promiscuous," that is, they are pollinated by pollinators of several different classes. It is remarkable that a large percentage, perhaps more than a quarter, of the orchids practice "false advertising" and do not reward their pollinators.

On present knowledge we may treat the orchids as a distinctive family in the superorder Liliiflorae (see Chapter 3). We could treat them as a distinct order, Orchidales, but the Orchidales becomes quite artificial if we include the Burmanniaceae or Corsiaceae with the orchids. My preference would be to treat the orchids as a family in a larger order, such as the Asparagales, if that relationship is confirmed. Further study of the anatomy and seed structure of primitive orchids and other monocotyledon groups would probably clarify the relationships of the Orchidaceae, but at that level molecular systematics may be our best hope.

The Orchidaceae have all the earmarks of a group in active evolution; species, genera, tribes, and subtribes are all difficult to delimit. As such, I believe it is an excellent group for studying evolution. One advantage of the Orchidaceae is that polarity is clear for many features. In floral features, for example, most transformation series go from simple, lilylike structures to highly derived conditions that are quite unique. Darwin devoted considerable time to a study of the structure and function of orchid flowers, but, strangely enough, the family has since been neglected by students of phylogeny and systematics. We need not understand phylogeny to appreciate the diversity of the orchids, but knowledge of the overall pattern of evolution lets us study anatomy, cytology, and other features in a more meaningful framework. We can make and test hypotheses about relationships, and that is what makes systematics exciting.

PHYLOGENY

To the lay person, the concept of organic evolution is counterintuitive, as species rarely change perceptibly during a human lifetime. Nonetheless, the evidence that species change through time is overwhelming. Evolution is a basic concept in biology. Almost any kind of data is most meaningful when related to the phylogeny of the organisms under study.

Phylogeny is essentially the geneology of species and groups of species. Traditionally, the branching of evolutionary lines (speciation) and their divergence are both considered as aspects of phylogeny. Some have tried to restrict the term "phylogeny" to branching pattern, alone, but the term "cladogenesis" is available for branching pattern, and there seems no good reason for limiting phylogeny in this way. At least four distinct processes are involved in producing the pattern we call phylogeny.

1. Separation. The division of one population or lineage into two is the usual first step. Geographic isolation is probably the major factor in separating populations, but not necessarily the only type of separation.

2. Speciation. Speciation and separation or isolation may occur simultaneously in special cases, as in allopolyploidy, the doubling of chromosome number in a hybrid. In most cases, though, geographic isolation is thought to precede speciation. Separation does not necessarily lead to speciation. Many isolated populations may merge again with other populations or become extinct. When long-isolated populations are markedly different, we treat them as distinct species, but this may be rather arbitrary (see Chapter 12). Separation and speciation, together, are often characterized as "cladogenesis," or evolutionary branching.

3. Change. Change, in itself, is not necessarily closely linked to speciation. The term "anagenesis" is sometimes used for change within lineages (independent of speciation).

4. Extinction. Extinction is the ultimate fate of most lineages and strongly shapes the patterns of variation that we study in living organisms.

Phylogeny occurs through geological time. A perfect fossil record with no gaps might be difficult to classify, but it would be wonderful for students of phylogeny. We may try to understand phylogeny by studying fossils, but the fossil record is inadequate for most plants and animals, though it is the only evidence we have for most extinctions. Fossils may give us hints, but in many cases we must try to understand the phylogeny of plants and animals by inference from the plants and animals of the present. We can, of course, know them in much greater detail than we do the fossils. In any case, phylogeny is traditionally shown as some sort of "tree." Many of the earlier attempts at phylogeny looked much like coniferous trees, each with a single trunk, a number of branches, and many twigs. Now, our phylogenies are more likely to be rather surrealistic shrubs.

In recent decades, several methods of phylogenetic analysis have been developed to infer phylogeny from the distribution of features in living plants or animals. There is still debate as to whether or not phylogenies should be directly converted into classifications, but there is general agreement, at least, that polyphyletic "groups" are not acceptable. Current methods of analysis work very well with highly derived groups, and in most cases their taxonomy will be straightforward. Groups with few derived features, however, are difficult.

PREFATORY COMMENTS

As in 1981, I list the genera under each tribe or subtribe, as many readers find this useful. I have tried to be as nearly correct as possible, but I cannot pretend to be a specialist on 700 genera. My generic lists may be quite accurate when I follow Eric Christenson or Carlyle Luer, but I deserve little credit for that accomplishment. Generic names are also listed in Appendix B, with authors and approximate numbers of species. In Appendix B, I include also some synonyms, probable synonyms, and names of doubtful validity. Chromosome numbers listed in Chapters 4 through 11 are diploid ($2n$); in Chapter 3, haploid ($1n$), diploid ($2n$), and the hypothetical "base number," x, will be specified.

Acknowledgments

I am very grateful to Drs. John T. Atwood, Jr., Eric Christenson, Mark Clements, Katharine Gregg, Carlyle Luer, Gustavo Romero, Mark Whitten, and Norris Williams for profitable discussion of many subjects. Dr. Wilhelm Barthlott has sent me much useful information on seed and velamen structure. Dr. Mark Chase has kindly kept me posted on his work on molecular systematics of the Orchidaceae, and both John A. Freudenstein and H. Kurzweil have kindly sent me important papers while they were still in press. Dr. Walter Judd has been especially helpful with stimulating and unbiased discussion of cladistic theory and practice. Dr. Finn Rasmussen sent photographs of the column of *Cephalanthera*. The staff of the Marie Selby Botanical Gardens, and especially Ron Strobel, have been generous in lending material for study or illustration. Dr. Kingsly Dixon sent material of *Epiblema*, Michael S. Owen sent fresh *Disa* flowers from Longwood Gardens, and Warren Stoutamire sent freshly preserved material of several Australian genera. David L. Jones and Joyce Stewart both sent me articles or journals that I was unable to find in the western hemisphere.

About a third of the illustrations are taken from the 1981 book. Other drawings of floral material were prepared by Wendy Zomlefer. I am responsible for diagrams that could be prepared from pre-packaged lines and symbols, and Paloma Ibarra has converted my crude sketches into clean drawings or diagrams in most other cases. A grant from the American Orchid Society supported the preparation of most of the new drawings.

My wife, Kerry, has been unfailingly supportive and understanding even when I shunned all human contact and spent hours at the computer organizing and writing down some imagined insight while it was still fresh in mind. Most of the photographs in the color plates were taken by Kerry; for others I am indebted to R. Escobar R. (47), J. P. Folsom (20), J. W. Green (18), P. Lavarack (31), C. A. Luer (7, 13, 19, 25, 26, 29, 35, 36, 55, 57, 94, 95), H. Page (62), E. A. Schelpe (22, 89, 90), FL Stevenson (61), J. Stewart (23, 24), P. Vaughan (1), and N. H. Williams (9).

I am grateful, too, to those who have disagreed with me (mentioning no names). Our differences and discussions have aided me to clarify my own thoughts and to understand why we may continue to disagree on some issues.

Addendum

The manuscript is now ready for the printer; new data have not appeared to undo the system, but important new data are expected in the near future. Mark Chase has shown me a preliminary analysis, based on the *rbcL* gene, that indicates no close relationship between the Cranichideae and either *Diceratostele* or the Tropidieae (Chapter 6). The analysis shows the Cranichideae to be more closely allied to the Orchidoideae than I had thought and suggests that the Sobraliinae should not be placed in the Epidendreae. This analysis may answer many of our questions when more genera have been compared.

2

Structure and Other Evidence of Relationship

Most of the following discussion of phyletic relationships is based on structural features, so these must be discussed in some detail. In most cases I try to indicate the polarity, or probable direction of evolution, for the features discussed.

GROWTH HABITS

There is a fundamental difference in growth between monocotyledons and dicotyledons. Dicots normally have a vascular cambium that lets the stems grow in diameter during the entire life of the plant. Most monocots lack such a tissue, though a few (none orchids) have something similar. Thus, once a section of orchid stem has matured, it may swell by absorbing water but it cannot grow in diameter. In discussing stem structure, we speak of nodes and internodes; the nodes are the points at which leaves are attached, and an internode is the section of stem between two nodes (Fig. 2-1B). The monocots usually show intercalary growth; that is, the basal part of the internode retains growth potential and continues to elongate for some time, supported by the leaf sheaths that surround it. This is especially obvious in orchids that flower from the top of a pseudobulb before the pseudobulb itself has matured, as in many species of *Coelogyne*. Once a

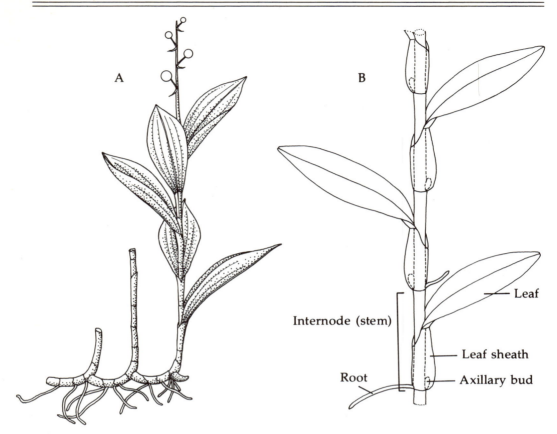

Figure 2-1. (A) The growth habit of a primitive orchid, with spirally arranged, plicate leaves and a terminal inflorescence. Each year a new shoot is produced from the base of a previous shoot. The basal portions of successive shoots together form the horizontal rhizome. (B) A diagrammatic sketch of three internodes of a monocot stem. (After Dressler 1981.)

stem has matured, however, new growth (except for roots) normally occurs only from a bud, either apical or axillary. A bud may form a new shoot, an inflorescence, or a solitary flower, depending on its position and the growth habit of the plant.

Holttum (1955) suggests that the basic growth form of the monocots is sympodial; each shoot has limited growth and later gives rise to a similar shoot from an axillary bud (Fig. 2-1A). This, then, is a modular habit of growth. The monopodial growth habit, in which the shoot has the potential for indefinite apical growth, has been derived from the sympodial habit in various groups of orchids (Fig. 2-2N–P). Sympodial plants may have shoots clustered together or spread out on a long rhizome, and new shoots may arise from any part of the older shoots where there is an axillary bud. Both sympodial and monopodial growth habits may be erect, creeping, or pendant.

POLARITY. The sympodial habit, with each shoot limited in growth, is surely the ancestral condition. The monopodial habit, with lateral inflorescences and continued apical growth, is typical of the Vandeae, but has evolved independently in several other groups.

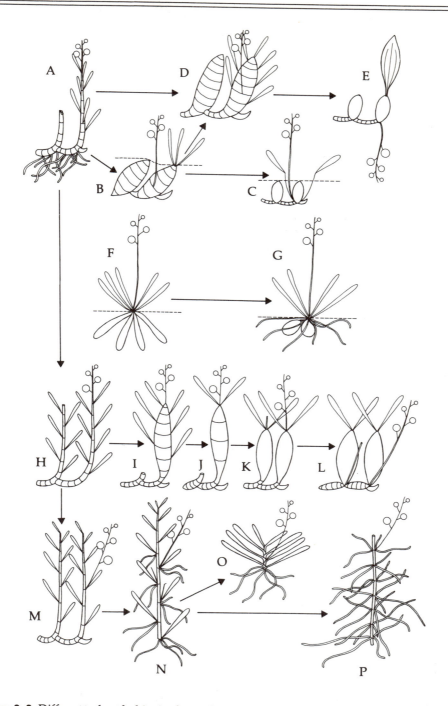

Figure 2-2. Different plant habits in the orchid family, showing possible patterns of evolution (highly diagrammatic). The growth habit is sympodial in A–M, monopodial in N–P. The leaves are spiral in A, distichous (two-ranked) in others. The inflorescence is terminal in A and F–K, lateral in all others. Corms of several internodes are shown in B, of a single internode in C. Pseudobulbs of several internodes are shown in D, I, and J; pseudobulbs of single internodes are shown in E, K, and L. Fleshy storage roots (tuberoids) are shown in F, and root–stem tuberoids in G. A leafless plant with photosynthetic roots is shown in P. (After Dressler 1981.)

ROOTS

As they are monocotyledons, orchids never have a carrotlike taproot, and, indeed, never have a primary root. The entire root system is made up of secondary roots that arise from the stem. Orchid roots vary greatly in thickness but are never thin and fibrous like those of grasses.

Velamen

One of the features found in most orchid roots is the velamen or, especially in the older literature, the *velamen radicum*. A recent review of the velamen by Pridgeon (1987) may be consulted for more detail. The velamen is morphologically homologous with the epidermis, but may have as many as 24 layers of cells. The velamen cells are dead at maturity, and the velamen is bordered within by the exodermis, a layer with long cells that are dead at maturity, usually with thickened walls, and shorter, living, passage cells that allow water and nutrients from the velamen to enter the cortex of the root.

The velamen is usually a spongy, whitish sheath around the root, and the cell walls usually have fibrous thickenings. There has been some controversy over the function of the velamen, but it now seems clear that most orchid roots can and do absorb water and nutrients through the velamen. The velamen is most obvious in epiphytes, but it is also present in many terrestrials, as well as in some Amaryllidaceae, Araceae, Dioscoreaceae, and Liliaceae. In epiphytic orchids, the growing apex of the root is commonly green when exposed to light, and there are often chloroplasts within the cortex of the mature root, though the velamen masks their green color. In some Vandeae, all photosynthesis takes place in the roots; the stem is short and bears only scale-leaves. In these leafless orchids, the velamen usually sloughs off of the upper surface, while a thin velamen remains on the lower surface.

In some orchids the velamen includes spongy, fibrous bodies in the cells adjacent to the passage cells. These have been termed "tilosomes," and seven types may be recognized (Pridgeon et al. 1983). These occur especially in the New World orchids, and most types appear to have evolved independently in different groups. Various functions have been proposed for these spongy bodies; they might be plugs to prevent drying or to prevent the entry of pathogens, or they might function in the condensation of water vapor.

Porembski and Barthlott (1988) have surveyed the structure of velamen in the Orchidaceae. They find that some terrestrial orchids have no velamen, though this may be either a primitive or a derived condition (or both in different groups). The *Calanthe* type of velamen structure is found in many orchid groups, and thus gives little indication of relationships. The *Cymbidium* type is general in the more advanced subtribes here treated as the cymbidioid phylad, however, and there are eight velamen types that are each characteristic of a tribe or subtribe. Some orchids have "unspecific" velamen, that is, a definite velamen that is not referable to any of the named types.

POLARITY. The primitive condition in the Orchidaceae may be either no velamen or velamen of the *Calanthe* type, but no attempt is made to polarize the derived types, most of which might have been derived from the *Calanthe* type.

Storage Roots

In many terrestrial orchids, the roots form storage organs, or tuberoids (a tuber must, by definition, be a stem). In some, the whole root may be very fleshy, as in many Spiranthinae. In others, as in *Cleistes* or some species of *Tropidia*, some segments of root are thick and the rest much thinner. We may call these nodular tuberoids to distinguish them from roots that are thick throughout. An unusual structure is the root–stem tuberoid of the subfamily Orchidoideae (Fig. 2-3). It is

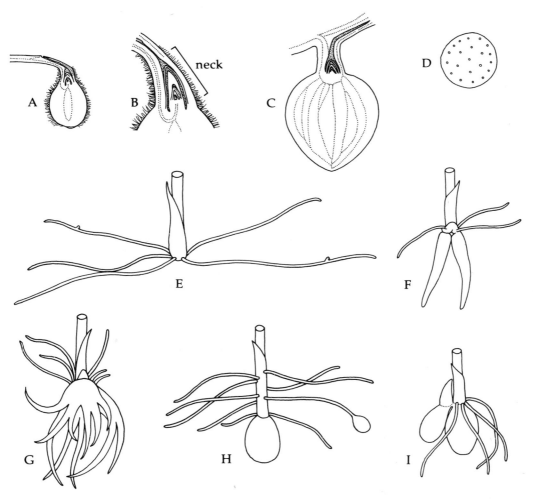

Figure 2-3. Structure of the root–stem tuberoid. (A) Longitudinal section through a globose tuberoid. (B) Part of the same at higher magnification. (C) Longitudinal section through a globose tuberoid, showing the many steles. (D) Cross-section of a globose tuberoid, showing the many steles. (E–I) Different forms of root–stem tuberoids. (A, B after Kumazawa 1956; C after Stojanow 1916, D–I after Ogura, 1954.)

largely a storage root, but the basal portion has a sheath of root structure around a core of stem structure with a bud. The tuberoid survives during the dormant season, and in the growing season the bud produces a new shoot; one of the axillary buds forms a new tuberoid that will be the next resting phase. During the growing season, the plant typically has two globose tuberoids, an old one from the previous season and a new one that will survive during the coming dormant season. These paired structures look rather like mammalian testicles, and the plants were thus called *orchis*, or testicle, by the ancient Greeks. This has become the generic name of an Old World genus and has given the family its name.

Fleshy storage roots are not limited to terrestrial orchids. Epiphytes and lithophytes, such as *Sobralia*, *Ponera*, *Isochilus*, and some species of *Epidendrum*, may have thick, fleshy roots. Plants of this sort usually have thin stems and leaves, and collectors find that such plants may survive shipment with their shoots cut off, but not without roots.

Roots normally produce only roots, but "adventitious" buds that give rise to new shoots may be produced on roots, as in *Listera*, *Pogonia*, *Psilochilus*, or *Phalaenopsis*. Such roots are superficially much like the slender root–stem tuberoids of some Orchidinae, but they are anatomically distinct, lacking the core of stem structure.

POLARITY. Slender, moderately fleshy roots are probably the ancestral state, though nodular tuberoids may have occurred with them. Very fleshy roots, as in many Cranichideae, are considered derived. The unusual, dimorphic roots of the Orchideae, Diseae, and Diurideae are clearly a distinct derived feature.

STEMS

Basically, the orchid stem is much like that of corn or lilies, or any other ordinary monocot stem. The vascular tissue is in many scattered bundles that are usually more numerous near the surface of the stem and embedded in softer parenchyma, or storage tissue. The stems may be thin and wiry, tough and "woody" (no true wood, of course), or soft and succulent, as in *Vanilla*.

Rhizome

The term "rhizome" may be used for any horizontal stem on or in the substrate. In sympodial orchids, the rhizome is a compound structure, in that it is made up of the basal sections of successive shoots. Typically, the first few internodes of the shoot are horizontal and may be somewhat thickened or hard and woody. The axillary buds at a few of the nodes are unusually well developed, and one or two of these grow to form the next shoot(s). In sympodial orchids with lateral inflorescences, the inflorescence also may arise from the rhizome. The creeping, horizontal stems of the Goodyerinae are called rhizomes, but there is usually not a sharp distinction between the rhizome and the aerial shoot in that group. There is no distinct rhizome in orchids with a monopodial habit of growth, and there may be none in sympodial orchids with creeping, climbing, or pendant growth if the new shoots do not arise from the bases of older shoots.

The term "secondary stem," found in many taxonomic descriptions, seems to refer to the vegetative shoot above the rhizome, but this use is inaccurate and confusing. Morphologically, the only primary stem is the first seedling shoot from a protocorm. All other stems are secondary, including the rhizome. Stern and Pridgeon (1984) have suggested the term "ramicaul" as a substitute for "secondary stem." This term has been criticized by H. Rasmussen (1985); Soto Arenas and Greenwood (1989) suggest that the older term "caulome" is available, but that it may be simpler to say "stem" in most cases.

Corms and Pseudobulbs

The term "corm" is used to refer to an underground storage stem like that of *Gladiolus*. Thus, *Bletia, Eulophia, Spathoglottis*, and some related genera have quite typical corms. There is no sharp distinction, however, between corm and pseudobulb. The term "bulb," as an English botanical term, refers only to a structure like that of the tulip or the onion, in which the bulb is largely made up of thickened leaf bases. The term "pseudobulb" was applied to the thickened stem structures of epiphytic orchids, which are neither bulb nor tuber, and the term has achieved general usage. A pseudobulb may be made up of a single thickened internode (heteroblastic) or several (homoblastic), and a pseudobulb of several internodes may bear leaves along its length or only at the apex. The sheaths that clasp the bases of pseudobulbs often bear leaf blades, and in a few orchids with leaf-bearing sheaths the pseudobulb has only a scale leaf at its apex. Pseudobulbs are quite incompatible with the monopodial growth habit, and monopodial orchids must use either leaves or roots as storage organs.

LEAVES

As noted above, each node of the stem bears some sort of leaflike organ, with an axillary bud at its base. In many orchids the rhizome bears mere scale leaves or leaf sheaths. Most orchid leaves are typical of the monocots, with many parallel veins and inconspicuous cross connections between them. In *Epistephium* and *Clematepistephium*, however, it is hard to call the leaf anything but net-veined (Fig. 2-4A). The arrowhead-shaped leaves of *Pachyplectron* (B) also look a bit out of place among the orchids, and the deeply lobed leaves of some *Acianthus* species (F) would seem more at home on a buttercup.

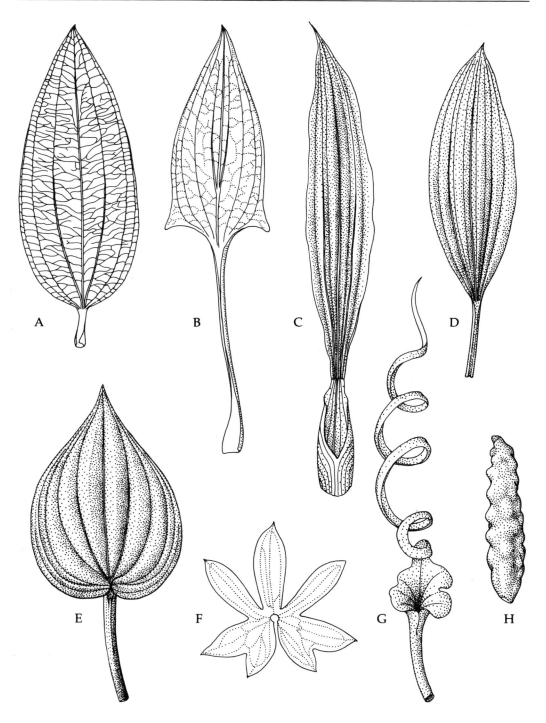

Figure 2-4. Various types of orchid leaf. (A) *Clematepistephium*, net-veined. (B) *Pachyplectron arifolium*, net-veined and hastate. (C) *Catasetum*, plicate with a sheathing base. (D) *Stanhopea*, plicate with a distinct petiole. (E) *Monophyllorchis maculata*, plicate and cordate. (F) *Acianthus bracteatus*, soft herbaceous and deeply lobed. (G) *Thelymitra spiralis*, twisted. (H) *Dendrobium cucumerinum*, fleshy. (After Dressler 1981.)

Arrangement

In most orchids the leaves are arranged distichously, or in two ranks, with the leaves alternating on opposite sides of the stem (Fig. 2-2H, N). In many cases the stem or pseudobulb bears only a single leaf, but if we check the orientation of scale leaves and sheaths, we find them to be distichous. The primitive condition, however, seems to be a spiral arrangement (A). In a few cases, by condensation of the internodes, there are two or more leaves arising at the same level (*Codonorchis*, *Isotria*).

Vernation and Folding

The way leaves develop has been considered important in orchid classification. Among the primitive groups the developing leaves are rolled, or convolute (Fig. 2-5A, B). In many, especially among the epiphytes, the leaves are duplicate during development (C), or folded once with each half flat. Such leaves always become conduplicate, that is, with a single fold at the midline and broadly V-shaped in cross-section, the veins of the leaf blade all being similar in size and not prominent (E). Leaves that are convolute in development may be either plicate or soft herbaceous. In the plicate, or pleated, condition, several veins are prominent,

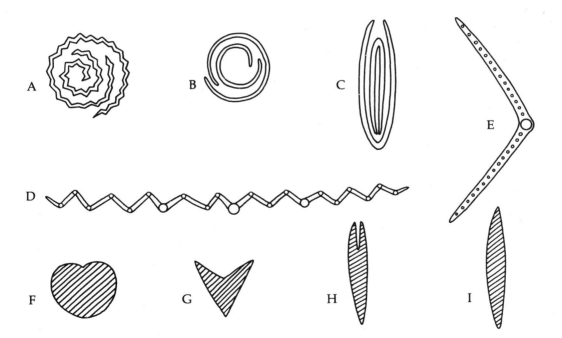

Figure 2-5. Diagrammatic cross sections of devloping and mature leaves. (A, B) Convolute development. (C) Duplicate development. (D) Plicate leaf. (E) Conduplicate leaf. (F–I) Fleshy leaves, cylindrical, triangular or laterally flattened. (After Dressler 1981.)

and the leaf is usually folded at each of these (D). Conduplicate leaves have evolved from plicate leaves in several orchid groups, and some groups have leaves that are not quite one thing or the other, as in *Cymbidium* and the *Chondrorhyncha* complex. *Bletia*, *Sobralia*, and *Spathoglottis* are genera with typically plicate leaves. Plicate leaves are always relatively thin, as are soft herbaceous leaves. Soft herbaceous leaves are typical of terrestrials with annual shoots, such as *Spiranthes* or *Habenaria*. The leaves may be weakly folded at the midline, and the leaves are thin, with all veins similar in size. Conduplicate leaves, however, are usually somewhat leathery or fleshy. In the extreme cases, fleshy leaves may be triangular or cylindrical in form (F, G). In a number of orchids we find the leaves to be laterally flattened, rather than dorsoventral (H, I). These are often called "equitant," as the base of each leaf "rides" on the one above it. The leaves of *Vanilla* fit none of these categories, as they are convolute in development but very fleshy.

POLARITY. Both soft herbaceous leaves and conduplicate leaves evolved from plicate leaves, and conduplicate leaves have evolved independently many times.

Sheaths and Petioles

In many cases, the basal portion of the leaf forms a sheath around the stem, a feature without which there could be little intercalary growth, as the soft, growing portion of the internode is too soft to support itself alone. In other cases, such sheaths are formed without leaf blades, and especially on rhizomes and inflorescences the sheaths (bracts) may be small. In the genus *Teuscheria*, the spotted sheath about the pseudobulb is hard and keeps its shape even after the pseudobulb has shriveled within. In some orchids the base of the leaf forms a narrow, subcylindric petiole, as in some Stanhopeinae (Fig. 2-4).

Articulations

We usually take the falling of leaves for granted, especially in autumn, but not all leaves are shed. A special abscission layer of breakaway cells near the base of the leaf, commonly called a joint or articulation, is lacking in many terrestrial orchids, so the leaves simply rot in place. An abscission layer is quite lacking in other subfamilies, but it is found in most members of the Epidendroideae, though it may be lost again in some highy evolved micro-orchids, as in some species of *Dichaea*, *Epidendrum*, and *Macroclinium*. The abscission layer is usually at the base of the blade or the petiole. In a few cases, as in *Teuscheria* and some *Oeceoclades* species, the abscission layer is well above the base of the petiole. One might, at first glance, expect the structure below the joint to be the tip of the pseudobulb, but in some species of *Oeceoclades* there are two leaves on a single pseudobulb, each with a joint in the middle of the petiole (Summerhayes 1957). Similarly, in the Pleurothallidinae the base of the leaf and the abscission layer may be independent; the base of the leaf is termed an "annulus" when it is below the abscission layer (Pridgeon 1982a).

Stomata and Subsidiary Cells

Land plants normally have small openings, or stomata, on one or both leaf surfaces that permit the passage of gases, and especially the entry of carbon dioxide, into the leaf. Each opening is bordered by two kidney-shaped "guard cells" that open and close the stomata by changing their shape. The guard cells are very different from other epidermal cells. In many cases the stomata and guard cells are accompanied by two or more distinctive subsidiary cells that are structurally unlike the surrounding epidermal cells. The presence or absence of subsidiary cells and their ontogeny are considered important in plant classification. There are, unfortunately, many systems of classifying stomatal types, each with its own abstruse terminology. It is easy to see the form of the mature stomatal complexes, but similar structure may evolve in different ways.

Most orchid groups have distinct subsidiary cells associated with their stomata in the majority of their members. The presence of four subsidiary cells with each stomate may be a derived condition, as suggested by N. H. Williams (1979), but outgroup comparison suggests the presence of subsidiary cells as the ancestral condition for the Orchidaceae. Recognizable subsidiary cells are consistently lacking only in the Orchidoideae, Neottieae, and Pogoniinae. The pattern in the Cranichideae is distinctive and may be characterized by the mesoperigenous development of the subsidiary cells (Williams 1975). H. Rasmussen (1981, 1987), however, indicates that some cells in the Orchidoideae may be mesoperigenous in origin, though not recognizable as subsidiary cells at maturity. Our limited observations suggest that the pattern in the Tropidieae and *Diceratostele* resembles that in some primitive Epidendroideae rather than that of the Cranichideae. As is too often the case, we have good information on the pattern and ontogeny of the epidermis in the advanced Epidendroideae, but more study of the primitive groups is needed.

I tentatively recognize three main patterns: the epidendroid pattern, usually with recognizable subsidiary cells that are perigenous in development, with trapezoid cells (Fig. 2-6E, F); the cranichid pattern, usually with recognizable subsidiary cells that are mesoperigenous in development (D); and the orchidoid pattern, without recognizable subsidiary cells at maturity (A–C). The epidendroid pattern appears to be the primitive condition, and the orchidoid pattern may have evolved independently two or three times.

Stegmata, or Silica Cells

Møller and Rasmussen (1984) have brought together information on an interesting feature that had been known but little appreciated. The orchids are one of the few families that possess stegmata, or silica cells, sheathing vascular bundle sheaths and fiber bundles in longitudinal rows. These are presumably a sort of structural reinforcement. These silica cells are found in all major orchid groups except the soft-leaved Orchidoideae, thus supporting the idea that the Orchidaceae are a very natural family. Silica cells are not found in delicate or soft,

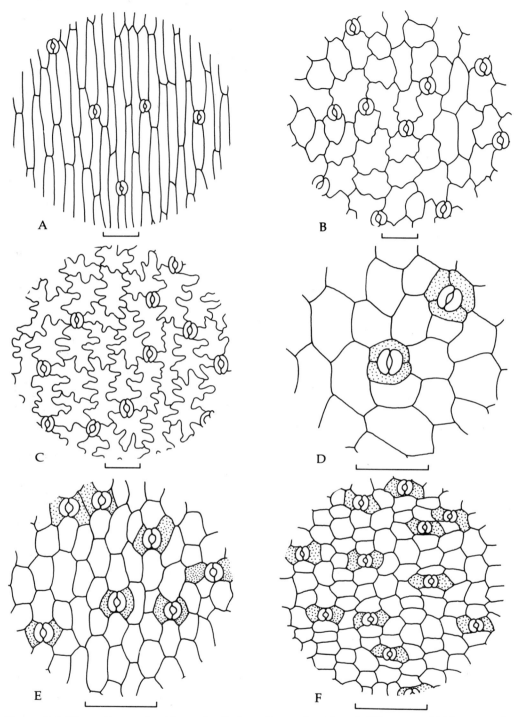

Figure 2-6. The patterns of epidermal cells from the undersides of different orchid leaves; subsidiary cells stippled. (A) *Calochilus herbaceus* (Diurideae). (B) *Habenaria petalodes* (Orchideae). (C) *Isotria verticillata* (Pogoniinae). (D) *Cyclopogon elatus* (Cranichideae). (E) *Sobralia fragrans* (Epidendreae). (F) *Neomoorea wallisii* (Maxillarieae). (A, G, and C) Orchidoid pattern. (D) Cranichid pattern. (E, F) Epidendroid pattern. Scale 0.1 mm. (Courtesy of N. H. Williams.)

herbaceous plants, and they have not been found in certain subtribes, such as the Bulbophyllinae. In most orchids the silica cells and the included silica bodies are roughly conical in shape, but they are spherical and rather lumpy in the Vandeae, Eriinae, Podochilinae, and Dendrobiinae (Fig. 2-7). Møller and Rasmussen (1984) suggest that the change from conical to spherical probably occurred only once within the Orchidaceae. The silica cells supply the clearest evidence that the "vandoid" orchids are a polyphyletic grade, and they indicate a phyletic relationship between the Vandeae, Eriinae, Podochilinae, and Dendrobiinae, a relationship supported by independent lines of evidence. There is one frustrating problem with the silica cells. If a group lacks silica cells, one cannot know which type of silica cell the group (or its ancestor) has lost.

POLARITY. As the spherical silica bodies correlate well with other features, they are considered derived.

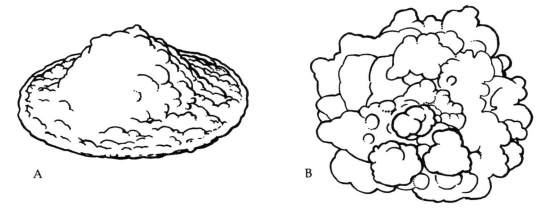

Figure 2-7. Silica bodies. (A) Of *Agrostophyllum stipulatum* (conical). (B) Of *Cleisostoma subulatum* (spherical). (Based on Møller and Rasmussen 1984.)

INFLORESCENCE

The orchid inflorescence is normally racemose, or indeterminate; the flowers are axillary on the rachis and usually flower from the base upward (Fig. 2-8A). Inflorescences may be branched (paniculate), in which case the ultimate branches are usually racemose (B). There are also orchids that normally bear one-flowered inflorescences, such as *Lycaste*, *Maxillaria*, *Dichaea*, and *Chondrorhyncha* (D). Some of these inflorescences are probably more complex, but the main axis of the inflorescence is completely hidden by sheaths and leaf bases, so that the individual flowers each appear on a separate stem.

In all cases, the flower is subtended by a bract. The bract is usually inconspicuous, but may be large or colored, as in *Cyrtopodium* or *Lockhartia*. The flowers

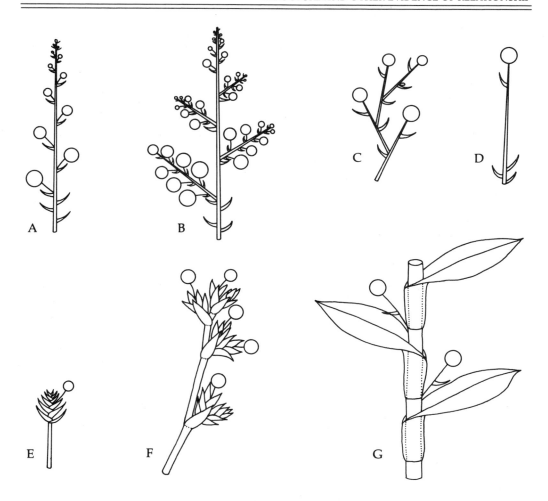

Figure 2-8. Inflorescence types (diagrammatic). (A) Raceme. (B) Panicle. (C) Cymose inflorescence. (D) One-flowered inflorescence. (E) Condensed inflorescence with successive flowers. (F) The inflorescence of *Sigmatostalix*; each flower arises from a cluster of bracts. (G) Leaf-opposed inflorescence, as in *Dichaea*. (After Dressler 1981.)

are often spirally arranged on the rachis, even when the leaves are distichous, but the bracts and flowers are distichous in some groups, and the flowers are whorled in a few cases, as in *Chamaeangis* and some *Oberonia* species. The inflorescence may arise from any part of the stem. In the primitive condition, the inflorescence is terminal and is simply a continuation of the shoot axis (Fig. 2-2A, F–K). In other cases the inflorescence is lateral from the side or base of the shoot, or from the rhizome. In the monopodial growth habit, of course, it is always lateral. Sometimes the inflorescence is condensed and the flowers are produced one by one from a very short rachis, as in *Epidendrum nocturnum*, *Systeloglossum*, *Bromheadia*, and *Thrixspermum* (Fig. 2-8E). The flowers may also be produced simultaneously, or nearly so, in a very dense cluster, as in *Elleanthus* or *Glomera*. Each flower of the genus *Sigmatostalix* arises not from a single bract but from a cluster of bracts, which may represent a branch of a more complex inflorescence that has been condensed (F).

The unusual inflorescence found in *Lockhartia* appears to be cymose, or determinate (Fig. 2-8C), a type common among the dicots but infrequent among monocots. The first flower to open is terminal on the inflorescence, then one or two subtending buds produce shoots that end in flowers, and each of these produces lateral shoots, and so on. Presumably, an ancestor of this orchid had a single-flowered inflorescence, and when selection favored an increase in the size of the inflorescence, this was achieved by cymose branching. In *Macroclinium* and *Notylia*, new clusters of flowers are commonly produced from the basal portions of old inflorescences. If the flower cluster of *Notylia* were reduced to a single flower, the inflorescence would be comparable to that of *Lockhartia*.

Although the normal inflorescence is produced from the axil of a leaf or bract, there are some notable exceptions. In *Dichaea* (Fig. 2-8G) the one-flowered inflorescence is produced directly opposite the base of a leaf (Wirth 1964). There are two possible explanations for such an inflorescence, and we have, at present, little data that will help us choose. In some cases a leaf-opposed inflorescence represents the ultimate in sympodial branching; that is, each flower is terminal and the next stem internode is produced from the axil of the subtending leaf. The alternate explanation is that the axillary bud "rides up" as the internode elongates and comes to lie opposite the leaf above the leaf to which it is actually axillary. The inflorescence of *Luisia teretifolia* is supraaxillary (Wirth 1964), but it is borne on the middle of the internode, so that it seems clearly to be derived from the lower axil. I have seen a similar condition in an unidentified *Epidendrum*.

POLARITY. A terminal inflorescence is clearly the ancestral condition. An upper lateral inflorescence occurs occasionally in other groups, but it occurs consistently in most of the dendrobioid subclade. A basal lateral inflorescence is also considered derived, but basal inflorescences are occasional in most groups characterized by terminal inflorescences, as in the Laeliinae and Pleurothallidinae. Terminal inflorescences also occur in groups characterized by lateral inflorescences, and these may be either ancestral features or secondary "reversals."

FLOWERS

The orchid flower typically shows bilateral symmetry; that is, one may draw one, and only one imaginary line down the middle of the flower to separate halves that are mirror images (Fig. 2-9B). This at once distinguishes the orchid flower from flowers like the iris, but there are many other flowers, both monocot and dicot, with bilateral symmetry. In the flowers of *Mormodes*, both the column and the lip are twisted to one side, thus breaking the normal orchid symmetry. Similar asymmetry is to be found in the flowers of *Ludisia*, *Macodes*, *Macradenia*, and *Tipularia*, though it is less striking in some cases.

Pedicel and Ovary

In the orchids, as in the amaryllids and several other families, the ovary is inferior—the bases of the other flower parts are completely united with the ovary so that these parts appear to arise above the ovary (Fig. 2-9A). Especially in the epiphytes, the ovary is not fully differentiated at flowering, and only continues to develop if the flower is pollinated. Thus, the plant invests little energy in the ovaries of unpollinated flowers. The ovary is usually slender at flowering, and it may be difficult to see any distinction between pedicel and ovary. The subtribe Pleurothallidinae is an exception, as there is always a joint, or abscission layer, between the pedicel and the ovary. In other groups the pedicel is jointed at its base and falls off with the flower if the flower is not pollinated. The conduplicate-leaved lady's slippers and most Vanilleae, however, have a joint between the ovary and the rest of the flower; if the flower is pollinated, the column and perianth fall off after fertilization has occurred.

As is typical of many monocots, the ovary is made up of three carpels. In a few primitive genera, such as *Apostasia* and *Selenipedium*, the ovary is divided into three locules, or chambers, but in most genera these divisions are lacking.

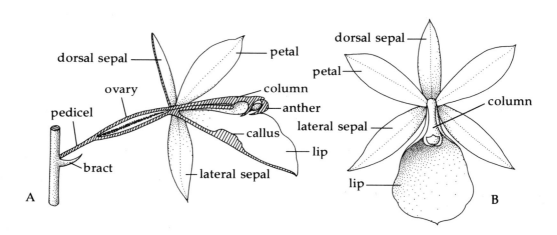

Figure 2-9. Flower parts. (A) Longitudinal section. (B) Front view of flower. (After Dressler 1981.)

Resupination

We are so accustomed to seeing the lip on the lower side of the orchid flower that we use the terms dorsal, ventral, and lateral with reference to this orientation. If the rachis of the inflorescence is erect, the lip is on the upper (adaxial) side of the young bud, but the pedicel usually twists as the bud develops, so that the lip comes to be on the lower side. The term "resupinate" is used for any orchid flower that has the lip on the lower side. I have suggested (Dressler 1983b) that resupination might have arisen in plants with bracteate inflorescences, in which the pedicels bent away from the bracts rather than away from the stem. In most cases the twisting is definitely oriented with respect to gravity and occurs regardless of the position of the plant or the rachis. Similarly, most orchids that are nonresupinate may also orient their flowers in relation to gravity so as to maintain their normal posture. In some cases the pedicel may simply bend down beside the peduncle, so that the flower is resupinate with little or no real twisting. The same sort of bending usually occurs in solitary flowers, such as *Lycaste,* in which case the pedicel bends over the apex of the stem rather than beside it. There may be some orchids with erect inflorescences that have passively nonresupinate flowers, just as orchids with pendant inflorescences could have passively resupinate flowers. In either case, this could be checked by turning the plant or the inflorescence upside-down before the buds mature.

In *Catasetum,* the pistillate flowers are always nonresupinate, but the male flowers are resupinate in some species and nonresupinate in others. When both pistillate and staminate flowers occur on the same inflorescence, we may observe the pistillate flowers orienting to gravity so as to have the lip uppermost, and the staminate flowers on the same inflorescence doing exactly the opposite.

POLARITY. Resupination occurs in all orchid subfamilies, and there can be little doubt that this is a basic feature of the family, though it has been lost or modified in many groups.

Perianth

In the Liliiflorae, the inner and outer perianth segments are usually similar in color and texture, but when speaking of orchids, one customarily uses the term "sepals" for the outer segments and "petals" for the inner segments. In a few primitive genera, notably *Epistephium* and *Lecanorchis,* there is a cupule or calyculus borne at the base of the perianth. A similar structure is found in *Bulbophyllum pachyrhachis* and its allies, but little is known of its origin or significance. Probably it is merely a sort of outgrowth, and does not represent the reduction of an extra circle of bracts or perianth parts inherited from some distant ancestor. The sepals have a protective function while the bud is developing, and they are usually valvate, with the edges meeting but not overlapping. In a number of cases, though, they do overlap. The two lateral sepals may be partially or completely united, in which case the term "synsepal" may be used. In other cases, all three sepals may be united, or the dorsal sepal may be united with the column. The petals are commonly thinner than the sepals and usually overlap in bud. They may be united with the sepals or with the column, and may be greatly reduced. In *Lepanthes* and *Habenaria,* the petals are often deeply lobed and may be much wider than long.

The median petal (opposite the fertile anther) is virtually always differentiated from the other two and is called the lip or labellum. It is usually larger and more complex than the petals and is one of the main elements that makes orchid flowers recognizable as such. The lip is often partially united with the column, rarely with other perianth segments. The lip or part of it may be hinged and movable; in a few cases the movements are active, rather than passive. Fleshy lumps, ridges, keels, or plates, usually referred to as the "callus," commonly appear on the lip. Often divided into three or more lobes, the lip is so complex in some orchids that the terms "hypochile" (basal portion), "mesochile" (midportion), and "epichile" (distal portion) are used to aid in description. The structures so named may not be morphologically homologous in different groups, however. Darwin tried to explain the complexity of the orchid lip and, at the same time, to account for two of the missing stamens by suggesting that the lip is a compound structure, made up of a petal and two staminodes or sterile stamens. This ingenious hypothesis has had some proponents, but it has fallen out of favor for lack of evidence. Neither the vascularization nor any other aspect of the lip anatomy indicates that it is a compound structure (Swamy 1948).

Nectaries, Elaiophores, and Osmophores

Since orchid pollen is rarely used as food by bees, the principal reward offered by orchid flowers is nectar. The lilylike ancestors of the orchids may have had shallow nectar glands between the perianth and the ovary. The most obvious sort of nectary in the orchids is the spur, a slender tubular or sacklike extension from one of the perianth segments, usually the lip (Fig. 2-10A). Spurs are well

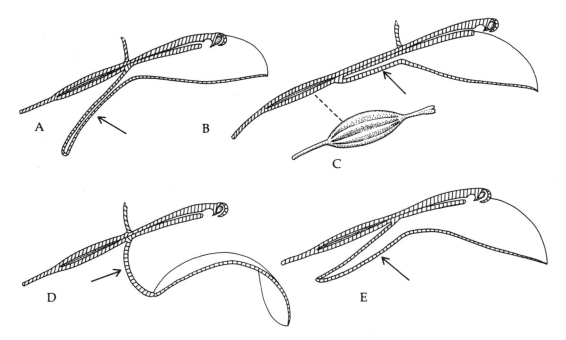

Figure 2-10. Various floral structures. (A) Spur at base of lip. (B) Cuniculus. (C) Beaked fruit (the beak represents the cuniculus of the flower). (D) Column foot. (E) Spur formed by column foot and lip. (After Dressler 1981.)

known in *Angraecum* and its allies, *Tipularia*, *Calanthe*, and many other genera. In *Comparettia* we find a spur formed by the united lateral sepals, but the nectar is supplied by two slender, solid projections from the base of the lip that extend into the sepaline spur. We find that a spur is not necessarily a nectary, as some orchids with a spur at the base of the lip actually offer no reward to the pollinators.

In many Laeliinae, the spurlike structure is less obvious, and the nectary runs through the "stalk" of the flower, so that it is usually evident only when we split the flower with a razor blade. This type of nectary has been called a cuniculus (Fig. 2-10B). In these genera, there is a floral tube between the ovary and the base of the perianth. This is most obvious in *Rhyncholaelia digbyana* or *Brassavola cucullata*, which have their flowers on long "stems," but the seeming stem is a floral tube between the ovary and the perianth. In these orchids, the capsule has a very long beak that corresponds to the floral tube (C). In *Epidendrum*, the claw of the lip is normally united with the column to form a tube that is continuous with the cuniculus. In some species, the nectary is swollen and externally obvious. Some Spiranthinae also have nectaries below the base of the perianth, but their structure appears to be different, as these nectaries occur alongside the ovary, rather than between the ovary and the perianth. Possibly the nectary of *Sarcoglottis*, etc. (Fig. 6-8) is actually an "external" spur that is completely united with the ovary. In many orchids there is a relatively shallow nectary on the lip or between the column and the lip. This is typical of *Listera*, *Stelis* and many species of *Pleurothallis*, for example. When there is a "column foot," there is often a shallow nectary on that structure, as in many *Dendrobium* and in *Scaphyglottis* (Fig. 2-10D). In *Hexisea* and *Systeloglossum*, we find deep, spurlike nectaries derived from the column foot or the column foot and the base of the lip (E). In *Dendrobium bigibbum* and its allies there is a spur formed in the middle of the column foot, and in *Thecostele* the column foot, itself, is hollow and spurlike.

The shallow, open "nectaries" of *Ornithocephalus*, *Sigmatostalix* and some groups of *Oncidium* have seemed incongruous, since such structures are typical of "promiscuous" pollination systems, not of advanced orchids that are visited by only a few kinds of pollinators. Vogel (1974) has shown that these are not nectaries but "oileries" or elaiophores. As only a few bees gather oil, an open gland is rarely subject to robbery by inappropriate insects. Most of the African Coryciinae also produce oil and attract bees of the genus *Rediviva*.

A third class of floral glands are the osmophores, or scent glands. Perfumes are not produced by all flower parts, or even by the whole perianth, but by specialized areas (Vogel 1962). The osmophores may be borne on the sepals, the petals, or part of the lip. The characteristic club-shaped tips of the petals or sepals in many Pleurothallidinae are osmophores. In many cases, the base of the lip produces perfume.

Column

In all orchids there is some degree of union between the style and the staminal filaments. In most cases these structures are so completely united that we cannot distinguish them. The combined structure is called the column or gynostemium (Fig. 2-11). In the Cypripedioideae and Apostasioideae, the stamens and staminodes are only partly united with the style. Also, in the Spiranthoideae and in the Diurideae, flowers with relatively short columns have only a slight degree of union. Of *Diuris* itself, one can almost say that it does not have a column (Fig. 2-11C–E). In most orchids, however, the column is quite obvious, and a good deal thicker than is usually the case with styles or filaments. The column often has lateral wings. These are frequently interpreted as staminodia, and in most cases they probably are. In the above-mentioned case of *Diuris*, the wings are virtually free from the column, and they are clearly sterile stamens. In the Cypripedioideae and in *Apostasia*, the median stamen may be represented by a staminode.

In most orchids the anther is seated in a clearly defined area at or near the end of the column. This is termed the "clinandrium." In some cases, the edges of the clinandrium are winglike and very prominent, and in a few cases longer than the column itself, as in some species of *Epidendrum* and *Oerstedella*.

Column Foot

The base of the column in many orchids forms a ventral extension to which the lip is attached (Fig. 2-10D). Sometimes this column foot is longer than the body of the column. The bases of the lateral sepals are usually attached to the column foot, and sometimes the bases of the petals as well. When there is a prominent column foot, the flower usually has a spurlike "chin" or mentum when seen from the side (E). The column foot is especially well developed in the *Scaphyglottis* complex and in the Dendrobiinae, where the column foot and the lip together sometimes form a complex structure.

Staminodia

In the apostasioids and lady's slippers, the lateral stamens are both fertile. In the monandrous orchids, only the median stamen is normally fertile, and the lateral stamens are often represented as sterile stamens (staminodia), or column wings. Kurzweil (1987a,b, 1988) finds that the staminodia of most Epidendroideae, the Limodorinae, and Tropidieae initiate from massive primordia that are developed shortly after the petals and just above them. Such large primordia may be present even in species that do not have obvious staminodia in the mature column. The staminodia of the Orchideae are smaller than those of the Epidendroideae and appear later in development, and the staminode primordia of the Listerinae and Cranichideae are obscure or lacking.

Figure 2-11. Columns. (A, B) *Corymborkis*, side view and ventral view of apex. (C, D, E) *Diuris*, ventral, side, and dorsal views. (F, G) *Serapias*, ventral, and side views. (H, I, J) *Schomburgkia*, side and ventral views with anther in place and ventral view with anther tipped away. (K, L) *Maxillaria*, ventral and side views. (M, N) *Oncidium*, ventral and side views. (O, P) *Systeloglossum*, ventral and side views. An, anther; Cl, hooded clinandrium; CW, column wing (or staminodium); Ro, rostellum; Stg, stigma; Vi, viscidium. Scale 2 mm. (After Dressler 1981.)

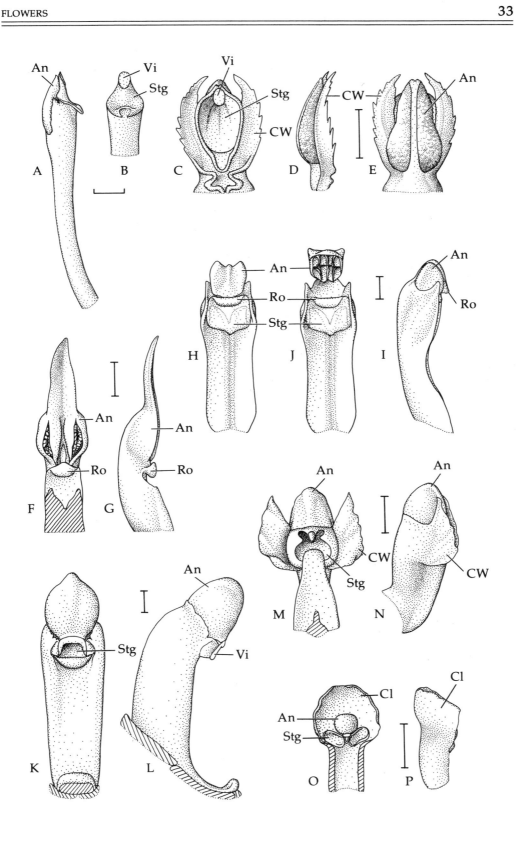

The large primordia of the Epidendreae would appear to be the primitive condition, and a similar condition would be expected in many Diurideae.

Fingerlike staminodia are taken as a derived feature of *Cephalanthera* by Burns-Balogh et al. (1987). Leaflike staminodia are frequent in the monandrous orchids, but it may be that fingerlike staminodia (like that of *Apostasia*) are primitive, even though poorly represented among living taxa.

Anther

The anther is much the same in most flowering plants, whether dicot or monocot. Basically, it is an oblong structure with four longitudinal sporogenous sacks, or locules. In *Neuwiedia*, each of the three anthers fits the classical pattern very well, though the lateral anthers are basally asymmetrical. In *Apostasia*, the median stamen is sterile or lacking. In the lady's slippers, too, the lateral anthers are fertile and the median anther is sterile or (rarely) lacking. All other orchids normally have only the median anther, and the two lateral stamens are sterile or lacking. In most Spiranthoideae, the anther is fairly narrow, but in other orchids it is usually relatively wide. In the Epidendroideae, the anther may be as wide as long or even a bit wider. In the Spiranthoideae and Diurideae, the anther is usually markedly pointed at the apex, and a similar condition occurs in some Orchideae. In the Cypripedioideae and Epidendroideae, the anther apex is normally wide and blunt. Both blunt and acute apices occur in *Neuwiedia*, so the polarity of this feature is unclear.

In the advanced orchid groups the anther may be modified in various ways. Each anther locule of the Epidendreae and related tribes may be transversely divided into two, so that eight pollen masses are formed, a condition also found in some species of *Caladenia* and *Eriochilus* (Diurideae). In this and other groups, some of the chambers may become united or suppressed, thus producing fewer pollen masses. Among primitive orchids, the longitudinal partitions that divide the anther into four cells are usually obvious. In *Coelogyne* and in the vandoid tribes, the number and size of the partitions are reduced, and their orientation within the anther quite different (Fig. 2-12). In most orchids, the anther opens toward the center of the flower or toward the surface of the column, though the anther of *Prescottia* may open laterally.

In the Spiranthoideae, Neottieae, and Diurideae, the anther is usually firmly attached to a thick filament and it remains in place after the pollinia are removed (Fig. 2-12C, E). In the Orchideae and Diseae, the base of the anther is so completely united with the column that there is no clear boundary between column and anther (Fig. 2-11G). In most Epidendroideae, the anther is lightly attached to a very short filament (the rest of the filament is part of the column) and falls away when the pollinia are removed (Fig. 2-12G–P). In the Orchideae and Diseae, the connective is often very broad and the anther cells widely separated, and in many Diurideae the bases of the anther cells are markedly divergent.

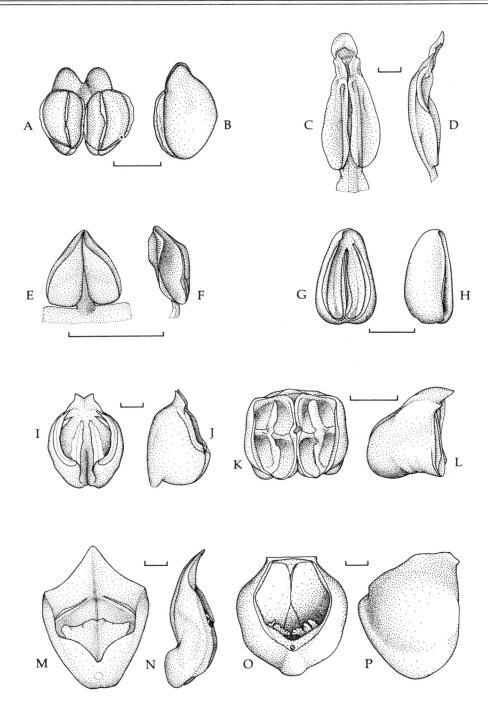

Figure 2-12. Anthers of various orchids. In each case, ventral (clinandrial) and side views are shown. (A, B) *Phragmipedium*. (C, D) *Sarcoglottis*. (E, F) *Prescottia*. (G, H) *Epipactis*. (I, J) *Sobralia*. (K, L) *Schomburgkia*. (M, N) *Cochleanthes*. (O, P) *Maxillaria*. Scale 1 mm. (After Dressler 1981.)

Anther Position

The position of the anther is traditionally considered an important feature, though there is much variation, especially in more advanced groups. In the primitive condition, the anther is erect and parallel to the axis of the column. This is the case in Spiranthoideae and most Orchidoideae. In most Diseae, the anther bends over backward, as in *Disa* and *Satyrium*. In *Satyrium* the anther actually comes to have the base uppermost and the apex pointed down toward the base of the flower. In most Epidendroideae, the anther is erect in the early bud, and then bends downward for 90 to 120 degrees (Fig. 2-13), so that it comes to rest, like a cap, on the apex of the column, or somewhat ventral in its position (*Vanilla, Coelogyne*). The terms "incumbent" and "operculate" are used for this condition (Fig. 2-11H–J). In the advanced Epidendroideae, the anther may be secondarily erect, as in *Stelis, Malaxis, Notylia, Rodriguezia*, and the Podochilinae.

In most orchids of the vandoid grade (Cymbidieae, Maxillarieae, Vandeae), the bending of the anther during development is less obvious. In 1981, I saw this as a fundamental difference between the epidendroid and vandoid grades and treated them as distinct subfamilies. Now we have more information on column development, and it is clear that Hirmer (1920) was correct in interpreting the vandoid anther as being bent like that of the epidendroid anther but bending earlier in its development. Kurzweil (1987a) has shown that there is much variation

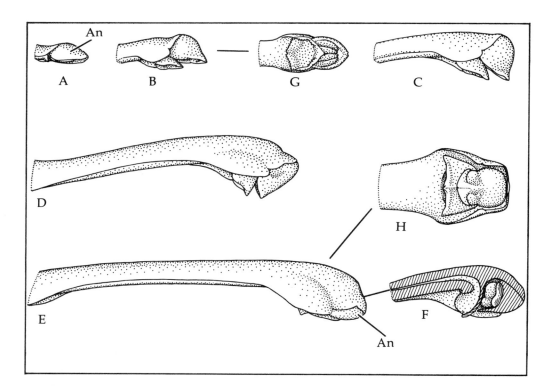

Figure 2-13. Development of the anther (An) in *Arundina graminifolia*. The anther is erect in the early stages and bends downward over the apex of the column. (After Dressler 1981.)

in the timing and that anther development cannot be taken to separate the epidendroid and vandoid grades.

The term "versatile" is usually applied to anthers that are attached in the middle and can swivel in various ways. The term has also been used for the anther of the epidendroid orchids. In these orchids the attachment of the filament is basal, or nearly so, but the anther pivots readily on its point of attachment, bringing the pollinia, or their caudicles, around to touch whatever has just brushed the rostellum (Fig. 8-1).

Acrotony and Basitony

The base of the anther is next to the stigma in the Orchideae and Diseae, and the apex of the anther is closest to the stigma (rostellum) in both Spiranthoideae and Epidendroideae. Some authors, generalizing from this, have separated the Orchideae and Diseae from all other orchids as the Basitonae (as opposed to the Acrotonae). In the Australian Diurideae, however, we find all degrees between acrotony and basitony (Mansfeld 1954); the terms "pleurotony" and "mesotony" have been used for the intermediate condition. Though both Spiranthoideae and Epidendroideae may be characterized as "acrotonic," the development of the column is quite different in the two groups.

Anther-rostellum Relationship

A median stigma lobe ventral to the anther is considered the primitive state, as the rostellum could easily evolve in this position, and both basal and terminal positions may be derived from a ventral one. A terminal relationship with an erect anther occurs only in orchids with distinct viscidia. Such a terminal relationship could scarcely function in cross-pollination without a viscidium, so it is improbable as an ancestral condition.

Burns-Balogh and Funk (1986) stress the position of the anther base relative to the stigma. Anther base at or near stigma base is associated with a terminal viscidium in every case, including several epidendroid genera (Fig. 6-1). The lower position of the anther base relative to the stigma appears to be a developmental concomitant of the terminal viscidium. When the tip of the stigma shifts upward near the tip of the anther (in both ontogeny and phylogeny), similar shifts occur in their basal relationships.

Anther Appendages

Auricles, or lateral appendages of the column, are typical of the Orchideae; these have been much debated, and some authors have considered them to be staminodia. Developmental studies show that they are dorsal appendages of the filament or the base of the anther (Kurzweil 1987b, 1990). In *Bonatea* and some other Orchideae, however, the lateral appendages appear to be compound structures, involving staminodial tissue in addition to the auricles (Kurzweil 1989b, 1991). The auricles include large cells full of needlelike crystals, or raphides, and may have a protective function. Superficially similar structures in the

Cranichideae and Diurideae are apparently appendages of the staminodia, rather than the anther, but none has been carefully studied.

Brieger (in Schlechter 1970–) notes that the epidendroid orchids often have a rooflike structure over the anther cells (Fig. 2-14B). In the Epidendroideae, a beaklike extension of the connective is prominent at the tip. This may well be phyletically homologous with the thick beak on the anther of *Triphora* and the Neottieae, which apparently functions in much the same way, loosening or exposing the pollinia when touched.

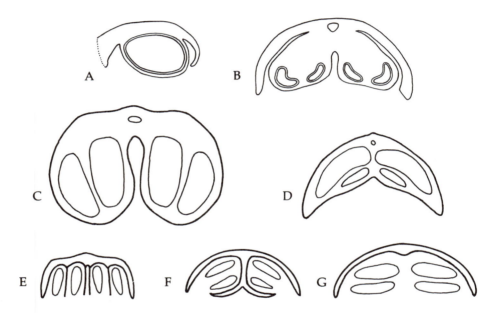

Figure 2-14. Anther structure. (A, B) Longitudinal and cross sections of young *Vanilla* anther). (C) Cross section of developing anther of *Encyclia* (caudicles not shown). (D) Cross section of developing anther of *Angraecum*. (E–G) Diagrams showing the divergence of anther cells in "vandoid" orchids. (E) Section of "epidendroid" anther with anther cells (and pollinia) parallel. (F) Section of anther of *Coelogyne* or *Zygopetalum*; the anther cells are divergent (pollinia superposed), and the inner anther walls are to be seen beneath the pollinia (parallel with the clinandrial surface in the flower). (G) Section of fully vandoid anther, such as *Vanda* or *Oncidium*, with superposed pollinia and only rudimentary walls within the anther. (A–D after Hirmer 1920; E–G after Dressler 1986a.)

Endothecial Structure

The innermost layers of the anther wall, or endothecium, usually have conspicuous thickenings in their cell walls, and their structure may be of systematic value. A survey of the orchids by Freudenstein (1991a) shows four main types of thickenings. The principal types of endothecial thickenings are as follows: Type I—tightly packed loops or helices; type II—scattered loops; type III—a circular arrangement of long anticlinal bars; and type IV—scattered bars.

Type I is found in the Apostasioideae, Tropidieae, Diurideae, and most primitive Epidendroideae. This would appear to be the ancestral state for the Orchidaceae. Type II is found in the Orchideae and Diseae and may be a uniquely derived feature of this clade. The presence of intermediates between type I and type II in some Diurideae is quite compatible with the Orchideae/Diseae as either a sister group or a derivative of the Diurideae. Type III, though distinctive, has a curious distribution, ocurring in the Cypripedioideae, Vanillinae, many Cranichideae, and some advanced Epidendroideae. It is likely that this type has evolved more than once. Type IV occurs primarily in the advanced Epidendroideae, though it occurs also in a few Cranichideae. Types I to III and the more heavily thickened variants of type IV occur primarily in terrestrial groups, and the less heavily thickened variants of type IV occur mainly in epiphytes; this last condition may be correlated primarily with firm or hard pollinia, rather than with an epiphytic habit.

Pollen and Pollinia

The predominant theme in orchid evolution may be the union of the pollen grains into pollinia. In the Apostasioideae, the pollen is powdery and no more coherent than in other Liliiflorae. In most other orchids, the pollen is at least sticky, as in *Vanilla* or *Cypripedium*. This stickiness may be due to abundant *Pollenkitt*, a sticky, semiliquid substance associated with the surfaces of all pollen grains. In the majority of monandrous orchids, and in some species of *Phragmipedium*, the pollen grains cohere into very definite masses. This aggregation of the pollen grains is adaptively related to the large number of ovules to be fertilized in the orchid ovary. The pollinia are soft in most Spiranthoideae and Orchidoideae and the primitive Epidendroideae. In the most advanced Epidendroideae, they are so hard that they may be termed "bony," and all intermediate steps between very soft and very hard pollinia can be found within the Epidendroideae. When the pollen grains are united into tetrads or larger units, the grains usually have thick exine on the outer surface of the pollinium, though there may be little or no exine developed on the surfaces between the grains within the aggregation.

When the grains are aggregated into distinct pollinia, there are several patterns. The relative abundance of different pollinia types are shown in Table 2-1. Relatively soft pollinia may be uniform in structure, as in the Cranichideae and most Diurideae, or they may be "sectile," aggregated into packets, or "massulae," as in the Diseae, Orchideae, Tropidieae, Goodyerinae, Gastrodieae, and Prasophyllinae (Fig. 7-10). Sectile pollinia may have a central core of elastoviscin

Table 2-1. Relative frequence of different pollen types. Larger percentages are rounded to whole numbers.

Free monads	0.08%
Sticky monads/tetrads	2.0%
Soft pollinia	6.6%
Brittle pollinia	0.53%
Sectile pollinia	11.0%
Firm/hard pollinia	80.0%

with wedge-shaped massulae attached at their smaller ends, as in the Orchideae or *Ludisia*. In some Goodyerinae there is a scooplike "skin" on one side of the pollinium, and the massulae are attached to the skin by a layer of elastovicin. A similar structure is shown by some Spiranthinae, though the pollinia are not sectile. This scooplike skin, possibly derived from the endothecium (P. Burns-Balogh, personal communication), is analogous to the translator of the periplocoid Asclepiadaceae, though by no means homologous. In the case of sectile pollinia, a single pollinium may leave several massulae in each of several different stigmas, thus pollinating a number of flowers.

We find the pollen grains as monads, or single grains, in the Apostasioideae, Cypripedioideae, the Vanilleae, some Diurideae, and a few Neottieae. In all other orchids, the pollen grains either remain in tetrads or are united into larger masses. Though monads are surely the primitive condition, it is quite likely that this state is secondary in some cases. Study of pollen grain structure has been neglected in the past, but the scanning electron microscope is now being used for this purpose with excellent results (Williams and Broome 1976; Schill and Pfeiffer 1977; Newton and Williams 1978; Schill 1978; Ackerman and Williams 1981; Burns-Balogh and Hesse 1988; Hesse et al. 1989). The work on the development of pollen and pollinia by Wolter and Schill (1986) is intriguing; further work along the same lines could give us features of real value in orchid systematics.

Sculpturing of Exine

In the Apostasioideae, Orchidoideae, and Spiranthoideae, the pollen grains are often heavily sculptured. Some Epidendroideae with soft pollinia also have more or less heavily sculptured pollen grains, but the more advanced members of the Epidendroideae have thick but relatively smooth exine on the outer walls of the pollinia. The pollen grains of some Vanillinae and Cypripedioideae are relatively thin-walled and have little sculpturing. The study of the exine patterns in the more primitive orchids will undoubtedly contribute useful data for classification, but it will not solve all our problems. For example, Schill and Pfeiffer (1977) consider the Neottioideae in the sense of Brieger (in Schlechter 1970-) to have rather uniform pollen structure, though that "group" may involve three distinct clades. At the same time, they find the sculpturing to be exceedingly diverse in the Orchideae, virtually the only natural group on which most authors agree.

Ultrastructure

Scanning electron microscopy gives us excellent three-dimensional images of the pollen grain surface, but transmission electron microscopy (TEM) can show the internal structure of the pollen wall. Some primitive orchids have monads, each with a single sulcus, and have a tectate-columellate or perforate exine structure with a distinct foot layer. In addition to the trend to adherence of pollen grains into tetrads, massulae, and pollinia, Zavada (1990) indicates two other important trends. One is from sulcate to ulcerate to inaperturate pollen grains, and the other shows reduction in the exine, from tectate collumellar to tectate granular or to atectate. In some of the more highly derived pollen types, there is little or no exine, even on the outer surfaces of the tetrads or pollinia. At the same time, there is a tendency to lose the foot layer and to elaborate the intine.

Elastoviscin

We read that the soft orchid pollinia are held together by "viscin" and that viscin is derived from the tapetum, but the term "viscin" has been used in several ways. Most orchid pollinia, and especially the caudicles, contain a clear, very elastic substance remarkably like rubber. With reference to the Ericaceae and the Onagraceae, the term "viscin thread" is used for strands of sporopollenin that are extensions of the outer wall (exine) of the pollen grain. As "viscin" has been used for different substances, some of them not very viscid, it might be better to abandon the term altogether. The term "elastoviscin" was suggested for the clear, elastic material (Dressler 1981). Schill and Wolter (1986) find this substance to be present in all orchid subfamilies and suggest that it is homologous with the *Pollenkitt* of other plant groups.

Cohesion Strands

Balogh (1982) reports finding sporopollenin strands in the pollinia of the Spiranthinae, but these apparently are not attached to the pollen grains. Such "cohesion strands" are seen only when the pollinia are broken open, and they are very different from the clear, elastic "viscin" of the caudicles. They may function as reinforcement in the soft pollen mass. Ackerman and Williams (1981) find straplike bands of sporopollenin connecting sister grains within the tetrads of some species of *Chloraea* and *Caladenia*.

Number of Pollinia

For the most part, the pollinia are molded by the form of the anther, and the number and shape of the pollinia reflect the form and partitions of the anther itself. In many orchids, there are four pollinia, representing the four anther cells. In many other cases, the four are more or less united into two. The two pollinia of *Cymbidium rectum* are partially united into a single pollen mass (Du Puy and Cribb 1988). The primitive number in the epidendroid phylad is apparently eight, and these may be globose, laterally flattened, or club-shaped (clavate). Within the Epidendreae, reduction to six, four, or two pollinia occurs independently in the Laeliinae and the Pleurothallidinae. In many cases, two or four rudimentary pollinia are so small as to be virtually nonfunctional. Perhaps the most improbable pollinia are those of *Brachionidium kuhniarum*, in which a pair of clavate pollinia have become united at one end in each half of the anther to form a set of hairpin-shaped pollinia (Fig. 10-1P).

Though the pollinia are usually either club-shaped or laterally flattened in the epidendroid grade, in the vandoid grade they are (when four) usually "superposed," that is, flattened parallel to the face of the clinandrium. This same orientation is found in *Coelogyne* and some species of *Sobralia*. This orientation probably evolved by a rotation of the anther cells (Fig. 2-14E–G). In the vandoid orchids, fusion of four pollinia into two is frequent, and reduction may have occurred also by the loss of one pair, as suggested by Holttum (1959). The number of pollinia is normally quite clear in orchids with hard pollinia, but in soft pollen masses, distinguishing between two deeply notched pollinia and four distinct pollinia that are in contact along one side may be difficult.

POLARITY. Powdery pollen is clearly the primitive condition and is found only in the Apostasioideae. Most living orchids have pollen that is at least sticky or coherent. Soft pollinia may be either mealy or sectile, and sectile pollinia represent a derived condition. Brittle pollinia, found only in the Cranichidinae, are an independent derivation from mealy. In the Epidendroideae, most pollinia are firm or hard, another derived condition. Firm or hard pollinia may have been derived from sectile pollinia in the Arethuseae, for example, but other hard pollinia may have been derived directly from a mealy condition.

Caudicles

When the pollinia are relatively hard or compact, there is usually a softer extension or tail by which the pollinia are attached to the viscidium or to pollinators. The caudicle functions both as a stalk and as a weak point, permitting the pollinium to break away from a pollinator and stay in the stigma. The caudicles are produced within the anther and are thus very different from the stipe, which is discussed below. There is a good deal of structural variation in the caudicles, and the terminology has been a bit confused (see especially Mansfeld 1935). In a very few cases the caudicles may be hard and bony, like the pollinia themselves. The caudicles may be granular, or made up primarily of pollen grains (fertile or abortive), or they may be clear and "hyaline," made up primarily of elastoviscin. In the latter case, the caudicles are translucent, very elastic, and lacking in cellular detail. Among the Orchideae, at least, the caudicles may be partially derived from the wall between adjacent anther cells. Elongate caudicles may be produced at either end of the anther cells. Some Arethuseae and Epidendreae produce caudicles on the ventral side of the anther cell with pollinia attached at each end. Others develop only the basal pollinia (see Fig. 8-4), so that the caudicles may be considered terminal. Elongate caudicles may be as many as the anther cells (intralocular) or may be produced between adjacent cells (interlocular). In a few cases, the caudicles are large and irregular in shape, or "massive," as in *Coelogyne*. In many other cases, they are small and are formed only where the pollinia are attached to the viscidium or stipe, as in most vandoid orchids, where the caudicle may be hidden in a slit in the pollinium. Caudicles are lacking in the Malaxideae and Dendrobieae; such pollinia are termed "naked", though viscidia occur in *Malaxis* and some Bulbophyllinae.

POLARITY. The primitive condition is clearly mealy pollinia without distinct caudicles, but caudicles may be secondarily lost, as in the Dendrobieae. The massive caudicles of *Coelogyne* are considered derived.

Stigma and Rostellum

The stigma is, as expected, three-lobed, though in most cases the median (dorsal) stigma lobe is much larger than the lateral lobes. One of the most distinctive features of the monandrous orchids is that part of the median stigma lobe, the rostellum, has become involved in pollen transfer. In the ancestral monandrous orchid, an insect backing out of the flower might brush the median stigma lobe and receive some sticky stigmatic fluid on its back, so that clumps of pollen would stick to the insect. In *Cephalanthera* and *Vanilla*, a narrow band of sticky material (distinct from the stigmatic fluid) is found along the margin of the median stigma lobe, and this may be the most primitive sort of rostellum. In genera such as *Sobralia, Cattleya*, or *Epipactis*, the distal part of the stigma projects downward, and is different in form and texture from the rest of the stigma. Here a clearly defined part of the median stigma lobe functions in pollen transfer.

Variation in stigma structure has been studied by Dannenbaum et al. (1989), who recognize three basic types. Type I is known only in the Cypripedioideae; type II occurs in the Cranichideae, Orchideae, and *Epipactis*; and type III occurs in *Cephalanthera* and the Epidendroideae. Primitive monandrous groups remain poorly sampled, so interpreting the distribution of stigma types is difficult. Type III may be a derived feature of the Epidendroideae, and type II may be the ancestral type for the monandrous orchids.

The homology of the rostellum was first recognized by Robert Brown. Darwin suggested that the entire median stigma lobe was modified to form the rostellum, an interpretation that is quite correct in some advanced Orchideae. For many years, most authors defined rostellum as the median stigma lobe, but *used* the term as "the modified portion of the median stigma lobe." Determining the limits of the stigma lobes is not always easy, and many doubtless failed to see the inconsistency. It was generally accepted that the median stigma lobe functioned only in pollen transfer, and that the fertile (receptive) surface involved only the lateral lobes. Recent workers have noted that the median stigma lobe usually includes receptive surface, and is often much larger than either lateral lobe (Mansfeld 1954; Dressler 1961; Vermeulen 1966), but Darwin's idea continues to crop up. I have preferred to redefine the term, so as to keep the traditional usage, but F. N. Rasmussen (1982) prefers to maintain the traditional definition. Given the confusion that exists, I think it better to specify "median stigma lobe" when referring to that portion of the stigma. The traditional concept of rostellum may be quite applicable in some orchids. In *Habenaria* and *Bonatea*, for example, the two fertile stigmatic surfaces are borne on long stalks that surely represent lateral stigma lobes, and the median lobe apparently has no receptive area. In the Cranichideae, there are often two stigmatic surfaces, usually assumed to be the lateral lobes. Developmental studies show, however, that these are both part of the median stigma lobe and that the lateral lobes are rudimentary or lacking (F. N. Rasmussen 1982; Kurzweil 1988). Kurzweil (1991) finds a similar condition in the Coryciinae and some related genera.

The fertile portion of the stigma usually forms a depression, or in some cases a flat or convex surface. Many Vanilleae, Arethuseae, and Coelogyneae have an "emergent" stigma—a membranous structure that projects well away from the body of the column. In *Vanilla*, the emergent stigma is clearly three-lobed; in other cases it may be an asymmetrical bowl, with the forward edge projecting much more than the rest. The emergent stigma is probably a primitive condition within the Epidendroideae. Even when not clearly emergent, the stigma of the Epidendroideae is usually deeply concave, and the forward edge, at least, projects downward from the column. Burns-Balogh and Bernhardt (1985) point out an important corollary of the deeply concave stigma. In orchids with soft pollen and shallow or convex stigmas, a portion of a pollen mass is generally sufficient to fertilize a flower, and a single pollen mass may pollinate a number of flowers. In the Epidendroideae the scoop-like structure of the stigma permits it to remove whole pollen masses from the pollinator. This is the general pattern in the subfamily, and this feature probably facilitated the evolution of hard pollinia. In the other subfamilies, the ovary has many ovules, but the number is minor, compared to that of most epidendroid orchids. In most Epidendroideae, each pollination delivers enough pollen grains to fertilize an enormous number of ovules.

Rostellar Membrane

The rostellar glue of *Epipactis* is covered by a membrane that breaks easily to expose fresh glue. In *Dendrobium*, also, the rostellar glue is usually covered by a thin membrane. In some species this has evolved into a very elegant mechanism for placing pollinia on the pollinator (see Fig. 10-9).

Scraper

In many species of *Dendrobium* there is a "scraper," that is, a nonsticky portion of the rostellum, between the stigma, proper, and the rostellar glue (see Figs. 10–8, 10–9). The scraper forms a backward-pointing shelf that scoops pollinia off the backs of departing pollinators.

Sensitive Rostellum

In *Listera*, the rostellum includes a sack, or pocket, of glue under pressure. When a pollinator touches the rostellum this sack breaks open, squirting out fresh glue onto the intruder. At the same time, the margins of the rostellum reflex, releasing the pollinia to fall onto the fresh glue droplet (Ackerman and Mesler 1979; Schick et al. 1987). The sensitive rostellum of *Listera* may have evolved from a membrane-covered rostellum like that of *Epipactis* (Dressler 1990a).

Accessory Structures

More than the pollen mass, itself, is involved in the transfer of pollinia from one flower to another. The term "accessory" is used for any structure formed outside the anther that becomes attached to the pollinia.

Viscidium

Some modified stigmatic glue is usually involved in attaching the pollinia to the pollinator. In many orchids there is a discrete "viscidium," or sticky pad, formed by the rostellum. Typically, this structure is sticky on one side and already has the pollinia attached to its other side when the flower opens. Thus, the viscidium sticks to the pollinator and is removed with the pollinia as a unit. This portion of the rostellum has been called gland, proscolla, retinaculum, or viscidium, and in recent years the last term has dominated. F. N. Rasmussen has used this term to apply to any rostellar glue, contrary to traditional usage (Dressler 1989a). Rasmussen would use the term "detachable viscidium" for the structure that I describe here (F. N. Rasmussen 1982). Schick (1988, 1989) follows Rasmussen and uses "retinaculum" for the detachable viscidium. Dressler and Salazar (1991) have suggested the term "viscarium" for rostellar glue that is not detachable as a unit.

In its development, the viscidium is at first made up of cells, but at least part of it breaks down to form a sticky glue. When the viscidium is relatively small, the entire structure may become glue at maturity, but it is clearly defined and leaves a distinct slit or cavity upon removal from the rostellum or stigma. In most cases there is a cellular, nonsticky layer between the glue and the pollinia (the scutellum of Schick 1988, 1989).

The viscidium in the tribe Orchideae and Diseae is basically in two parts, a condition occasionally found in other groups. This led Vermeulen (1959) to suggest that the rostellum of the Orchideae might be derived from the lateral stigma lobes. Developmental studies show, however, that the viscidia of the Orchideae are derived from the median stigma lobe (Kurzweil 1987b). The viscidia of the Orchideae probably evolved from an ancestor with the anther locules widely separated at the base and with a tongue-like extension of the rostellum between the anther locules. Rather than evolving a single median viscidium, this group evolved two distinct viscidia, one attached to each of the pollinia, with the rostellar extension between them. Some Orchideae have a single viscidium, but even these usually have the rostellar extension above the viscidium, thus suggesting that the single viscidium is secondary in this group. In *Orchis* and *Dactylorhiza* each viscidium is covered by a sheath, or bursicle, that is easily broken when touched, thus exposing the fresh, sticky viscidium. More delicate and less obvious membranes may occur on the viscidia of other orchids.

POLARITY. The ancestral condition was clearly an undifferentiated median stigma lobe, or no rostellum, and the primitive rostellum is simple with exposed glue. From this type several other types have been derived: rostellar glue covered by membrane, sensitive rostellum (probably derived from membrane-covered glue), and rostellum with a clearly defined viscidium (or viscidia) that is removed as a unit. This last type has evolved independently many times within the Epidendroideae.

Stipe

In many orchids one finds a nonsticky strap or stalk between the pollinia and the viscidium. When this is not formed within the anther (the caudicle), it is termed a "stipe" (the Latin *stipes, stipites* is often used). Recent work by F. N. Rasmussen (1986a) has shown that this term applies to rather different structures. In the best known case, a thin strap is formed by the epidermis of the rostellum and clinandrium (anther bed). This is often said to be rostellar in origin, but it may be stylar as often as stigmatic, so "columnar" may be a better term. This is the stipe that is readily seen in *Vanda, Oncidium,* and other advanced Epidendroideae. It is usually a single layer of cells, and Rasmussen has coined the term "tegula," or tegular stipe, for this structure. Rasmussen (1986a) notes, however, that the stipe of *Stanhopea* is several cell layers thick.

In some other orchids, a similar structure develops in a different way. In the Tropidieae, Prasophyllinae, and *Bulbophyllum ecornutum,* for example, an extension of the rostellum forms a connection between the pollinia and the viscidium. This is termed a "hamulus," or hamular stipe.

POLARITY. Each type of stipe has evolved independently in several groups, and caudicles are usually a prerequisite for the evolution of stipes (not in the Bulbophyllinae). One finds viscidia without stipes in *Cymbidium, Cryptarrhena,* and some species of *Maxillaria,* thus suggesting that the stipe may have evolved several times within the cymbidioid phylad. A hamular stipe with a sclerenchyma core is a specialization known only in the Tropidieae.

Pollinaria

The actual package carried by the pollinator includes the pollinia, the viscidium, and often a stipe as well. The term "pollinarium" has been coined for this unit: the pollinia from a single anther and all the structures that are removed with them. In many Orchideae and in some Vandeae there are two separate viscidia, each with a separate stipe or caudicle and one or more pollinia. I had suggested the term "hemipollinarium" (Dressler 1981), but Ed Greenwood (personal communication) suggests that it would be simpler to say that these orchids have two pollinaria in each flower. The pollinarium is a useful and clearly definable concept, especially with respect to orchids that have viscidia. In orchids without viscidia, however, a pollinator may remove either some or all of the pollinia.

The features of the pollinaria have long been considered of taxonomic importance; indeed, one can often identify an isolated pollinarium to genus or even to species (Figs. 9–9, 9–10). With the number and form of the pollinia themselves, and the shape and texture of both the viscidium and the stipe, the pollinarium is a fairly complex structure and includes many different features. Some workers now mount fresh pollinaria on paper triangles, like small insects, so that they may be compared without the distortion that occurs in pressed flowers.

In material preserved in liquid, unfortunately, the viscidium and the caudicles often dissolve.

The Orchidaceae are not unique in the formation of pollinia. One other plant family, the Asclepiadaceae, or milkweeds, shows an analogous structure. In the milkweeds, the pollen grains are held together in a hard, bony matrix, and the retinaculum (or translator) and corpusculum (or "gland") of the milkweeds are formed by a stigmatic secretion molded between the stigmatic head and the anthers. Further, the two connected pollinia of an asclepiad are derived from the adjacent halves of different anthers. Thus, though the two families both have their pollen grains united into pollinia, this union is achieved in quite different ways, and the associated features are very different in their origin and structure. This, and the fact that asclepiad flowers are radially symmetrical, has led to very different patterns of floral evolution in these families.

Some of the features discussed here show not only clear polarity but nearly irreversible trends. Some orchids may lose viscidia or even stipes in the case of obligate self-pollination, but this is surely an evolutionary dead end. Especially in the more advanced groups with complicated pollinary apparatus and hard pollinia, any reversal would be nonfunctional in cross-pollination.

Unisexual Flowers

Catasetum and *Cycnoches* are well known for producing male (staminate) and female (pistillate) flowers, sometimes in the same inflorescence. Nonfunctional intermediate flowers are occasionally produced as well. Taxonomists describe the related genus *Mormodes* as having "perfect," or hermaphroditic, flowers, but most species of *Mormodes* also show distinct dimorphism in the flowers (Allen 1959; Dressler 1968). The pistilloid flowers are larger, have a much broader stigma, and do not have hairs on the lip. The staminoid flowers have hairs on the lip in some species and appear to be more short-lived than the pistilloid flowers. Both these types of flower are technically, and perhaps functionally, hermaphroditic, but they would seem to be on the way to becoming unisexual. In the Catasetinae, bisexual flowers are effectively protandrous, as the stigma cannot function until the viscidium has been removed (Romero 1990). This feature, in conjunction with the somewhat sensitive rostellum, probably contributed to the evolution of unisexual flowers in this group.

We usually say that the Catasetinae are the only orchids with unisexual flowers, but Chen (1979) finds that different plants of *Satyrium ciliatum* may have staminate, pistillate, or bisexual flowers. Whether or not other species of *Satyrium* show a similar pattern is not known.

FRUIT

The structure of orchid fruits has been neglected by botanists. Fruiting specimens without flowers may be difficult to identify to species, but there are several features in the fruit that may be of value in classification. Hallé (1977), in his book on the orchid flora of New Caledonia, illustrates fruit structure for most species.

Though the ovary is basically made up of three carpels (and the receptacular tissues surrounding them), this is not obvious in the orchid fruit. Rather than splitting between carpels, the orchid capsule usually splits down the midline of each carpel; and in most orchids the midvein of the carpel, with a little additional tissue, separates from each half-carpel, so that the fruit splits into three wide valves and three narrow ones, though in some cases there are only three wide valves. The fruit usually splits from near the apex, and the edges of the wide valves may be connected by transverse fibers through which the seeds must sift. In most orchids, the valves remain attached apically, but they may separate completely and spread widely, as in *Lockhartia* and some *Maxillaria* species. Some orchid capsules split along a single line, as in *Angraecum*; in others, as in *Dichaea*, *Vanilla*, and some *Pleurothallis*, the fruit forms two unequal valves.

In *Cattleya* the wider valves each bear two prominent ridges, so that the fruit is nine-ribbed. In *Encyclia* subgenus *Osmophyta*, this same valve usually forms a single high keel, making the fruit triangular in section. Fruits may also be spiny or warty, and the fruits of many orchids are distinctly beaked, the beak often representing a floral tube or cuniculus. Beer (1863) and Malguth (1901) note that terrestrial orchids usually have thin-walled, rather dry and papery fruits, but epiphytes usually have much thicker, rather fleshy fruit walls. These authors also point out that the fruits of terrestrial orchids are almost always erect, but those of epiphytes are often pendant. In a few terrestrials, as in *Nervilia* and *Corybas*, the peduncle elongates greatly as the capsule nears maturity. The increased height doubtless aids in seed dispersal.

Orchid fruits often have long hairs interspersed among the seeds, termed "spring hairs" (*Schleuderhaare*) on the theory that they are hygroscopic and that their movements aid in dispersing the orchid seeds. Malguth (1901) compares the hairs in orchid fruits to fungal capilletia, and uses the term "capilletium" for the orchid hairs. The term "elater" is also used. Malguth finds that the hairs in the fruit of the Vandeae and a few other orchids are strongly hygroscopic, but that those of most other orchids are not markedly hygroscopic. Malguth finds the hairs to be characteristic of epiphytic orchids and nearly lacking in terrestrials, even in the terrestrial members of otherwise epiphytic genera. Hallé (1986) discusses these hairs, as elaters, and suggests that their distribution may have systematic significance.

Though 99.9 percent of the orchids have capsular fruits and shed dry seeds, a moist pulp is associated with the seeds of *Vanilla*, *Palmorchis*, and *Cyrtosia*; some species of *Neuwiedia* also have fleshy fruits. The fruits of *Selenipedium* do not split open at maturity, but whether or not they are fleshy is not clear. The presence of fleshy fruits in members of these primitive groups suggests that the group from which the orchids evolved may have had a fleshy fruit rather than a dry capsule. Also, very long, slender fruits are often found in relatively primitive orchid groups, such as *Apostasia, Selenipedium, Sobralia,* and most Vanillinae.

SEEDS

Seed structure, like fruit structure, has been much neglected. The early work of Beer (1863) gives some idea of the diversity of orchid seeds. More recently, Barthlott and his co-workers (Barthlott 1976; Barthlott and Ziegler 1981; Rauh et al. 1975) have used the scanning electron microscope to survey a larger selection of orchid seeds. Barthlott (1976) concludes that seed structure will be especially useful at the subtribal and tribal levels.

Vanilla, Apostasia, Cyrtosia, Palmorchis, Selenipedium, and some species of *Neuwiedia* have seeds with hard seed coats. Other Vanilleae, such as *Epistephium* and *Galeola,* have a hard seed coat over the embryo and a more or less developed wing around the seed. Some species of *Neuwiedia* have small seeds with sacklike appendages at each end. Most other orchids have a loose, rather papery seed coat around the embryo. The seeds range from 0.15 mm to 6 mm in length and vary a great deal in width and in structural details. Aside from the size and shape of the seeds, the size and shape of the cells in the seed coat offer features of taxonomic interest, especially in the structure of the cell walls (Figs. 2–15, 2–16). The outer cell wall, for instance, may have longitudinal thickenings, transverse thickenings, or netlike thickenings, or it may be covered with wax deposits, as in many Epidendreae.

Cotyledons

The orchids are considered members of the Monocotyledonae, but in fact the embryos and seedlings usually lack cotyledons. The early seedlings of seven genera, *Arundina, Bletilla, Dendrochilum, Encyclia, Polystachya, Sobralia,* and *Thunia,* have been reported to have at least rudimentary cotyledons (Nishimura 1991). Nishimura has studied the seedlings of four genera and finds that *Bletilla* has a definite, vascularized cotyledon. The cotyledon of *Sobralia macrantha* is somewhat smaller and lacks vascular tissues, and the cotyledons of *Arundina* and *Bletia purpurea* are present but rudimentary. Nishimura emphasizes that all of these are members of the Epidendroideae (*Polystachya* of the vandoid grade), but that no cotyledons have been reported from supposedly more primitive groups.

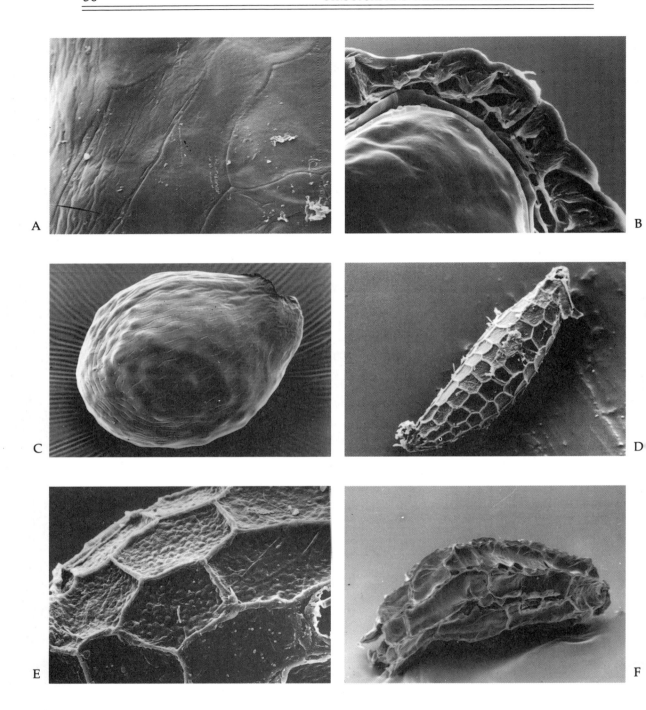

Figure 2-15. Orchid seeds with crustose seed coats. (A, B) *Selenipedium palmifolium*, surface detail (× 550) and section showing the broken testa (× 1000), note inner layer of testa cells. (C) *Vanilla planifolia*, whole seed. (D, E) *Palmorchis silvicola*, complete seed (× 200) and magnification of surface (× 550). (F) *Diuris longifolia* (*Diuris* type). Photographs courtesy W. Barthlott.

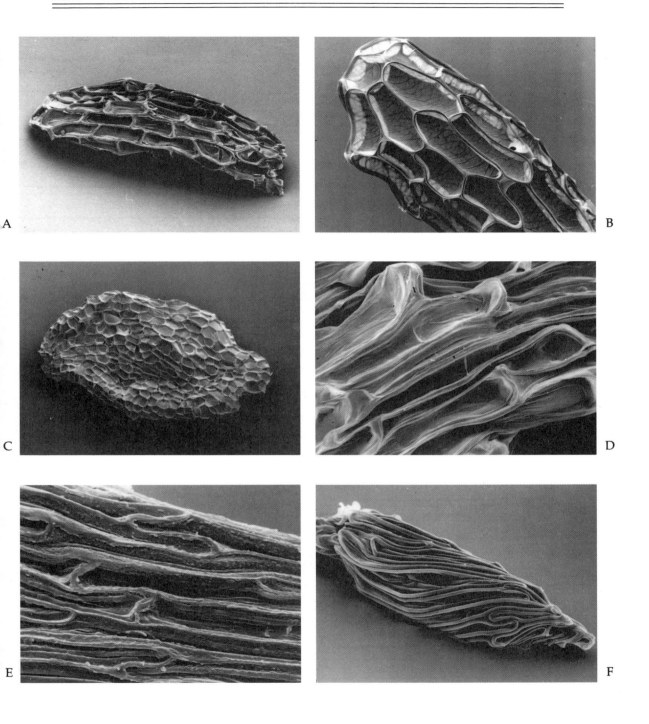

Figure 2-16. Orchid seeds: some dust seed types. (A) *Dactylorhiza iberica* (*Dactylorhiza* variant of *Orchis* type). (B) *Corallorhiza trifida* (*Eulophia* type). (C) *Stanhopea* sp. (*Stanhopea* type). (D) *Bletia purpurea*, surface detail (*Bletia* type). (E) *Encyclia sceptra*, surface detail (*Epidendrum* type). (F) *Vanda testacea* (*Vanda* type). Photographs courtesy W. Barthlott.

Summary of Seed Structure

A good survey of orchid seed structure has been done by B. Ziegler (unpublished), and Dr. Wilhelm Barthlott has kindly sent me information on a number of genera that were not available during the thesis work. As this work is not yet published, I summarize the seed types recognized by Ziegler, with a few modifications. I treat his *Chondrorhyncha* type as a variant of the *Maxillaria* type. The *Vanilla*, *Apostasia*, *Neuwiedia*, and *Elleanthus* types were not treated as distinct types by Ziegler, nor assigned to any of the types he described. They are treated as distinct types here, to facilitate discussion. When possible, published illustrations of each type are cited.

1. *Vanilla* type. Seeds slightly flattened or lenticular; 400–600 μm long, about 350 μm wide; brownish black; outer periclinal and anticlinal walls thickened and crustose; cells thus tabular (Fig. 2-15C; Barthlott 1976, Fig. 4).

2. *Galeola* type. Seeds winged, 1.0–3.5 mm across; brown or dark brown; testa of central sector of seed coat appressed to embryo, like *Vanilla* type in structure; wing cells like *Limodorum* type (Barthlott 1976, Fig. 1c; Barthlott and Ziegler 1981, Fig. 2; Dressler 1981, Fig. 3.31A; Rauh et al. 1975, Fig. 6).

3. *Apostasia* type. Seeds globular or ovoid; about 300 μm long; about 6 cells long and wide; dark brown; inner periclinal and anticlinal walls thick and crustose; outer periclinal walls collapsed, the seed coat thus pitted (Barthlott 1976, Fig. 1g; de Vogel 1969, Figs. 4o, 7h).

4. *Neuwiedia* type. Medial portion of seed coat resembling that of *Apostasia*, base and apex of seed coat inflated, balloonlike, shiny, and light brown; seed about 600 μm long; cell walls of balloon sectors thin (de Vogel 1969, Fig. 4r).

5. *Disa* type. Kernel or dust seeds 200–500 μm long; brown or blackish brown; the medial testa cells are elongate, the basal and apical cells isodiametric; the cells are troughlike in cross section (the outer periclinal wall completely collapsed), the inner and anticlinal walls are crustose (Barthlott 1976, Fig. 1d).

6. *Diuris* type. Very similar to the *Disa* type (they are combined by Ziegler), but with prominent intercellular spaces (Fig. 2-15F).

7. *Orchis* type. Dust seeds, light to dark brown; the medial cells distinctly elongate, the medial sector 1–2 cells long; the terminal cells more or less isodiametric; distinct parallel or reticulate thickenings are present on the anticlinal walls (Barthlott 1976, Fig. 1b; Tohda 1983).

7a. *Habenaria* variant. Similar to the *Orchis* type, but without thickenings on the walls. May be considered a transition to the *Disa* type.

7b. *Dactylorhiza* variant. Without thickenings on the walls; the testa cells very shallowly troughlike in cross section (Fig. 2-16A).

8. *Goodyera* type. Generally dust seeds; seed about 5 cells wide; cells about the same size throughout, isodiametric or slightly elongate; intercellular spaces are prominent, especially at the cell corners, which may be rounded (Tohda 1985).

9. *Limodorum* type. Colorless or brownish balloon seeds about 1.0 mm long and 0.1 mm wide; the cells are of the same size throughout, isodiametric or slightly elongate; the seeds are about 15–30 cells long and 10 cells wide, thus relatively many-celled (Barthlott 1976, Fig. 5; Tohda 1986).

10. *Gastrodia* type. Dust seeds, or usually thread seeds, about 1.5 mm long, 10–100 μm wide; nearly white to light brown; cells similar and more or less isodiametric; reticulate thickenings are always present; intercellular spaces or covered cell-border (as in *Limodorum* type) are never present.

11. *Elleanthus* type. Dust-like kernel seeds about 200 μm long. Basal and apical testa cells slightly elongate, medial cells strongly elongate and twisted, deeply troughlike in cross section, with clear cell-border ridges; periclinal walls with strong, longitudinally reticulate thickenings.

12. *Epidendrum* type. Seeds oblong to elongate dust-like, 500–1000 μm long, usually pale yellow; all cells similar and distinctly elongate; cell corners are not rounded, but prominently acute-angled; anticlinals high, narrow, and sharp-angled; cell border not visible; with or without anticlinal thickenings; may be verrucose (Fig. 2-16E).

12a. *Epidendrum secundum* type. A balloon-thread seed up to 6.0 mm long, thus among the largest known orchid seed; the medial sector inflated, about 300–400 μm wide, the cells of the medial sector isodiametric.

13. *Pleurothallis* type. Dust seeds from 150 to 300 μm long; yellowish or brownish; the testa cells are all of the same size and distinctly elongate; seeds almost always 2–3 cells long; flat marginal ridges are present and always over-topped by a distinct cell-border ridge; the anticlinal walls have prominent thickenings; the seeds may be covered by warts or scales that are easily soluble (Rauh et al. 1975, Fig. 2).

14. *Dendrobium* type (*Coelogyne-Dendrobium* type of Ziegler). Short or oblong dust seeds, with a total length of 300–500 μm; yellow to orange in color, some-times yellowish green or red-orange, rarely brownish; the testa cells all of the same size and quite elongate; the transverse anticlinal walls are strongly bowed, so that the ends of the cells are broadly rounded; the surface of the seed is always dull and velvety, that is, covered by very fine warts; larger warts may also be present.

15. *Eulophia* type (*Corallorhiza-Eulophia* type of Ziegler). Club-shaped balloon seeds to thread seeds; whitish to light brown; the testa cells are distinctly elongate; intercellular spaces are never present; the cell border is never covered by a folded "cuticle," and may be both raised and sunken in the same seed; it may be present only as a small three-rayed star at the cell corners; definite thickenings are present on both anticlinal and periclinal walls; the marginal ridge may be warty (Fig. 2-16B; Barthlott 1976, Figs. 3, 6; Barthlott and Ziegler 1981, Fig. 4).

16. *Bletia* type. Dust seeds to long balloon seeds, sometimes almost threadlike; the testa cells are flat trough-shaped in cross section, show no thickenings, and are always elongate and compressed laterally, so that the trans-verse anticlinals are bowed and project outward; resembles the *Limodorum* type (Fig. 2-16D; Barthlott 1976, Fig. 1e).

17. *Cymbidium* type. A very distinct seed type; always small balloon seeds, from 500 to 1000 μm long; white, with the yellow embryo visible through the testa; the testa cells are polygonal or slightly elongate; the cell border is not visible; the cell corners are strongly raised, and each is covered with a small or large wax hood; the anticlinal and periclinal walls are densely covered by parallel, longi-tudinal thickenings (Dressler 1981, Fig. 3.31E; Rauh et al. 1975, Fig. 3).

18. *Maxillaria* type. Dust seeds or oblong dust seeds, 250–500 μm long; yellowish to brownish; the terminal testa cells are isodiametric or nearly so; though the cells of the medial sector are strongly elongate, this sector usually not more than 2 cells long; a marginal ridge is present and is round in section; the cell-border ridge is fine and may be concealed by the marginal ridges; the periclinal walls have prominent reticulate or longitudinal thickenings; fine warts or micropapillae may be present (Chase and Pippen 1988, Figs. 5, 6; Dressler 1981, Fig. 3.31D).

18a. *Chondrorhyncha* variant. Substantially smaller than the typical *Maxillaria* type, about 250 μm long, or only 3–4 cells long; the marginal ridges are weakly developed, and the periclinal walls show dense, reticulate thickenings.

19. *Stanhopea* type. White or brownish balloon seeds more than 500 μm in diameter; both the basal and apical sectors may project as tiny stalklets; the testa cells in the balloon-like medial sector are all of the same size and isodiametric-polygonal, but the terminal cells are more or less strongly elongate; the testa cells are not or are incompletely collapsed, so that the outer periclinal wall stretches over the cell lumen; the anticlinal walls are reticulate; a smooth marginal ridge is present, as are cell border ridges (Fig. 2-16C).

20. *Vanda* type. Dust seeds or oblong dust seeds; may be yellowish, but usually brown or blackish brown; length from 300 to 500 μm; the testa cells are always so strongly elongate that the longitudinal anticlinal walls are in contact; the surface of the seed, then, is made up of marginal ridges; a cell border ridge is present but may be covered by the marginal ridges; periclinal thickenings are not seen; the surface may bear either micropapillae or warts. The three variants listed below all appear to be adaptations found in twig epiphytes (Chase 1987a, Chase and Pippen 1988) (Fig. 2-16F; Barthlott 1976, Fig. 1a; Chase and Pippen 1988, Figs. 23, 25).

20a. *Gomesa* variant. Similar to the *Vanda* type, about 400 μm long; the transverse anticlinals are strongly arched and project out from the seed surface.

20b. *Thrixspermum* variant. Substantially larger than the *Vanda* type, 600–1200 μm, with unicellular trichomes, especially at the ends of the seed; these may be capped by micropapillae that may be hooklike (Barthlott 1976, Fig. 2; Rauh et al. 1975, Fig. 7).

20c. *Oncidium variegatum* variant. A distinctive type or subtype; yellow; a cell border ridge clearly visible; the strong, rounded marginal ridges are covered by verrucose sculpturing; this seed type also bears stalked micropapillae and biseriate trichomes (Barthlott and Ziegler 1981, Fig. 3; Chase and Pippen 1988, Figs. 7, 8, 10).

21. *Vanda-Maxillaria* transition form. This category includes seeds that combine the features of the *Vanda* and *Maxillaria* types. Some seeds approach one or the other type more strongly, thus supplying a complete series of intermediates.

CHEMICAL DATA

The chemical compounds produced by plants may be used as evidence concerning their relationships. A number of different classes of compounds may be studied. Hegnauer (1963) and Gibbs (1974) summarize our chemical knowledge of the orchids. It is clear that the family has not been very well sampled.

Alkaloids

In the orchids, the alkaloids are probably better known than any other class of compounds. Reviews of our knowledge of orchid alkaloids are offered by Lüning (1974) and Slaytor (1977). Alkaloids are especially frequent in the Cryptostylidinae, Dendrobiinae, Malaxideae and Vandeae, though significant alkaloids are also found in at least a few species of most other groups. To a nonchemist, it appears that each group has its own distinctive alkaloids, but there are also a few alkaloids that are found in other families. On present knowledge, then, alkaloids might give us useful information on relationships within tribes or subtribes but do not seem to tell us much about the relationships between major groups.

Flavonoids

The leaf flavonoids of a number of orchids have been surveyed by C. A. Williams (1979). She finds that the Neottieae, Spiranthoideae and Orchidoideae usually have flavonol glycosides, and the Vanillinae and advanced Epidendroideae usually have flavone C-glycosides, with some exceptions in each group. Williams indicates that the flavonoids of the orchids do not suggest a relationship with the Liliaceae, but point rather to a possible relationship with the Commelinaceae, Iridaceae, or Bromeliaceae.

Other Classes

A good deal is known about the aromatic and terpenoid constituents of orchid fragrance, especially in the orchids that are pollinated by male euglossine bees (Williams and Whitten 1983), but these compounds clearly do not reflect phylogenetic relationships. Rather, distantly related species that are pollinated by the same bee species may have nearly identical fragrances.

Some work has been done on the anthocyanins of orchids (Arditti and Fisch 1977). Again, these compounds may sometimes reflect pollination syndromes rather than phylogenetic relationships.

Gibbs (1974) found no evidence of cyanogenesis in orchids. Pinto and Costa (1988) give evidence of seasonal cyanogenesis in a number of orchids, but its significance may be more ecological than systematic.

MOLECULAR EVIDENCE IN SYSTEMATICS

Technology now permits the literal reading of genetic code. Molecular data may be used in systematics in several ways. In general, these techniques are costly in time and equipment, but they may give very valuable data. They are especially valuable when there is much parallelism, or when the data from the analysis of structure is inadequate or conflicting. In such cases, data on molecular details may greatly improve our hypotheses about phylogeny or may help to choose between conflicting ideas. In many cases, data from molecular studies may agree with one of the hypotheses based on morphology. When the two kinds of evidence conflict, neither one should be considered "right," but data of other types should be sought before any reasonable hypothesis is rejected (Doyle 1987). Donoghue and Sanderson (1992) emphasize that all available information should be used in determining relationships, and that the current popularity of molecular systematics should not be allowed to result in one-character taxonomy.

Chloroplast DNA

Genetic sequences may be studied in nuclei, ribosomes, or mitochondria, but in plants, the genetic material of chloroplasts is especially favorable for such study (Palmer 1987; Palmer et al. 1988). Chloroplasts of land plants have their genetic material arranged in a circular genome (sometimes incorrectly called a chromosome) with about 120 genes and made up of 120 to 217 kb (kilobase, or 1000 units) of genetic material. In most plants, there is an "inverted repeat" segment in the ring and, curiously enough, the repeated segments do not change independently. Chloroplasts are easily extracted in quantity, and the chloroplast DNA is relatively conservative (especially in the inverted repeats), that is, it does not evolve rapidly, so distantly related plants may still be compared. In practice, separate lots of the ring genome are broken into pieces enzymatically, by each of 10 to 20 different "restriction endonucleases." The sample from each of these is separated by gel electrophoresis, and may be transferred to a filter and "hybridized" with purified chloroplast DNA of known structure marked with radioactive phosphorus. Thus, the homology of the fragments under study may be determined, and the restriction sites (breaking points) may be mapped. Certain genes are especially well known, and are often used for comparison. One of the most important is *rbcL*, a large subunit of *rabisco*, that produces the carbon dioxide fixing enzyme.

Changes in the DNA include both point mutations and rearrangements, including inversions, deletions, and insertions of genetic material. Noncoding deletions or insertions in the "spacers" between genes are generally not used in such analyses, as they are highly variable and homology is difficult to determine. Aside from this, the mutations may be compared in different taxa and are analyzed by cladistic procedures, using parsimony. Major rearrangements of the genetic material are rare, but where they occur, as in the Asteraceae, they can be quite important in determining phylogeny (Palmer et al. 1988).

Mark Chase and his colleagues are using chloroplast DNA to study relationships within the Orchidaceae and to determine the relationships between the orchids and other monocotyledon families (Albert et al., in press; Chase and Palmer 1988; Palmer et al. 1988). Already, they have clarified the relationships of and within the Oncidiinae and have clarified relationships among the lady's slipper genera. One hopes that their work will clarify relationships at the subfamily level.

Enzyme Electrophoresis

Enzyme electrophoresis allows the treatment of many samples in a short time and is especially useful in comparing species and closely related genera (Buth 1984). Enzymes are extracted and separated by starch gel electrophoresis, and the number and types of enzymes may be compared between different samples. Enzyme electrophoresis has been used by Schlegel et al. (1989) to compare several species of *Orchis*, *Dactylorhiza*, and *Gymnadenia*. Further use of this technique should lead to a better classification for this confusing group.

Chase and Olmstead (1988) have used enzyme electrophoresis to show that chromosome evolution in the Oncidiinae has been primarily through aneuploidy, rather than polyploidy. Polyploids should have a greater number of isozymes than diploids, but the number of isozymes is quite stable in the subtribe, thus indicating that plants with higher numbers are not polyploids.

3

Orchid Phylogeny

S everal aspects of orchid phylogeny, the probable evolution of some features, and the evolution and relationships of the subfamilies are discussed here; evolution within subfamilies is discussed in the corresponding chapters.

The fossil record of the orchids is so scanty and brief that it tells us nothing significant about orchid evolution (Schmid and Schmid 1977; Mehl, in Senghas and Sundermann 1986). Wolter and Schill (1985) find that many orchid pollinia are destroyed by acetolysis, and suggest that fossils of sectile pollinia might be recognized as orchidaceous, but that other types of pollen grains might show no clearly orchidaceous features even if they survived preparation.

WHENCE THE ORCHIDS?

One of the first questions under phylogeny must be, "To what other group of plants are the orchids most closely related?" or, "What is the proper outgroup to compare with the Orchidaceae?" The orchids seem most closely related to something that would once have been classified in the Liliaceae, or possibly the Amaryllidaceae. These two families have been broken up into many, however, and the former Liliaceae are now placed in several orders or superorders. In part, this may be rampant splitting, but it also reflects the real diversity of the lilylike plants that were classed together because all have six tepals and six stamens. In some classifications, all are treated as the superorder Liliiflorae, and this inclusive group should be the logical place to seek the orchids' closest relatives. Unfortunately, there is little agreement on the proper classification of these plants, and this makes our task no easier.

Huber (1969) outlined two major complexes in the lilylike plants: the colchicoid group, including the Liliaceae in the narrow sense, and the asparagoid group. These are treated as the orders Liliales and Asparagales by Dahlgren and his co-workers. There are few if any clear derived features that consistently separate the Liliales and Asparagales, and there is little agreement on the exact delimitation of these orders.

Dahlgren and Clifford consider the orchids most closely allied to the Liliales and cite a number of features in support of this view (Dahlgren and Clifford 1982, p. 307). Several of these features are secondary features within the Orchidaceae; they are features found only in quite derived groups, and they almost certainly evolved independently within the orchids. The other features cited are all quite as prevalent in the Asparagales as in the Liliales. Dressler (1983b), using Huber's comparison of asparagoid and colchicoid groups (1969), argued that the features of the more primitive orchid groups point to an alliance with the Asparagales (see Table 3-1). More recently Dahlgren et al. (1985) include the orchids (as three distinct families) in the Liliales. They also include the Alstroemeriaceae in the Liliales, though this family shows most of the asparagoid features that occur in primitive orchids.

Of the extant plant families that might be compared with the orchids, the Alstroemeriaceae, Philesiaceae, and Convallariaceae (the latter both included in Asparagales by Dahlgren and co-workers), especially, show a number of features suggestive of a phyletic relationship with the Orchidaceae (Table 3-2). If the remote ancestor from which the orchids evolved were living today, one might expect it to show simultaneous microsporogenesis and a bracteate raceme, similar to that of *Palmorchis* or *Selenipedium*. Frölich and Barthlott (1988) find characteristic *Convallaria*-type epicuticular waxes to occur in most Liliales and Asparagales, but to be absent in all of the orchids sampled. They suggest that this might be taken to support treating the orchids as a separate order, the Orchidales, but this gives no clue to the orchids' closest relatives.

This is the sort of problem that is difficult to solve by morphological analysis, alone. It is likely that molecular systematics will give us more dependable data on

Table 3-1. Comparison of the Orchidaceae with the Asperagales and the Liliales.[a]

Feature	Orchids	Asparagales	Lilales
Shrubby habit	+P	+	−
Fleshy roots	+P	+	−
Velamen	+	+	*Lilium*
Alisma type leaf	+P	+	−
Reticulate venation	+P	+	−
Abundant raphides	+	+	few
Subsidiary cells by oblique division	+	+	−
Anther introrse	−	−	some
Simple style	+	+	few
Abscission at tepal base	+	+	−
Fleshy fruit	+P	+	few
Inner seed coat collapsed	+P	+	−

[a]+, present; −, absent; P, features primitive within the Orchidaceae.

Table 3-2. Distribution of features considered primitive for the Orchidaceae in Alstroemeriaceae, Philesiaceae, and Convallariaceae.[a]

Feature	Alstroemeriaceae	Philesiaceae	Convallariaceae
Fleshy roots	+		+
Velamen			+
Raphides	+	+	+
Shrubby or viny	+	+	+
Alisma type leaf	−*	+*	+
Reticulate venation	+	+	+
Filaments united	−	+	1
Anthers basifixed	+	+	+
Anthers introrse	+	+	+
Ovary inferior	+	−	−
Ovary three-locular	+	−	+
Stigma simple	+	+	+
Stigma wet	+	+	−
Fruit fleshy	−	+	+
Perianth caducous	+	−	+

[a]+, present; −, absent; 1, in one genus only; * refers to leaves that are twisted at the base (not found in Orchidaceae).

the relationships of the monocot orders and families. A preliminary analysis by Chase et al. (in press) suggests that the Liliiflorae are paraphyletic, but that the Asparagales and Liliales may, with some revision, become monophyletic groups. Their analysis indicates that *Alstroemeria* is a member of the Liliales, but that the orchids are a subgroup of the Asparagales. Of the nonorchid genera sampled in their study, *Hypoxis* is the most closely allied to the Orchidaceae. It must be emphasized, though, that more groups need to be sampled.

The Burmanniaceae and Corsiaceae, both largely saprophytic, are often placed near the Orchidaceae, partly because of their very small seeds, but this feature is to be found in virtually all saprophytic groups, whether monocotyledons or dicotyledons. Rübsamen (1986) concludes that these families may be related to the orchids, but not to the Apostasioideae, a conclusion that seems to be based on features that are general in saprophytic (or mycotrophic) groups and are probably parallelisms. Chase et al. (in press) find *Burmannia* to be allied to *Vellozia*, but not closely allied to the orchids.

Though the exact relationships of the Orchidaceae within the Liliiflorae may require further study, a generalized Liliiflorean flower may be taken as the outgroup to polarize most floral features. For vegetative features, plants such as *Cypripedium*, *Epipactis*, and *Palmorchis* are taken to represent the primitive condition (Dressler 1983b).

Some Primitive Features in the Orchidaceae

Some features are clearly primitive by out-group comparison or by comparative anatomy. Other features are strongly correlated with these in several different groups, and we may reasonably suspect these other features of being primitive. There are several such features in the primitive orchid groups (see Table 3-3; Dressler 1983b).

1. Branched stems. Irregular branching above the rhizome is not common in sympodial orchids, though the stems may be regularly superposed in a few genera, such as *Otochilus* and *Scaphyglottis*. Among relatively primitive groups, branching occurs in *Apostasia, Selenipedium, Tropidia,* the Vanillinae, *Elleanthus, Helleriella,* and *Xerorchis*. Such branching may have been a feature of the ancestral orchids. At the same time, such branching is much more probable in plants with long, slender stems.

2. Net-veined leaves. We find this feature in *Epistephium* and *Clematepistephium,* of the Vanillinae. Such a feature is considered relatively primitive in *Smilax* and its relatives, and it might be a primitive feature in the Vanillinae. Similar veining is also to be found in a few Cranichideae, such as *Pachyplectron* and some Goodyerinae, and a few Diurideae, especially *Acianthus* and *Corybas*.

3. Climbing habit. A truly vinelike habit is restricted to the Vanilleae among the orchids, and may or may not be primitive for the family.

4. Three-locular ovary. This is clearly a primitive feature in the orchids.

5. Fleshy fruit. Fleshy, indehiscent fruits are found in *Cyrtosia, Palmorchis,* and *Rhizanthella* and in some species of *Neuwiedia*. Fruits of *Vanilla* and *Selenipedium* may be leathery rather than fleshy, and the fruit of *Vanilla* is usually dehiscent.

6. Crustose seed coat. A hard, or crustose, seed coat is found in *Apostasia, Cyrtosia, Palmorchis, Selenipedium, Vanilla,* in some Orchideae and Diurideae, and in some species of *Neuwiedia*. This is discussed at greater length under seed structure (Chapter 2).

7. Abscission layer below perianth. The perianth and column absciss, or drop off the ovary in *Neuwiedia, Paphiopedilum, Phragmipedium,* the Vanilleae, and Pogoniinae. This may be comparable to the abscission layer that is frequent at the base of the perianth in asparagoid Liliiflorae. Such a feature is not usually associated with an inferior ovary, but does occur with an inferior ovary in the Alstroemeriaceae.

8. Gregarious flowering? Short-lived flowers occur in various orchid groups, but gregarious flowering (short-lived flowers all opening on the same day) occurs in the Epidendroideae, including some quite primitive genera. This may well be a primitive feature in the monandrous orchids. It occurs in the Palmorchideae and Triphoreae and sporadically in other tribes and subtribes: *Bromheadia,* some species of *Nervilia,* Dendrobiinae, Sobraliinae, and Vandeae.

Table 3-3. Distribution of primitive features in different orchid groups.[a]

	1	2	3	4	5	6	7	8	9	10	11	12	13
Apostasioideae	+	+	+	+	+	+	+	+	+	+	+	+	+
Cypripedioideae	+	+	+	+			+	+	+	+		+	+
Spiranthoideae	+	+	+										
Diurideae					+		+	+				+	+
Neottieae	+	+	+	+			+	+				+	+
Vanilleae	+	+	+	+			+		+	+		+	+

[a] The features are: 1. slender, but fleshy roots; 2. stems slender and rather woody; 3. leaves spiral, convolute and plicate; 4. inflorescence terminal, with leafy bracts; 5. perianth parts free and similar; 6. three fertile anthers; 7. pollen as free monads; 8. filaments united to each other (staminal column); 9. abscission layer between ovary and perianth; 10. ovary three-locular; 11. stigma with equal and similar lobes; 12. fruit fleshy; 13. seed with a hard crustose coat.

What Sort of Plants Were the Ancestral Orchids?

Based on the features discussed above, as well as more extensive discussions by Rolfe (1909–1912), Dressler and Dodson (1960), and others, we may make a fairly good estimate as to the appearance of the primitive orchids.

Primitive orchids were surely sympodial, that is, they had rhizomes in or on the soil; erect shoots arose from the rhizome, the basal portion of each new shoot adding a new segment to the rhizome. The roots were surely fleshy, but probably not very thick, though they may have had some thickened nodules, as in *Apostasia* or *Tropidia*. The stems were relatively slender, may have been branched, and may have been a bit woody (hard, or tough, but without secondary thickening). They were probably erect, though some may have been climbing.

The leaves were probably spiral and plicate and might have been petiolate and net-veined. The inflorescence was usually terminal and may have been a bracteate raceme. The flowers had inferior ovaries that were divided into three chambers, or locules. At some point in their evolution, the three stamens on the lower (actually adaxial with respect to the inflorescence axis) side of the flower were lost. The filaments of the remaining anthers may have been partly united among themselves, but were not united with the perianth. The anthers were more or less basifixed and introrse (opening toward the center of the flower), with powdery or slightly sticky pollen. The fruits were fleshy, and the seeds, though small, had a hard seed coat.

THE PROBABLE EVOLUTION OF SOME FEATURES

Corm Versus Reed Stem

Intuitively, one would expect a slender stem to be the ancestral condition. Nevertheless, the "reed-stem" state of the Epidendreae and of *Bromheadia* is considered a secondary condition, possibly derived from the cormous pattern (but see also Figure 5-1 and discussion in Chapter 8).

Thickened stems are rare in other subfamilies but are frequent in the Epidendroideae. Thickened stems occur in some saprophytic orchids, and such thickenings (in seedling stages) may have been the starting point for the evolution of corms and pseudobulbs. In the reed-stem habit of the advanced Epidendroideae, the segments of the rhizome are commonly thickened, unlike the reed-stem habit in other subfamilies (or in the Neottieae, Palmorchideae, Triphoreae, and Vanilleae). One could visualize the evolution of the advanced reed-stem habit directly from the primitive reed-stem habit, from ancestors with the less clearly defined thickenings of some saprophytes, or from cormous ancestors (Fig. 3-1). It is also possible that the reed-stem state evolves secondarily from the pseudobulbous state. John Atwood (personal communication) suggests that this might occur through the retention of a seedling growth form. The hypothesis that the reed-stem habit of the Epidendreae, for example, is secondary or derived is based on: (1) the seeming close relationships of reed-stem and cormous groups that share both eight pollinia and the *Bletia* seed type; (2) the presence of corms in several primitive groups, such as *Nervilia* and *Arethusa*; and (3) the repeated evolution of pseudobulbous habit from reed-stem habit (or vice versa).

The secondary reed-stem habit is thought to be ancestral within the epidendroid phylad, except for the Arethuseae and Coelogyneae. If this is correct, the evolution of pseudobulbs occurs independently in all of these tribes except the Vandeae. Among highly derived groups, the thickened rhizome of the primitive Epidendreae may be secondarily lost in creepers, climbers, or twig epiphytes without pseudobulbs.

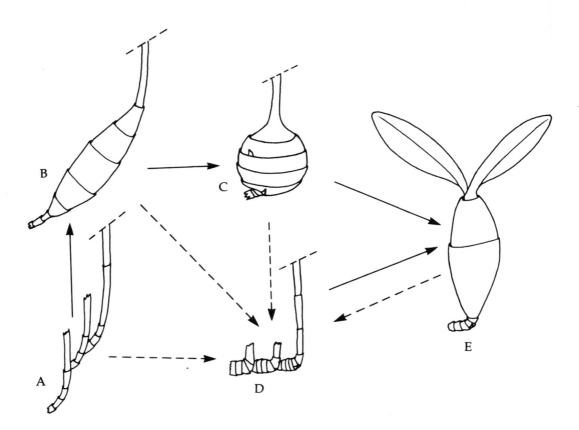

Figure 3-1. Hypothetical derivation of the reed–stem habit from either a cormous habit or a saprophytic stage with the stem basally thickened. (A) Slender stem without thickening. (B) Irregular basal thickening of saprophyte stem, as in Gastrodieae. (C) Cormous stem, as in Bletiinae, etc. (D) Reed–stem with thickened rhizome, as in Epidendreae. (E) Pseudobulb. (After Dressler 1990c.)

Nectaries

Nectaries are often found on the lip, but may also occur on sepals or petals. This has been cited to support a phyletic relationship between the orchids and the Liliales in the strict sense, for the Asparagales usually have septal nectaries, nectaries borne in the ovary wall. Note, however, that tepaline nectaries do occur in a few Asparagales. The real question, "did the common ancestor of the Orchidaceae have septal nectaries, tepaline nectaries, or no nectaries?" is not easily answered. Nectaries appear to be quite lacking in both the apostasioids and the lady's slippers. Among the Australian Diurideae, Beardsell and Bernhardt (1983) find only three genera, *Acianthus*, *Microtis*, and *Prasophyllum* (then including *Genoplesium*), that offer nectar on the lip. All other genera achieve pollination through some sort of false advertisement. Some of the Neottieae have nectaries, but others attract pollinators through deceit. We have, unfortunately, little information about the distribution of nectaries in the other primitive Epidendroideae.

The frequent occurrence of primitive orchids without nectaries might be taken as evidence that nectaries are secondarily derived within the orchids. If the ancestral orchids lacked nectaries, this could explain in large part the prevalence of pollination based on deceit throughout the family (Ackerman 1986a,b; Dafni 1984). Once the orchids evolved very sticky or coherent pollen masses, they could not use pollen to attract foraging bees. If they had no nectaries, they could survive only by developing deceitful methods of pollination, "reinventing" the nectary, or evolving entirely new means of attraction, such as "pseudopollen," stigmatic fluid, waxes or resins produced on the lip, or the use of perfume as both advertisement and reward, as in euglossine pollination. A colleague argues (in personal communication) that flowers without nectar are usually derived from nectariferous ancestors, but that does not tell us the status of the ancestral orchids. It is clear, I think, that nectariferous flowers may evolve from ancestors without nectaries, even if this is not the most frequent pattern. Current analysis of orchid relationships based on chloroplast DNA (see Chapter 2) indicates that the family most closely allied to the orchids may be the Hypoxidaceae, which apparently lacks nectaries (Dahlgren and Clifford 1982). Thus, outgroup comparison favors the primitive absence of nectar in the Orchidaceae, though the pattern of relationships may change when more asparagoid families have been analyzed.

Evolution of Bilateral Symmetry

The ventral stamens are not recognizable in normal orchid flowers, though staminodia or even fertile stamens may be present in self-pollinating forms and other "abnormalities." Kurzweil (1987a,b, 1988, 1990) finds apparent primordia of the ventral stamens in the buds of several orchids, and these may form part of the column or the column foot. We can only speculate about the earlier stages in orchid evolution. At one time there may have been orchid-like Liliiflorae with five stamens or with three large and two small stamens. I have suggested (Dressler 1981) that the loss of the ventral anthers was analogous to the loss of one or three anthers in the dicotyledonous Scrophulariales. With bees or wasps regularly entering the horizontal flowers to seek nectar near the base, the lower anthers would be ineffective in pollination, and even interfere with the function of the upper anthers. Thus, any flower with the lower anthers reduced would be more effective, and selection would favor the loss of these anthers. This scenario assumes that the early orchids had a nectary of some sort near the base of the flower.

Whatever the factors that led to the loss of the lower anthers, the early orchids with three anthers on the upper side of the flower differed from their dicotyledonous analogs in an important feature. One single anther was directly over the stigma; in the dicotyledons, on the contrary, a pair of anthers share that position, each a bit to one side. Thus, in the orchids it was easy for the median anther to take on the major role in depositing pollen on the pollinator, just where it would be most effective in pollinating other flowers. Most modern orchids are "monandrous," that is, have only the single, median anther.

Evolution of the Column

In all orchids there is some degree of union between the style and the filaments or staminodia, forming a "column" or "gynostemium." In most orchids, these structures are united up to the base of the anther. It is generally accepted that among the monandrous orchids, the most primitive column structure is to be found in the "neottioid" grade, the orchids with the anther erect (parallel with the axis of the column) or nearly so and extending beyond the stigma (see Fig. 3-2). In an orchid with the anther extending beyond the stigma, it would be easy for pollinators to receive some stigmatic liquid when leaving the flower and for this sticky liquid to attach clumps of pollen to the pollinator. This makes a logical and functional starting point for much of orchid evolution. Though such a flower would be functional, it might not be very efficient. Part of the anther would be concealed behind the stigma; with very soft pollen masses, part of the pollen might remain in the anther, unavailable to function in pollination. There are several ways that selection might improve such a pattern. The slightly sticky pollen might become more coherent, forming distinct pollinia. The anther could move upward relative to the stigma, so that most of the pollen mass would be exposed and more easily removed by pollinators. This is the pattern found in most orchids of the neottioid grade (see Fig. 3-2).

The evolution of a viscidium, that is, a sticky portion of the rostellum that becomes attached to the pollinia and is removed with them as a unit, would be an obvious improvement in efficiency. If the viscidium evolved near the middle of the anther, selection might favor shifting the viscidium toward the apex of the anther. The apical viscidium readily touches any potential pollinator that approaches the flower closely enough. This is the spiranthoid condition. In the more advanced neottioid pattern, with the anther largely above the level of the stigma, one or two viscidia might evolve near the base of the anther. This pattern occurs in the Diurideae and could give rise to the pattern of the Orchideae.

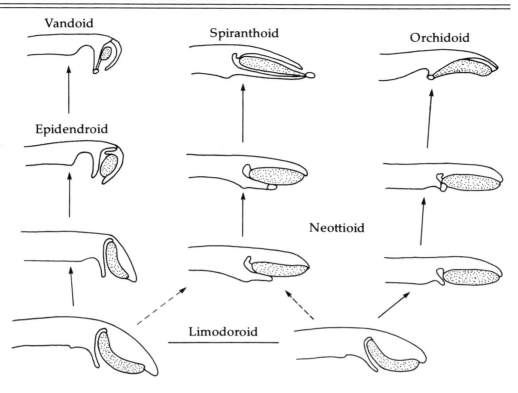

Figure 3-2. The overall pattern of column evolution in the monandrous orchids. It is hypothesized that all monandrous orchids evolved from a limodoroid grade (not necessarily the same ancestor in each case). The advanced Orchidoideae evolved basal viscidia and complete union of anther and column. Most Spiranthoideae (and some other groups) evolved a terminal viscidium and a beaklike column. A fully incumbent anther has evolved within the Epidendroideae (possibly more than once), and different epidendroid groups have independently evolved both viscidium and tegular stipe (the vandoid grade).

Finally, the function of the neottioid column could be improved in quite another way. If the anther and part of the stigma were bent sharply downward, this would greatly increase efficiency, as pollinators leaving the flower would almost certainly touch the edge of the stigma and, at the same time, lift up the anther and touch the pollen mass with the same part of the body that had just touched the stigma. This, in very simple terms, is the epidendroid pattern (see Figs. 3-2, 2-13).

The bending of the anther during development might suggest that the Epidendroideae had evolved from an ancestor with the anther at right angles to the column, as was suggested by Dressler (1983b, Fig. 3). The variation in anther shape, however, suggests another interpretation (Dressler 1990a). The anthers of some Neottiae, Triphoreae, and Gastrodieae have the locules strongly convex (Fig. 3-3A). The anther locule is straight and erect in its early development, and then bends so that part of the locule parallels the median stigma lobe, which is markedly emergent in these genera. I suggest, then, that the incumbent anther of the advanced Epidendroideae evolved through the reduction of the distal part of the anther, so that the remaining portion of the anther locules would be parallel

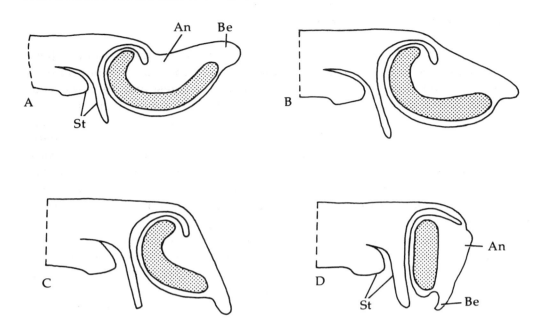

Figure 3-3. Hypothetical derivation of the incumbent anther (D) from an anther with convex locules, as in *Cephalanthera* or *Triphora* (A). An, anther; Be, anther beak; St, stigma. Anther locule (pollen) stippled. (A) Based on *Cephalanthera damasonianum* (after F. N. Rasmussen 1982; Figs. 5, 6, Dressler 1990a.)

with the stigma lobe (clinandrium). The anther of *Wullschlaegelia* is very close to an intermediate stage, such as Fig. 3-3C.

The convex anther locules of some primitive Epidendroideae and the emergent median stigma lobes, typical of the same groups, may be adaptations that evolved early in the monandrous orchids and improved the removal and deposition of pollen masses by pollinators (Dressler 1990a). With the evolution of viscidia and other more efficient mechanisms, the emergent stigma lobes and the convex anther locules become less critical. They have been lost in most orchid groups, and occur especially in the Neottieae and Triphoreae. Primitive members of both the Spiranthoideae and Orchidoideae, however, approach this condition.

POLARITY. Erect anthers are surely the primitive condition, though they might be secondarily derived in the Spiranthoideae or Orchidoideae. Incumbent and recumbent anthers are considered independent derivations. Anthers with strongly convex cells are interpreted as derived, relative to straight anther cells, and ancestral to fully incumbent anthers. A distinct beak at the anther apex is clearly derived, and the thin beak typical of the advanced Epidendroideae may be derived from a thick beak such as that of the Neottieae or Triphoreae.

Evolution of the Orchid Seed

Most orchid seeds are minute, with a thin, more or less transparent, papery seed coat loosely surrounding the small, undifferentiated embryo. Some authors have been so orchid-centric in their viewpoint as to suggest that the few orchid seeds with hard coats are some sort of abnormal development from the more usual type of orchid seed. Comparison with other Liliiflorae would suggest that the ancestral condition must have been a crustose seed coat. Further, some of the seeds with hard seed coat have layers of tissue that are lacking in other orchids, and the relatively few seeds with hard coats are diverse in structure and found in different, but always relatively primitive, orchid groups (see Fig. 2-15). Information on seed type, largely from Ziegler's unpublished thesis and data supplied by W. Barthlott, is given in more detail in Chapter 2.

There can be little doubt that the crustose seed coat is a primitive condition for the orchid family. One would guess that the primitive condition is a more or less globose seed, though whether it should have the outer wall of the testa cells thick, as in *Vanilla*, or the outer wall thin and collapsed against a thick inner wall, as in *Apostasia*, it is difficult to say. In any case, the testa cells are about as wide as long in *Palmorchis, Selenipedium*, and the Apostasioideae. Dust seeds with thin cell walls in the testa have evolved independently in several different clades, but we may guess that a seed coat like that of *Limodorum*, with the testa cells isodiametric and thin, would be relatively primitive among dust seed types. From this simple type, the cells may all become elongate, or only the cells around the embryo may become elongate, with the basal and apical cells remaining short, as in *Maxillaria*. The Diurideae and Diseae, however, have a crustose seed coat with elongate medial cells and shorter terminal cells, and the thin-walled testa cells of the *Orchis* type show the same pattern. Seed coats with the longitudinal anticlinal walls densely pressed together, as in the *Vanda* type, are highly derived, and there is evidence that the *Vanda* type has evolved independently several times. Balloon seeds, such as those of the *Stanhopea* type, go against the evolutionary trend, in that all cells tend to be isodiametric, and some seeds may have the medial cells isodiametric and the basal and apical cells somewhat elongate. Various types of ornamentation of the cell wall appear to be derived conditions: reticulate and longitudinal thickenings, warts, papillae, etc.

Of the existing seed types with crustose coat, the *Disa* and *Diuris* types could be secondarily derived. Each is centered in a relatively dry area, and they could be adaptations to a dry, seasonal climate. This, however, does not explain the close resemblances between these two types, or the intermediates between them.

Though it is clear that seeds with a crustose coat were the ancestral condition, more work is needed before we can assign clear polarity to most seed types. The *Orchis* type appears to be derived from the *Disa* type (or vice versa) and differs from that type primarily in lacking thickenings of the testa cell walls. The relationship between the *Diuris* type and the *Goodyera* type is less intuitive, but there are numerous intermediates between them, and both types occur in some genera of the Caladeniinae. The *Goodyera* type may have been derived from the *Diuris* type, at least in the Diurideae. In the Epidendroideae, the *Limodorum* and *Gastrodia* types are both relatively primitive, and the *Eulophia* type also is considered relatively primitive, as it is found in the Triphoreae and in both the Cymbidioid and Epidendroid phylads. Thus, the *Bletia*, *Cymbidium*, and *Maxillaria* types all may be derived from the *Eulophia* type. The *Dendrobium* seed type probably evolved at least twice, and the *Vanda* type appears to have evolved at least four times.

Cytology

Chromosome Size

Unusually large chromosomes are known in a few groups. These are considered primitive by Chatterji (1986), though the chromosomes of the Apostasioideae are very small (Okada 1988). I prefer to withhold judgment on the polarity of this feature, but large chromosomes may be primitive, at least within the Epidendroideae.

Resting Nucleus

Tanaka (1971) has reviewed the types of resting nuclei in the orchids. This preliminary work is promising, and more extensive sampling would be desirable. Most orchid groups show either the simple or the complex chromocenter type. The round prochromosome type and the similar rod prochromosome type are largely limited to the Cranichideae, though Sera's recent study (1990) shows a few Goodyerinae with the chromocenter type. The diffuse type occurs in the Neottieae and the Pogoniinae, and these groups show no other type of resting nucleus. The diffuse and prochromosome types are tentatively considered as derived features for their respective groups. Okada (1988) treats three types of resting nucleus: chromocenter, diffuse, and prochromosome.

Chromosome Number

In 1981 I found the cytological diversity of the orchids bewildering, and felt that the recurrence of $2n = 40$ in most groups obscured any overall pattern. There have been at least three recent attempts to outline the chromosomal pattern in the Orchidaceae: Chatterji (1986), Mehra (1983), and Vij (1989). When faced with an array of chromosome numbers, cytologists almost always seek the base number in their study group. This assumes that polyploidy has been more important than aneuploidy, and often postulates a base number between 6 and 11.

In the Oncidiinae, there are some taxa with $n = 14$ and many with $n = 28$, and *Psygmorchis* has been reported to have $n = 7$. Thus it has been suggested that 7 is the base number (x) for *Oncidium* (or for the subtribe), with 28 considered tetraploid ($4x$), 42 hexaploid ($6x$), and the common 56 octoploid ($8x$) (Sinoto 1962). Dodson and Dressler (1972), however, argue that *Psygmorchis* is highly derived, and that both $n = 5$ and $n = 7$ probably represent aneuploid reduction from a higher number, a pattern that is especially frequent in short-lived plants. Chase and Olmstead (1988) have examined enzyme polymorphism in the

Oncidiinae and find no evidence of polyploidy, thus suggesting that aneuploidy accounts for most of the chromosome number diversity in that subtribe (numbers above 80 are doubtless polyploid).

Jones (1970) stresses that there are no clear laws governing chromosomal evolution. I now argue, for reasons more phylogenetic than cytological, that a base number of about 20 may be the ancestral condition for some orchid groups. The crosses between *Cymbidium* and Maxillarieae reported by Tanaka et al. (1987) are interesting in this respect. Tanaka (1971) stresses the similar resting stage morphology of the nuclei in both groups. I would suggest that this is, indeed, a case in which crossability is a primitive feature. That is, I suggest that both groups retain the ancestral chromosome number ($n =$ about 20) and something very close to the ancestral chromosome structure. This, in turn, suggests that a base number near 20 is the ancestral condition for both the Cymbidieae and the Maxillarieae. A base number of 20 is also frequent in the Arethuseae, and base numbers of 19, 20, and 21 occur in a high percentage of the epidendroid phylad.

In the Cypripedioideae, aneuploidy appears to have gone upward from a base number between 10 and 13 (Atwood 1984; Karasawa 1979). Ten is the probable base number in both *Cypripedium* (Karasawa and Aoyama 1986) and *Phragmipedium*, and this may be taken as the probable base number for the sub-family (Atwood 1984). One could play number games much like those of the cytologists; with 10 as the base number of the lady's slippers, and 20 a major number in the Epidendroideae, the Epidendroideae might seem to be tetraploids. On the other hand, Yokota (1987) suggests that $n = 20$ in *Epipactis* has been derived by aneuploidy from $n = 17$ in *Cephalanthera*. Indeed, aneuploidy through centric fission could lead from a base number of 10 to a base number of 20. If the ancestral condition was a set of 10 metacentric chromosomes, then a set of 20 telocentric chromosomes would be the logical product of continued centric fission.

Some very early reports indicate chromosome numbers quite different from more recent reports, and especially much lower numbers. Early techniques and early microscopes probably were inadequate and these discordant chromosome numbers should thus be distrusted. I have usually omitted such counts from the chromosome numbers reported. Also, in the case of counts based on mitotic divisions (vegetative material), knowing whether an aberrant number represents a diploid plant or a chance haploid may be difficult.

In the Orchidoideae, there is again a predominance of 20, 21, and 22 as haploid numbers. Yokota (1987) outlines the evolution of the Orchideae from a hypothetical ancestor with $n = 7$, but this case is reminiscent of the Oncidiinae. Most Orchidinae show $n = 20$ or 21, but the more derived *Habenaria* shows a confusing series of aneuploids. Though some Epidendroideae may have reached $n = 20$ by aneuploidy, the report of $2n = 20$ in *Orchis coriophora* (Cauwet-Marc and Balayer 1984b) suggests that the Orchidoideae may have reached the same number by polyploidy. This agrees with the suggestion of Heusser (1938) that the base number of the Orchideae is "2×10."

The pattern in the Spiranthoideae is less clear. A haploid number of 15 is frequent, especially in Old World groups (but also in American *Cyclopogon*), but recent counts by Martínez (1985) show $n = 23$ in a number of different Cranichideae. Both $n = 15$ and $n = 22$ occur within *Spiranthes* in the narrow sense. Sera's study of Asiatic Goodyerinae (1990) shows an aneuploid series of haploid numbers from 15 to 10. The most common numbers are 14 and 15, and some *Zeuxine*, with 10, are probably derived. How to relate these numbers to other genera with $n = 20, 21, 22,$ and 23 is not clear.

Figures 3-4 and 3-5 outline the variation in chromosome number in the orchid subfamilies and within the subfamily Epidendroideae. Within the Epidendroideae, we see, as indicated by Chatterji (1986), that chromosome number has been very stable in the epidendroid phylad, including the Vandeae, while there is much less stability in the cymbidioid phylad, with much aneuploidy. Polyploidy is especially prominent in the Gastrodieae. Most orchid chromosomes are very small, but improved techniques may yet permit us to test hypotheses about chromosomal evolution. In any case, a more thorough review of orchid cytology is needed, with careful attention to more than just chromosome number.

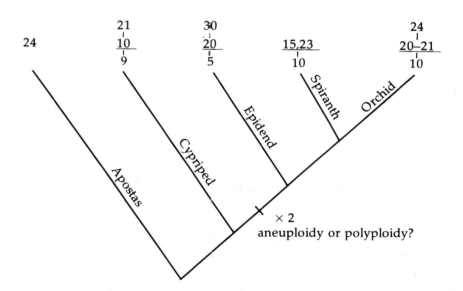

Figure 3-4. The range of haploid chromosome numbers in the orchid subfamilies. The base number is underlined in the Cypripedioideae, and the predominant numbers are underlined in those subfamilies for which there are sufficient data.

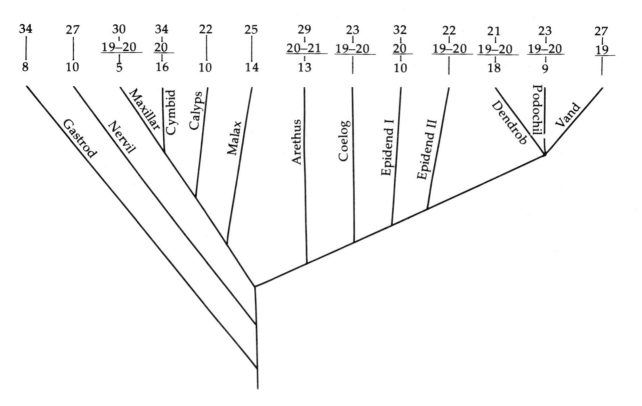

Figure 3-5. The range of haploid chromosome numbers in the advanced tribes of the Epidendroideae. The predominant numbers are underlined in those tribes for which there are sufficient data, if there are predominant numbers.

OVERALL PATTERN OF EVOLUTION

It is suggested in the introduction that the orchids might be of special interest in studying evolutionary patterns. There are many transformation series, and their polarity is generally clear. In comparing the orchids with other Liliiflorae, it is obvious that most of the transformation series go from an ordinary lilylike flower to something unique in the plant kingdom. In spite of the clear polarity, the classification of the orchids has been difficult because of the great amount of parallelism. There are relatively few uniquely derived features, and groups that look superficially quite natural have proven to be polyphyletic or paraphyletic grades.

The Apostasioideae are probably a relict group of tropical origin. *Neuwiedia*, especially, has almost diagrammatically primitive flower structure for the family. *Apostasia* is apparently pollinated by pollen-gathering female bees (when not self-pollinating) and parallels other vibration flowers, such as *Solanum*.

The lady's slippers are considered "primitive" in possessing fertile lateral stamens, but it might be more correct to consider them as simply specialized or derived in a different pattern from that of the monandrous orchids. The curious trap flowers seem to depend on deceit to attract the pollinators in all cases.

In the Spiranthoideae, the Cranichideae form a distinctive and relatively derived group. No unique features are known to tie the Diceratosteleae and Tropideae to each other or to the Cranichideae, though their juxtaposition seems reasonable on the information now available. There are no clear connections between the Spiranthoideae and Orchidoideae, but their common ancestor(s) may have resembled groups here treated as primitive Epidendroideae.

In the Orchidoideae, more information is needed to determine whether the *Diuris* complex is derived or ancestral relative to the *Caladenia* complex, and it may well combine primitive and derived features. In any case, the Diurideae appear to be a sister group to the Orchideae/Diseae, and their treatment in the same sub-family seems natural. The Orchideae and Diseae, together, are a distinctive and highly derived group that is surely monophyletic. Nevertheless, the tribes and subtribes within this group are poorly delimited and in need of revision.

The major innovation that characterizes the Epidendroideae is the fully incumbent anther with a projecting median stigma lobe, an elegantly functional mechanism that scrapes off entire pollinia while depositing glue and then fresh pollinia, as the pollinator backs out of the flower. This mechanism probably facilitated the evolution of really firm and then hard pollinia among the epidendroids. Many primitive epidendroid flowers function quite well without a viscidium. Though viscidia have evolved in many epidendroid groups, they appear to have evolved relatively late, as compared to the Spiranthoideae and Orchidoideae. The Neottieae are primitive in most features, though the Listerinae have a unique, sensitive rostellum that squirts glue out forcibly when touched. There are no clear ties between some of the primitive tribes and either the epidendroid or the cymbidioid phylad, and the relationships at that level are by no means as clear as one would prefer.

The advanced Epidendroideae show two more important innovations: thickened stems, including corms and pseudobulbs, and quite hard pollinia. Corm-like thickenings appear to be the ancestral condition in both major phylads, but the majority of the epidendroid phylad have either slender stems or elongate pseudobulbs. In terms of species diversity, the American Epidendreae and the Old World dendrobioid subclade are the major orchid groups, while the Vandeae and the Maxillarieae show the greatest generic diversity.

The epidendroid phylad shows a pattern of reduction from eight pollinia to six, four, or two in many groups, but there is no indication that the cymbidioid phylad was derived from an ancestor with eight pollinia. We find the evolution of the "vandoid" character complex in the Vandeae, with parallels in the Polystachyinae, Coelogyneae, and Bulbophyllinae. The Dendrobieae, though allies of the Vandeae, have achieved similarly elegant pollination mechanisms usually without either caudicles or stipe. The vandoid character complex is well developed in the cymbidioid phylad, and the major vandoid orchid groups of the Americas are members of this group. There is much parallelism between the

Vandeae and Maxillarieae. The reed-stem growth pattern of the epidendroid phylad seems to have favored the early or frequent evolution of distichous and conduplicate leaves, but the cormous growth pattern of the cymbidioid phylad has favored the early or frequent evolution of lateral inflorescences.

Starting with an asymmetrical lily flower, we can trace the evolution of the orchid flower. In the advanced orchids, the pollinary apparatus, with pollinia, caudicles, stipes, and viscidium, really represents a new level of organization. These flowers are as unlike buttercups as buttercups are unlike pine cones. Their pollination systems are often highly specific and may involve pseudocopulation and other forms of mimicry. In functional complexity and evolutionary level, both the Goodyerinae and the Orchideae parallel the vandoid tribes.

The orchids may be, as Gould (1977) indicates, "jerry-built of available parts used by ancestors for other purposes." At the same time, they offer "an almost endless diversity of beautiful adaptations," and contrivances "almost as perfect as any of the most beautiful adaptations in the animal kingdom" (Darwin 1888). Since these contrivances, like most biological systems, are built of the parts available to their ancestors, we find many cases of parallelism and convergence.

DIVERSITY AND SIZE OF GENERA

It is interesting to compare the species diversity of different grades and groups (Fig. 3-6). One would expect the primitive groups to be relatively small, and, effectively, the Apostasioideae, Cypripedioideae, Diceratosteleae, Tropidieae, and the primitive epidendroid groups, together, have only about 638 species (following Atwood 1986). The Cranichideae, alone, have about 1125 species, and the Orchidoideae are larger, with nearly 1900 species. For this purpose, I have divided the advanced Epidendroideae into three grades: the epidendroid grade with caudicles but no stipe, the dendrobioid grade with naked pollinia (excluding the few genera of Bulbophyllinae that independently evolved viscidia and stipes), and the vandoid grade with viscidia and usually stipes. The epidendroid grade has the greatest number of species (6612), but the vandoid grade is not far behind (5396). The dendrobioid grade (Malaxideae and most Dendrobieae) has 3114 species.

The primitive groups have relatively small genera, averaging about 16 species per genus. When I first prepared such a chart, the genera of the Cranichideae were appreciably larger than those of the primitive grade, but with the recent proliferation of genera in the American Goodyerinae and the Spiranthinae, the Cranichideae now average only about 13 species per genus. The Orchidoideae have somewhat larger genera, with an average of 17 species per genus. The genera of the epidendroid grade are yet larger, with an average of 44 species in each, but the vandoid genera average only 16 species. The genera of the dendrobioid grade, however, average about 183 species. I suspect that these figures represent something of an artifact, reflecting our traditional emphasis on characteristics of the pollinia and associated features. The pollinia of the dendrobioid grade have virtually no distinguishing features, and thus we cannot, without rethinking our concepts, distinguish many genera in the Malaxideae or Dendrobieae. The vandoid grade, on the other hand, offers a wealth of details in the pollinia, viscidia, and stipes, and they have been very finely split as compared to other advanced orchids. Overall, the genera are probably excessively split in the Vandeae. Some have tried to split the American Oncidiinae as finely as the Old World Vandeae, but this, by itself, will not improve our understanding of either group. In the Malaxideae, especially, the current classification is rather arbitrary, and a critical revision is much needed.

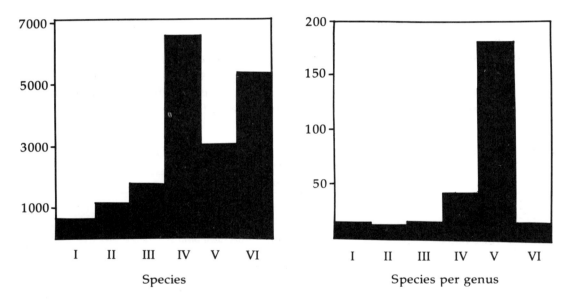

Figure 3-6. The numbers of species and of species per genus in different orchid groups and grades. The Roman numerals represent (I) the Apostasioideae, Cypripedioideae, Diceratosteleae, Tropidieae, and primitive epidendroid groups; (II) Cranichideae; (III) Orchidoideae; (IV) advanced Epidendroideae with caudicles but no stipe; (V) the Malaxideae and Dendrobieae (excluding the few genera of Bulbophyllinae with stipes); and (VI) the van-doid grade (Calypsoeae, Cymbidieae, Maxillarieae, Vandeae, and the Bulbophylline genera with stipes).

PHYLOGENETIC DIAGRAMS

This and some of the following chapters include phylogenetic diagrams representing my hypotheses of orchid relationships, with derived features shown on the diagrams. Norris Williams and I have analyzed some data sets with a computer program, but the diagrams shown here were basically made "by hand." I have tried to use or emphasize the features that are uniquely derived or have evolved only a few times, while giving little weight to those features that have evolved repeatedly. In philosophy, then, this is "Hennigian argumentation" and related to compatibility analysis. In practice, my method has been to make diagrams, over and over again, until I feel that I have the clearest and most consistent diagram for the available data. A number of apparently uniquely derived features are found in single, clearly defined groups (Table 3-4).

The features used in Figure 3-8 are listed in Table 3-5. Similar tables are given in each chapter with complex phylogenetic diagrams. In these lists, "0" represents the ancestral state. When there is significant doubt as to polarity, a "0" state is not shown. In two-state features with an ancestral state shown ("0→1"), state 1 is to be understood in the diagrams. Otherwise, the state is specified. A number on the diagram represents the appearance of the indicated feature and state, and that

Table 3-4. Probable uniquely derived features in the Orchidaceae, and the groups in which they occur. Losses or modifications occur within some groups.

Feature	Group(s)
Adaxial anthers lacking	All
Resupination	All
Silica bodies around vascular bundles	All
Spherical silica bodies	Dendrobioid subclade
Sticky pollen	All but apostasioids
Root–stem tuberoids	Orchidoideae
Corms	Advanced Epidendroideae
Cylindrical, herbaceous leaves	Prasophyllinae
Bulbophyllum velamen type	Bulbophyllinae
Coelogyne velamen type	Coelogyneae
Cymbidium velamen type	Cymbidieae, Maxillarieae
Dendrobium velamen type	Dendrobiinae
Epidendrum velamen type	Laeliinae, Meiracylliinae, Arpophyllinae
Malaxis velamen type	Malaxideae
Pleurothallis velamen type	Pleurothallidinae
Spiranthes velamen type	Prescottiinae, Spiranthinae
Vanda velamen type	Vandeae
Apostasia seed type	Apostasioideae
Cymbidium seed type	Advanced Cymbidieae
Disa seed type	Diseae
Diuris seed type	Diurideae
Epidendrum seed type	Laeliinae, Coeliinae
Gastrodia seed type	Gastrodieae
Maxillaria seed type	Maxillarieae
Orchis seed type	Orchideae
Pleurothallis seed type	Pleurothallidinae
Stanhopea seed type	Stanhopeinae

state should be present in the groups above that point, unless a reversal or another number in the same series is shown, indicating another state of the same feature. Parallelisms, features that have evolved more than once, are each indicated by two parallel lines on the diagram. Losses or reversals are shown by × preceding the number. Numbers that are shown next to group names, rather than along an axis of the diagram, represent features that vary within the corresponding groups. These are considered derived features within their groups and occur with the ancestral states in the same groups. Features characteristic of the basal elements, but lost at more advanced levels, would be treated as derived features for their respective groups. As an example, a phylogenetic diagram of computers is given in Figure 3-7.

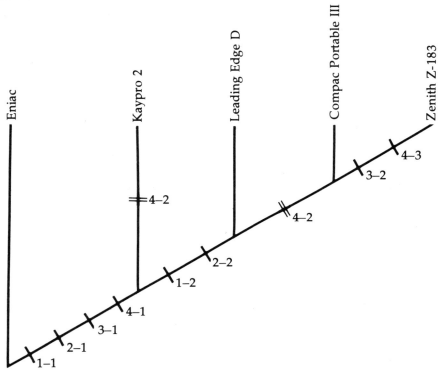

Figure 3-7. A phylogenetic diagram of selected computers. We have clear "transformation series" for the computers, and if we pretend that they evolve as plants and animals do, then phylogenetic analysis gives us the above diagram. There is parallelism in state 4-2, that is, it has evolved twice. To diagram this state as evolving only once we would have to show parallelisms or reversals in two other features. "Transformation series" as follows:
1. Central processing unit: 0. vacuum tube; 1. 8-bit microchip; 2. 16-bit microchip ($0 \to 1 \to 2$).
2. Operating system: 0. punch card; 1. CPM; 2. MSDos ($0 \to 1 \to 2$).
3. Storage: 0. punch tape or card; 1. floppy disk; 2. high density floppy disk (3½ inches) ($0 \to 1 \to 2$).
4. Size: 0. very large; 1. desk top; 2. portable; 3. lap top ($0 \to 1 \to 2 \to 3$).
     In each case the "evolution" has been from the primitive state (0) to successively more "derived" states, as indicated in parentheses after each feature.

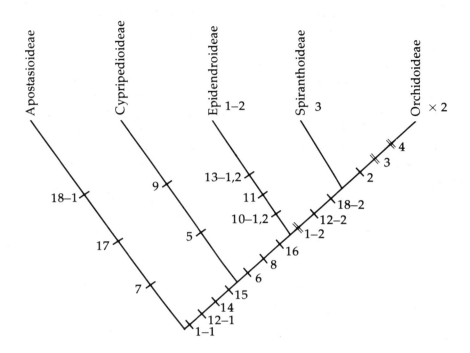

Figure 3-8. Phylogenetic diagram of orchid subfamilies. The features are given in Table 3-5.

Table 3-5. Features, their states and polarization, as used in Figure 3-8.[a]

1. Chromosome size: 1. some or all large; 2. all very small ($1 \to 2$?)
2. Roots: 0. monomorphic; 1. dimorphic (root–stem tuberoids) ($0 \to 1$)
3. Leaf type: 0. plicate; 1. soft herbaceous ($0 \to 1$)
4. Subsidiary cells: 0. present; 1. lacking ($0 \to 1$)
5. Lip: 0. not deeply saccate; 1. deeply saccate ($0 \to 1$)
6. Lateral anthers: 0. fertile; 1. sterile or lacking ($0 \to 1$)
7. Lateral anther bases: 0. symmetrical; 1. asymmetrical ($0 \to 1$)
8. Union of filaments with style: 0. partial; 1. complete ($0 \to 1$)
9. Median anther: 0. fertile; 1. shieldlike staminode ($0 \to 1$)
10. Anther posture: 0. erect; 1. subincumbent; 2. fully incumbent ($0 \to 1 \to 2$)
11. Anther cells: 0. straight or nearly so; 1. strongly convex ($0 \to 1$)
12. Anther apex 1. obtuse; 2. acute
13. Anther beak: 0. lacking; 1. thick; 2. thin ($0 \to 1 \to 2$)
14. Stigma: 0. three equal lobes; 1. broad and asymmetric ($0 \to 1$)
15. Pollen texture: 0. powdery; 1. sticky ($0 \to 1$)
16. Pollen grains: 0. colpate; 1. porate/ulcerate ($0 \to 1$)
17. Pollen grains: 0. exine moderately sculptured; 1. exine heavily sculptured ($0 \to 1$)
18. Seed type 1. *Apostasia*; 2. *Disa* or *Diuris*; 3. *Goodyera*

[a] When there is a significant doubt as to polarity, a "0" state is not shown.

PHYLOGENY OF THE SUBFAMILIES

The probable phylogeny of the orchid subfamilies is shown in Figure 3-8. The Apostasioideae is a straightforward group and is clearly separated from the rest of the family by a couple of important features. The Cypripedioideae are closer to the rest of the family in several features. The Epidendroideae are very diverse and in most cases have a clearly incumbent anther. Some of the primitive groups have merely a subincumbent anther, but these have markedly convex anther cells, a blunt anther beak and an emergent median stigma lobe. These features may be functionally linked but can occur independently. The blunt anther and the large chromosomes of the Neottieae and the Pogoniinae suggest that the primitive Epidendroideae are more similar to the Cypripedioideae than are the Orchidoideae or Spiranthoideae. The diversity of the primitive Epidendroideae also suggests that this may be the "core" group from which the other monandrous orchids have been derived. The Orchidoideae and Spiranthoideae share similar habit and seed type, but the habit evolved independently in the Spiranthoideae (assuming that the Tropidieae are properly placed in the Spiranthoideae), and the *Goodyera* seed type also may have evolved independently in the two groups, possibly from different ancestral types. From the diagram, alone, one could argue for including the Spiranthoideae in the Orchidoideae, and I have at times (in frustration) supported such a classification. The evidence suggests, however, that the Spiranthoideae and Orchidoideae are clearly separable, and the presence of acute anthers in both groups does not constitute very strong evidence that they are sister groups.

Subfamily	Division	Tribe	Series	Subseries

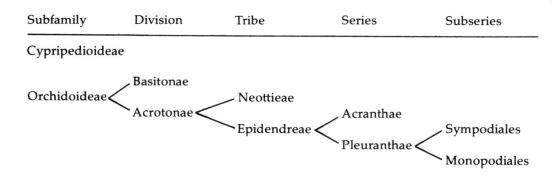

Figure 3-9. A diagram of Schlechter's orchid classification, with modern names for subfamilies and tribes.

PREVIOUS SYSTEMS OF CLASSIFICATION

I will not attempt to chart all previous systems of orchid classification, as in Burns-Balogh and Funk (1986) or Dressler (1981). Such charts necessarily become more complex and confusing with each generation. The most widely used system for the last few decades has been that of Schlechter (1926), which had the advantage of listing all genera known at that time. Schlechter's system had a curious hierarchic asymmetry (Fig. 3-9), and many of the names used are invalid under the current rules of nomenclature. Garay (1972) revised and modernized the Schlechter system and brought the names into line with the rules of nomenclature, as in Table 3-6. The system used here is diagrammed in Figure 3-10, and Figure 3-11 compares Garay's system with the one used here.

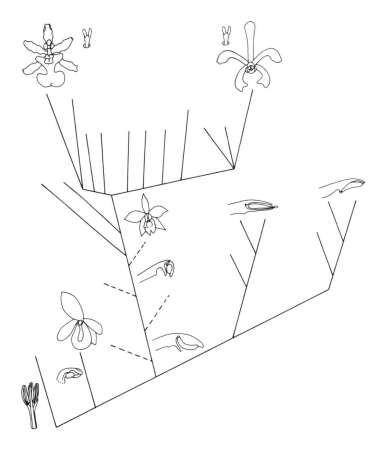

Figure 3-10. A diagram of the classification used here. The basal portion represents the diagram in Fig. 3-8, and the upper part represents Figure 8-3. The connection between the two is arbitrary, as the relationships of the primitive epidendroids are unclear. The groups are grouped into grades by level: the Apostasioideae and Cypripedioideae with fertile lateral anthers; the Spiranthoideae, Orchidoideae, and some primitive Epidendroideae with soft pollinia; the epidendroid grade and the vandoid grade.

Table 3-6. Comparision of Schlechter's classification (1926), with modern names, and Garay's classification (1972). Names on the same level represent the same groups. Garay's divisions of the Epidendroideae do not correspond exactly to those of Schlechter.

Schlechter	Garay
Apostasiaceae	Apostasioideae
Cypripedioideae	Cypripedioideae
Orchidoideae	
[Div. Basitonae]	Orchidoideae
Orchideae	Orchideae
	Diseae
	Disperideae
[Div. Acrotonea]	
Neottieae	Neottioideae
	Neottieae
	Epipogeae
	Cranichideae
Epidendreae	Epidendroideae
[Ser. Acranthae]	
[Ser. Pleuranthae]	Epidendreae
[Subser. Sympoidalies]	Vandeae
[Subser. Monopodiales]	

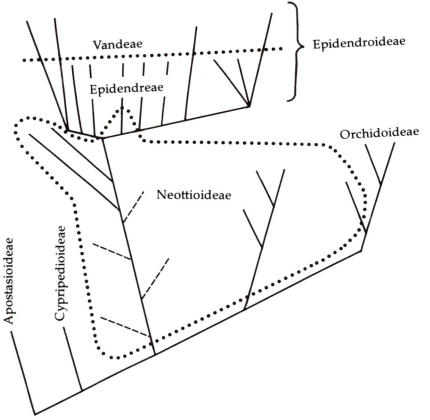

Figure 3-11. A diagram of the classification used here (see Fig. 3-10) with Garay's classification shown by dotted lines.

4

Orchids with Two or Three Fertile Anthers

The apostasioids and the cypripedioids are considered in the same chapter, simply because both groups are relatively small. They are only distantly related to each other. Among the apostasioids, the genus *Neuwiedia*, especially, has many primitive features. The lady's slippers are often considered primitive, but they are, rather, specialized in a way different from the other orchids.

Subfamily **Apostasioideae**

The Apostasioideae include less than 20 species; these are limited to tropical Asia and Australasia, where they could be relicts of a once wider distribution. They are of special interest as a link between the orchids and more ordinary lilylike plants. In this subfamily, the lip is similar to the petals, and the lateral anthers are fertile; the median anther may be fertile, sterile, or lacking.

The Apostasioideae appear to be a sister group of the remaining orchids, and their treatment either as a subfamily or as a distinct family is not unnatural. Though the two genera are quite distinct, I can find no reason to place them in separate tribes or subfamilies.

DESCRIPTION. Habit: forest floor terrestrials, with elongate stems; roots without velamen, nodular storage roots in *Apostasia*; may have stilt roots. Leaves: spiral, convolute, plicate, nonarticulate. Inflorescence: terminal, erect, or spreading, simple or branched, flowers spiral. Flowers: small, white, yellowish, or yellow, perianth with slight bilateral symmetry, resupinate or not; the rest of the flower may be shed from ovary after pollination; anthers 2 or 3, elongate, the

lateral anthers basally asymmetrical, the median stamen may be fertile, represented by a fingerlike staminode, or lacking; filaments partly united with each
other and the style; pollen grains as monads, powdery, with reticulate or
perforate-reticulate sculpturing; style slender, stigma lobes equal and similar or
slightly asymmetrical. Fruit: three-locular, fleshy or capsular. (See Fig. 4-1.)

DISTRIBUTION. Tropical Asia to northern Australia.

POLLINATION. Some of the apostasioids are self-pollinating. Vogel (1981)
suggests that both genera may be pollinated by bees that vibrate on the flower to
release the pollen from the anthers, as in *Solanum* or the Melastomataceae. The
flower of *Apostasia* resembles known vibration flowers.

CHROMOSOME NUMBER. 48 very small chromosomes (possibly tetraploid).

SEED STRUCTURE. *Apostasia* type, or in some *Neuwiedia* species with a balloon-
like appendage at each end, the central portion with seed coat of the *Apostasia* type.

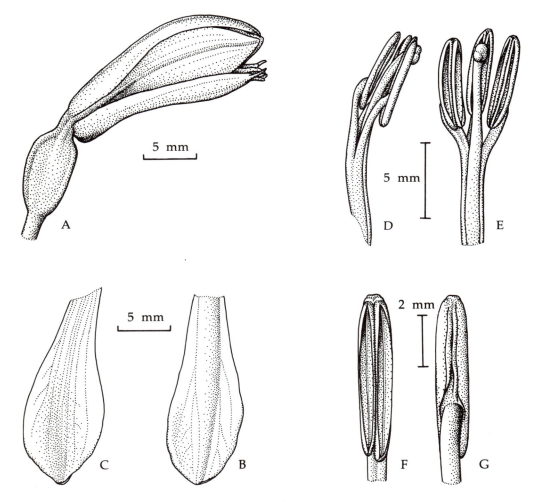

Figure 4-1. *Neuwiedia veratrifolia* (Apostasioideae). (A) Flower, side view. (B) Lip, flattened. (C)
Petal, flattened. (D) Column, side view. (E) Column, ventral view. (F) Anther, ventral view. (G)
Anther, dorsal view. Prepared from material preserved in liquid. (After Dressler 1981.)

SPECIES. About 16.

GENERA. 2: *Apostasia, Neuwiedia.*

DISCUSSION. *Neuwiedia* is very much like the *Ur*-orchid that one might hypothesize as the ancestor of the whole family. It has the basic orchid symmetry, with 3 fertile anthers on one side of the flower, and it shows resupination and some union of the filaments with each other and with the style. It also shows some primitive features that we find in several other orchid groups: crustose seed coat, fleshy fruit, three-locular ovary, and an abscission layer between ovary and perianth.

Okada's paper on chromosome number and morphology (1988) is most welcome. He finds diploid numbers of 48 in an *Apostasia* and 96 in a *Neuwiedia* and suggests that the base number may be 12. The resting nuclei are of the prochromosome sort, similar to those of the Cranichideae.

The traditional classification that included the apostasioids with the lady's slippers was quite artificial, and modern authors treat them as separate groups. Some separate the apostasioids as the family Apostasiaceae, but I see little profit in breaking up a family as natural as the Orchidaceae. A few authors have treated *Apostasia* and *Neuwiedia* as only remotely related, and Burns-Balogh and Funk (1986) place *Neuwiedia* in a separate subfamily. Still, the resemblances in seed, pollen, and anther form argue against such a separation, even though one anther has been lost by *Apostasia* in evolving a more efficient vibration flower. In that genus, the two remaining anthers tightly clasp the style, and together form a tube, functionally analogous to the tubular anthers of *Solanum*, for example. Note that the median anther of some *Neuwiedia* species is markedly smaller than the lateral anthers (de Vogel 1969).

The seed coat of *Apostasia* is distinctive, with inner and anticlinal cell walls very thick. The thin outer wall collapses, causing the seed surface to be pitted. The seed of *Neuwiedia* species with fleshy fruits is quite similar, while that of the capsule-bearing species has balloon-like appendages, but still has the seed coat of the central portion, around the embryo, similar to that of *Apostasia*. Schill and Pfeiffer (1977) report the presence of an operculum, or lid, on the furrows of the pollen grain in both genera. Such a structure has not been confirmed in *Neuwiedia*, but the sculpturing of the pollen wall is similar in the two genera, and very different from that of the lady's slippers (Newton and Williams 1978).

REFERENCES. Okada 1988; de Vogel 1969; Vogel 1981.

Subfamily **Cypripedioideae**

The lady's slippers are well-known wild flowers of the north temperate zone, and some of the tropical genera are popular greenhouse subjects, so that these have attracted the attention of amateur and botanist alike and are among the most studied of all orchids. The Cypripedioideae are clearly more closely allied to the monandrous orchids than are the Apostasioideae, and they, too, are a sister group rather than ancestral, and are quite specialized in their own way. Though fairly well represented in tropical Asia and tropical America, the cypripedioids may have dispersed by northern routes, when tropical climates were more wide-spread, as suggested by Atwood (1984).

DESCRIPTION. Habit: terrestrial, lithophytic, or epiphytic in humus; roots fleshy but slender, velamen lacking or of the *Calanthe* type; stems elongate or condensed, may be branched in *Selenipedium*, without pseudobulbs. Leaves: spiral or distichous, convolute and plicate, or conduplicate and leathery/fleshy, nonarticu-late, sometimes with pale blotches. Inflorescence: terminal, of one to many flowers; flowers spiral or distichous. Flowers: small to rather large, resupinate, with an articulation between ovary and perianth in the genera with conduplicate leaves; lateral sepals usually united; lip deeply saccate; lateral anthers fertile, median anther sterile and shieldlike; filaments largely united with style; pollen grains as monads, sulcate, laevigate (smooth) or slightly pitted or grooved, sticky and pastelike or united into pollinia (some *Phragmipedium*); style thick, the stigma large and convex, the median lobe larger than lateral lobes. Fruit: three-locular or one-locular, capsular or (in *Selenipedium*) fleshy or leathery and indehiscent. (See Fig. 4-2.)

DISTRIBUTION. Tropical America, North America, Eurasia, tropical Asia, and tropical Australasia.

POLLINATION. The slipper orchid flower is a one-way trap flower, with entrance into the lip easy from the front, but exit much easier to the rear, where the insect must first pass beneath the stigma and then an anther; as far as known, none offers any reward. *Cypripedium* is usually pollinated by small bees (Andrenidae, Halictidae, or Megachilidae), while the very large *C. acaule* is pollinated by queen bumblebees (*Bombus*). Though bee pollination has been reported for *C. reginae*, Vogt (1990) finds syrphid flies to be the main pollinators, with flower beetles as occasional pollinators. Most *Cypripedium* species, and presumably *Selenipedium* as well, simply offer attractive flowers that may be visited by naive bees. After a few visits the insects learn to avoid the flowers, and the percentage of visitation and pollination is relatively low. *Cypripedium acaule* is notable for the low percentage of pollination (Gill 1990), but that species is now unusually abundant in many areas of the northeastern United States in old second growth forest. The bumblebee pollinators are probably scarcer in the forests than they would be in more open habitats. Both Vogel (1962) and Stoutamire (1967) suggest that *Cypripedium debile* may be pollinated by fungus flies, attracted by the funguslike odor of the flower. Both halictid bees and syrphid flies have been reported as pollinators of *Phrag-mipedium longifolium*. Recent work suggests that the species of *Paphiopedilum* sect. *Coryopedilum* are pollinated by syrphid flies that lay their eggs especially on the staminode (Atwood 1985); these flies normally lay their eggs where there are aphids, and they are presumably duped into laying their eggs on the lady's slipper flowers.

CHROMOSOME NUMBERS. 18, 20, 24, 26, 28, 30, 32, 36, 38, 40, 42, 44. *Cypripedium* has 20 chromosomes, while 18, 20, 22, 28, and 30 are reported for *Phragmipedium*, and 26–44 chromosomes are known in *Paphiopedilum*. These are presumably aneuploid series in both tropical genera. The chromosomes are relatively large in this subfamily.

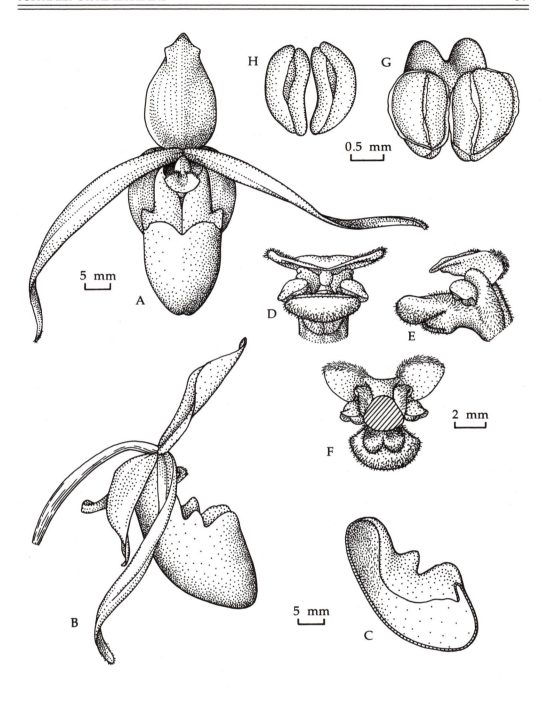

Figure 4-2. *Phragmipedium longifolium* (Cypripedioideae). (A) Flower, front view. (B) Flower, side view. (C) Longitudinal section of the lip. (D) Column, front view. (E) Column, side view. (F) Column, from base. (G) Anther. (H) Pollinia. (After Dressler 1981.)

SEED TYPE. Similar to the *Vanilla* type in *Selenipedium* (see Fig. 2-15A, B); *Limodorum* type in others.

SPECIES. 122.

GENERA. 4: *Cypripedium, Paphiopedilum, Phragmipedium, Selenipedium*.

ALTERNATIVE CLASSIFICATIONS. Some treat this as a separate family, Cypripediaceae. Atwood (1984) suggests that *Cypripedium arietinum* and *C. plectrochilum* might be better separated as the genus *Criosanthes*.

DISCUSSION. As the slipper orchids have caught the attention of both botanists and orchid growers, we have a great deal of published information, though we know relatively little about the tropical species in the field. I will refer especially to the article by Atwood (1984). It is not only one of the more recent papers, but it has a more botanical orientation and includes a careful phylogenetic analysis of the group. More is known about chromosome number and structure in the cypripedioids than in any other major orchid group.

Seed structure is distinctive in the cypripedioids, and fairly distinctive for each genus. In *Selenipedium* we find seeds with a crustose seed coat. Here, as in *Vanilla*, the anticlinal and outer walls of the cells are very thick, but the inner walls are thin and collapse at maturity. Indeed, the seeds of *Selenipedium* and *Vanilla* are similar, and the presence of vanillin in the fruits of both suggest a phyletic relationship (though distant). *Phragmipedium* and *Paphiopedilum* have seeds similar to the *Limodorum* type, though somewhat different from each other, while the seeds of *Cypripedium* are typical of the *Limodorum* type. B. Ziegler (unpublished) considers the seed of *Phragmipedium caudatum* to be a *Körnchensame* and compares it to that of *Selenipedium*, but does not specify a crustose seed coat.

Selenipedium is certainly rather primitive relative to the other genera in the three-locular ovary and the crustose seed coat, and its many-flowered bracteate raceme with one flower opening at a time is also likely to be a primitive feature (Atwood 1984). Its perianth is not deciduous, as in the conduplicate-leaved genera, and the sepal margins separate before the bud is mature, though the sepals are valvate in most other primitive orchids. Though *Selenipedium* is strikingly distinct in some obvious features, both *Cypripedium irapeanum* and *C. californicum* share some anatomical features with *Selenipedium*, thus suggesting relatively close ties between these genera. The recently described Asiatic *Cypripedium subtropicum* resembles *Selenipedium* in floral features (Chen and Lang 1986).

Atwood (1984) finds that *Cypripedium arietinum* and the Asiatic *C. plectrochilum* show some strikingly primitive features. Atwood cites the very antherlike staminode, the free lateral sepals, and the curious, spurlike appendage on the lip. The lip appendage seems quite unlike the spurs of other orchids, and could easily be a derived feature limited to this small complex. Similarly, free lateral sepals occur sometimes in other species, and could be a "reversion" related to the very narrow sepals of these species. The antherlike staminode, however, seems especially primitive.

The members of *Paphiopedilum* section *Parvisepalum* have flowers reminiscent of *Cypripedium*. Work on chloroplast DNA suggests that section *Parvisepalum* may retain primitive floral characteristics, but that similar features in *Phragmipedium schlimii* represent parallel evolution, probably associated with pollination by bees (Albert et al., in press). In the past we have interpreted *Paphiopedilum* and *Phragmipedium* as distinct lines derived from a more *Cypripedium*-like ancestor and showing several parallelisms (short stems, conduplicate leaves, etc.). There are enough resemblances between the two genera, however, to suggest that they may be sister groups; this close relationship is now confirmed by molecular systematics (V. Albert, personal communication).

Though the most primitive features are to be found in tropical species, Atwood (1984) suggests that the group may well have dispersed between continents in the north, with the more primitive groups moving to the south as climatic zones shifted. Thus, the conduplicate-leaved genera could also have arisen from a common northern ancestor and spread southward as tropical climates became more restricted.

The slipper orchids have fertile lateral anthers, doubtless a primitive feature for the orchid family (as is a fertile median anther), but Atwood stresses that the group is advanced in many features and is scarcely a "relict" group. The group early adopted a very different adaptive system; the median anther is sterile and helps to guide pollinators past the fertile anthers. Though fertile lateral anthers are shared with the apostasioids, the two groups are very different in most features, including even the forms of the fertile anthers and staminodia, and are not closely related.

REFERENCES. Albert et al., in press; Atwood 1984, 1985; Burns-Balogh and Hesse 1988; Catling and Knerer 1980; Chen and Lang 1986; Cribb 1987; Karasawa 1979; Karasawa and Aoyama 1986; Karasawa and Saito 1982; Nilsson 1979a; Sood and Mohana Rao 1988; Vogt 1990.

5

Primitive Orchids with a Single Fertile Anther

Here I treat those monandrous orchids with soft pollinia that fit neither the Spiranthoideae nor the Orchidoideae. These groups fit the Epidendroideae better than any other subfamily, but they have few derived features not shared by other monandrous orchids. The Neottieae and the Triphoreae fit the Epidendroideae less well than do the other groups treated here.

Current systematic theory emphasizes the importance of derived features, and groups with few derived features are thus problematic. They should not be grouped by ancestral features, yet their derived features are too few to determine their phylogeny with any confidence. For the present, I treat all these primitive epidendroids in alphabetic order. I discuss possible relationships between these tribes and subtribes, but I do not attempt a phylogenetic diagram. It is scarcely standard systematic procedure to treat the Spiranthoideae and Orchidoideae after the primitive Epidendroideae but before the advanced Epidendroideae; but some of these epidendroids are less advanced than either the spiranthoids or the orchidoids, while many other epidendroids are much more advanced. At the same time, the common ancestors of the spiranthoids and orchidoids may have been similar to some of these primitive epidendroids.

Though the other groups treated here are distinctly more "epidendroid" than either the Neottieae or the Triphoreae, the variation within the Gastrodieae and Triphoreae suggests that the incumbent anther may have evolved independently in two or more groups, probably from similar ancestors. Which, if any, of these groups might be better treated as a separate subfamily or subfamilies is not clear.

Table 5-1. Features, their states and polarization, as used in the phylogenetic diagram in Figure 5-5.[a]

---

1. Leaf type: 0. plicate; 1. soft herbaceous (0 → 1)
2. Subsidiary cells: 0. present; 1. lacking (0 → 1)
3. Hypochile formed by base of lip: 0. lacking; 1. present; 2. reduced (0 → 1 → 2)
4. Lip producing nectar: 1. yes, 2. no (1 → 2?)
5. Staminodial primordia: 0. massive and early in development; 1. lacking or not obvious in development (0 → 1)
6. Staminodia: 0. leaf-like; 1. finger-like; 2. lacking or obscure (0 → 1 → 2?)
7. Anther posture: 0. erect; 1. subincumbent (0 → 1?)
8. Rostellum: 0. simple with exposed glue; 1. covered by membrane; 2. sensitive (0 → 1 → 2?)
9. Anther/rostellum relationship: 0. ventral; 1. subterminal (0 → 1)
10. Column length: 0. elongate; 1. very short (0 → 1)
11. Stigma: 0. median lobe projecting from column axis; 1. projecting median lobe reduced (0 → 1?)
12. Pollen development: 0. central; 1. lateral (0 → 1)

---

[a] When there is a significant doubt as to polarity, a "0" state is not shown.

Hesse et al. (1989) survey the structure of the pollen grains in the primitive epidendroids and find a good deal of variation within the grade, with relatively primitive pollen structure in some Vanillinae, but with a mosaic of advanced and primitive features in most groups.

Possible Relationships

At this point, any attempt to determine relationships between the primitive epidendroid groups must be rather intuitive. We know a good deal about *Vanilla* and the European Neottieae, but we need to know much more about the other Vanilleae, the Gastrodieae, Palmorchideae, Pogoniinae, and Triphoreae.

The Vanilleae do not seem to be closely tied to any other group, and family status was once suggested by Lindley. I do not know of any clear derived feature to support even subfamily status, but this group does seem to be both primitive and quite isolated. The similarity to *Selenipedium* in fruit and seed structure and the presence of vanillin is noteworthy.

The Gastrodieae may be tied to the advanced Epidendroideae by the presence of thickened stems. They seem rather diverse and may represent the remnants of a once more extensive group. *Nervilia* resembles some Gastrodieae in the sectile pollinia, but in other respects the flowers seem quite different.

The Pogoniinae have been classified in the Vanilleae, but this placement is not supported by any derived features (F. N. Rasmussen 1982). At the same time, we know little about the morphology of *Galeola* and some other Vanilleae. One could treat the Pogoniinae as another distinct tribe, but I leave it as a subtribe of uncertain affinities for now. Though the Pogoniinae have fully incumbent anthers, they resemble the Neottieae in lacking subsidiary cells, in the sinuous epidermal cell walls, in the resting nucleus, and in having large chromosomes. These may well be sister groups.

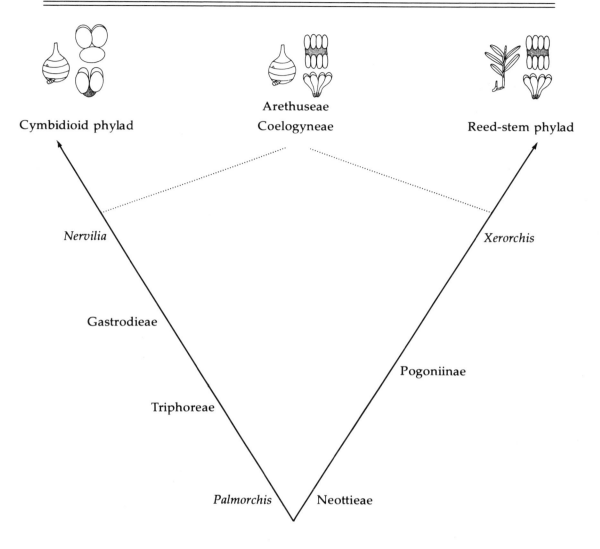

Figure 5-1. A scheme of possible relationships between the primitive and advanced Epidendroideae. This is simply an intuitive diagram based on resemblances between the various groups. The Arethuseae resemble both the cymbidioid phylad and the epidendroid phylad, but if the epidendroid phylad evolved independently of the cymbidioid phylad, as some resemblances suggest, then the Arethuseae can scarcely be closely related to both.

Even though the Triphoreae resemble the Neottieae in the form of the anther, I find no evidence of a close relationship between the two groups. It is possible, though, that the Triphoreae are more closely allied to the Palmorchideae, with which they share gregarious flowering and general similarity in flowers.

Nervilia has definite corms and may be more closely related to the advanced Epidendroideae than are any of the other groups treated in this chapter.

In Chapter 8, I discuss the cymbidioid and epidendroid phylads but recognize the possiblity that the Arethuseae and Coelogyneae could be considered a third phylad, rather than the basal members of the epidendroid phylad.

Comparing the primitive epidendroids suggests that the reed-stem epidendroid phylad and the cormous cymbidioid phylad might have evolved quite independently from different primitive groups (Fig. 5-1). Intuitively, the Arethuseae/Coelogyneae seem more closely linked to the cormous cymbidioid phylad than to the reed-stem phylad, but several different patterns could be hypothesized.

Tribe **Gastrodieae**

DISCUSSION. All the Gastrodieae are saprophytes, and the group gives the impression of being the remnants of a once more diverse group. Some genera are so distinctive that one is tempted to place each one in a separate tribe or subtribe. When we know more about the primitive "epidendroids," this might well be treated as a distinct subfamily.

Subtribe **Gastrodiinae**

DESCRIPTION. Habit: saprophytic, with fleshy tuber or coralloid underground stem (both may occur on the same plant); roots without velamen; leafless.

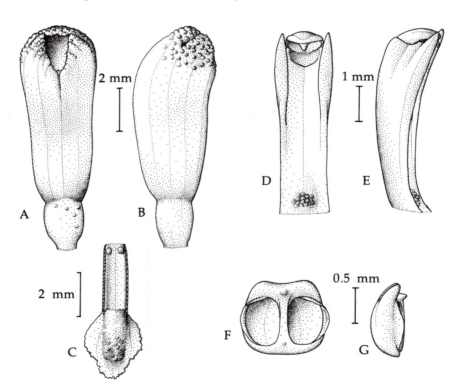

Figure 5-2. *Gastrodia siamensis* (Epidendroideae: Gastrodieae). (A) Flower, front view. (B) Flower, side view. (C) Lip, flattened. (D) Column, ventral view. (E) Column, side view. (F) Anther, ventral view. (G) Anther, side view. Drawn from material preserved in liquid. (After Dressler 1981.)

Inflorescence: terminal, simple, of one to many spiral flowers. Flowers: small or medium-small, fleshy, resupinate or not, sepals and petals may be united; column slender, often winged, usually with a prominent foot; anther terminal, incumbent; two pollinia, sectile or mealy, with or without a viscidium; exine of pollen grains coarsely reticulate to psilate-finely granulate; stigma entire, sometimes at base of column. Fruit: a capsule, often long-stalked at maturity. (See Fig. 5-2.)

DISTRIBUTION. Pantropical, but primarily in tropical Asia and Australasia.

POLLINATION. We have little information, but Jones (1981) reports that *Gastrodia sesamoides* is pollinated by a small xylocopid bee, *Exoneura*, that gathers pseudopollen from the lip.

CHROMOSOME NUMBERS. 16, 18, 22, 30, 36, 40.

SEED STRUCTURE. *Gastrodia* type.

SPECIES. 36.

GENERA. 6: *Auxopus, Didymoplexiella, Didymoplexis, Gastrodia, Neoclemensia, Uleiorchis.*

DISCUSSION. All of this group are saprophytes, many show some degree of union between sepals and petals, and all have the *Gastrodia* seed type, as far as known.

REFERENCES. Aoyama and Tanaka 1986; Jones 1985a; Liang 1984; Veyret 1980.

Subtribe **Epipogiinae**

DESCRIPTION. Habit: saprophytic from a coralloid underground stem or a fleshy tuber, leafless. Inflorescence: often thick and fleshy basally, simple, of few to many spiral flowers. Flowers: small to medium-large, resupinate or not; lip simple or concave with a prominent basal spur; column short, anther fleshy, incumbent or suberect; two pollinia, sectile, with caudicles and a viscidium; exine of pollen grains coarsely reticulate to psilate-finely granulate; stigma near base of column, entire. (See Fig. 5-3.)

DISTRIBUTION. Eurasia, tropical Africa, tropical Asia and Australasia.

POLLINATION. Bumblebee pollination has been reported for *Epipogium aphyllum*. Most populations are probably self-pollinating.

CHROMOSOME NUMBER. 68.

SEED STRUCTURE. *Gastrodia* or *Limodorum* type.

SPECIES. About 5.

GENERA. 3: *Epipogium, Silvorchis, Stereosandra.*

DISCUSSION. *Epipogium* parallels the Orchideae in several features, especially in the sectile pollinia with prominent caudicles, but there seems to be no direct phyletic relationship between these groups. The orchidoid features of *Epipogium* are probably parallelisms. *Silvorchis* has been assigned to the Orchideae, but according to Garay (1986) it is a member of the Epipogiinae. Both Vermeulen (1966) and F. N. Rasmussen (1982) agree that *Stereosandra* has a "viscidium" derived from the anther, a feature known in no other orchid. Aside from this unusual feature, *Stereosandra* could easily be a close relative of *Epipogium roseum* with a few mutations that would be viable only in a self-pollinating population.

REFERENCES. Arekal and Karanth 1981; Nageswara Rao 1987; Vermeulen 1965.

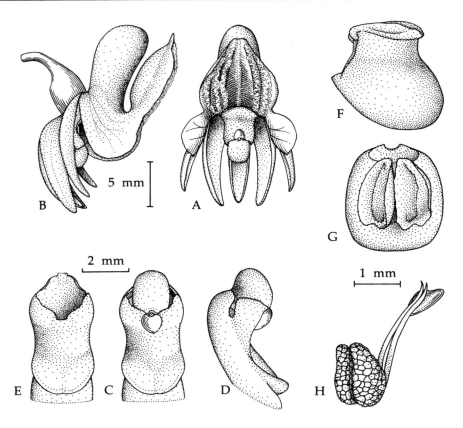

Figure 5-3. *Epipogium aphyllum* (Epidendroideae: Gastrodieae). (A) Flower, front view. (B) Flower, side view. (C) Column, ventral view. (D) Column, side view. (E) Column, after removal of anther. (F) Anther, side view. (G) Anther, ventral view. (H) Pollinarium. (A–G) Drawn from material preserved in liquid. (After Dressler 1981.)

Subtribe **Wullschlaegeliinae**

DESCRIPTION. Habit: saprophytic, stems slender, roots fusiform, with velamen of the *Calanthe* type, leafless, aerial parts clothed with bifurcate hairs. Inflorescence: terminal, of many spiral flowers, the apex nodding in bud. Flowers: tiny, resupinate or not, sepals free, lip simple, basally saccate; column short, the anther fleshy, with curved thecae, embedded in the column; two pollinia, sectile, with very slender massulae, exine of pollen grains reticulate to ornate; stigma entire, on apex of column, concave, with a prominent viscidium that is attached to the pollinia ventrally. (See Fig. 5-4.)

DISTRIBUTION. Tropical America.

POLLINATION. Self-pollinating.

CHROMOSOME NUMBER. Not known.

SEED STRUCTURE. *Gastrodia* type.

SPECIES. 2.

GENUS. *Wullschlaegelia*.

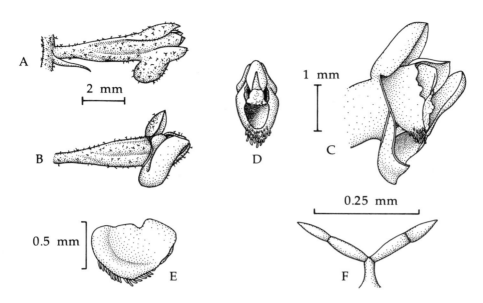

Figure 5-4. *Wullschlaegelia calcarata* (Epidendroideae: Gastrodieae). (A) Flower, side view. (B) Flower, side view, with lateral sepal removed. (C) Column, side view, with sepals and petals in place. (D) Column, front view. (E) Anther, side view. (F) Branched hair from inflorescence. (After Dressler 1981.)

DISCUSSION. *Wullschlaegelia* is distinctive in its fusiform roots, bifid hairs, and narrow massulae. Still, the seed type suggests an alliance with the Gastrodiinae, and the peculiar anther may be interpreted as incumbent. The genus certainly has nothing to do with the Cranichidinae, where it was long classified.

REFERENCE. Dressler 1980.

Tribe **Neottieae**

By the quirks of nomenclature, the European *Neottia* has been taken as the "type" of the monandrous orchids with soft pollinia. *Listera* and *Neottia* are undoubtedly very atypical. They are the most specialized of the Neottieae in the narrow sense, and even including *Cephalanthera* and *Epipactis*, the Neottieae remain a rather isolated group. They are not very closely tied to the other Epidendroideae, and, even less to the Orchidoideae or Spiranthoideae. *Cephalanthera*, especially, agrees with the Triphoreae in column structure and is relatively primitive in all of its features (see Fig. 5-7). The Neottieae are primarily northern and are more diverse in the Old World than in North America.

Chen (1982) and Chen and Tsi (1987) discuss several apparent Neottieae with more or less peloric (radially symmetrical) flowers, and interpret these as primitive genera (*Archineottia, Diplandrorchis, Sinorchis, Tangtsinia*). All of these appear to be self-pollinating forms, and I consider them to be "reversions" that are viable because of self-pollination. Chen argues that the orchids were ultimately derived from radially symmetrical flowers, so that these radially symmetrical orchids may be primitive. However, even *Neuwiedia* shows some bilateral symmetry, and it is difficult to see how some of these peloric forms could reproduce except through self-pollination. It would be interesting to cross these forms with normal *Neottia* or *Cephalanthera* and raise seedlings.

There are a number of features in which the Neottieae resemble the cypripedioids and, especially, *Cypripedium*. Both have similar fleshy but slender roots, and, indeed, the habits are so similar that it might be difficult to separate some species of *Cypripedium* and *Epipactis* without flowers. The habits of *Cypripedium* and *Epipactis* are primitive, but, at the same time, unlike those of other Epidendroideae. This similarity may reflect convergence, since both occur in similar temperate habitats, but the other primitive Epidendroideae are primarily tropical. Thus, the resemblances between the Neottieae and the lady's slippers may be a combination of primitive features and convergence.

PHYLOGENY. Probable relationships between the better known Neottieae are diagrammed in Figure 5-5. I have not included *Aphyllorchis* or the Chinese genera described by Chen. I consider *Cephalanthera* to be the basal member of the group, with the Listerinae the most derived.

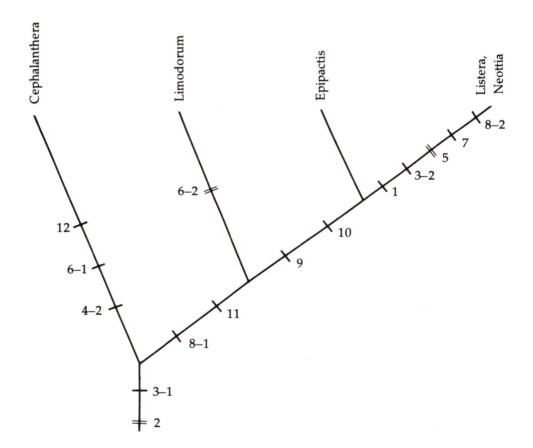

Figure 5-5. Phylogenetic diagram of the Neottieae. The features are given in Table 5-1.

ALTERNATE CLASSIFICATIONS. Some would include the Neottieae in a much wider Neottioideae, but this would be both polyphyletic and paraphyletic. F. N. Rasmussen (1982) refers *Listera* and *Neottia* to the spiranthoid complex, but seed structure and cytology both argue against such a placement (Dressler 1990a). If limited to this one tribe, the Neottioideae would seem to be a monophyletic and natural group, but it is not clear that it is more deserving of subfamily status than some of the other tribes treated in this chapter.

REFERENCES. Ackerman and Williams 1980; Burns-Balogh et al. 1987; Cauwet-Marc and Balayer 1984a; Mehra and Vij 1972; Senghas and Sundermann 1970; Tohda 1986.

Subtribe **Limodorinae**

DESCRIPTION. Habit: terrestrial or saprophytic, with elongate stems and more or less clustered roots, roots fleshy but relatively slender, without velamen. Leaves: spiral, scattered on stem, convolute, plicate, nonarticulate. Inflorescence: terminal, of several to many spiral flowers, simple. Flowers: small to medium; mid-lobe of lip often hinged to a more or less saccate base or lip with a spur; column short or long, the anther dorsal, tilted slightly downward, extending beyond the rostellum; pollinia two, soft and mealy, the pollen grains as monads or tetrads, porate or tenuate, usually reticulate; without a distinct viscidium; stigma entire. (See Fig. 5-6.)

DISTRIBUTION. Tropical Asia, northern hemisphere, and tropical Africa.

POLLINATION. *Epipactis* has relatively open nectaries near the base of the lip and attracts a variety of insects. Many species are pollinated by wasps; *E. atrorubens* and *E. palustris* are usually pollinated by bees, and *E. gigantea* is reported to be pollinated by syrphid flies. Ivri and Dafni (1977) report that *Epipactis consimilis* is pollinated by female syrphid flies, and they suggest that warts on the lip of this species mimic aphids and help attract flies that normally lay their eggs among aphids. Self-pollination is also frequent in this group. *Limodorum* and *Cephalanthera* are reportedly pollinated by bees. The flowers of *Cephalanthera* offer no nectar, and they may mimic other flowers. Dafni and Ivri (1981b) consider *C. longifolia* a facultative mimic of *Cistus*, and Nilsson (1983c) considers *C. rubra* a mimic of *Campanula*. Curiously enough, *C. rubra* flowers before the *Campanula* and is pollinated by the early-emerging male bees of the genus *Chelostoma* (Megachilidae). The female *Chelostoma* gather pollen almost exclusively from *Campanula*.

CHROMOSOME NUMBERS. 20, 32, 34, 36, 38, 40, 48, 56; chromosomes of two size classes, some quite large.

SEED STRUCTURE. *Limodorum* type.

SPECIES. 51.

GENERA. 4: *Aphyllorchis, Cephalanthera, Epipactis, Limodorum.*

DISCUSSION. The Limodorinae include a number of saprophytes. Some species assigned to *Aphyllorchis* may prove to be members of other groups.

The column structures of *Epipactis* and *Cephalanthera* are strikingly different, and a few authors have suggested that they are only distantly related. The two genera are very similar in other structural details, however, and natural

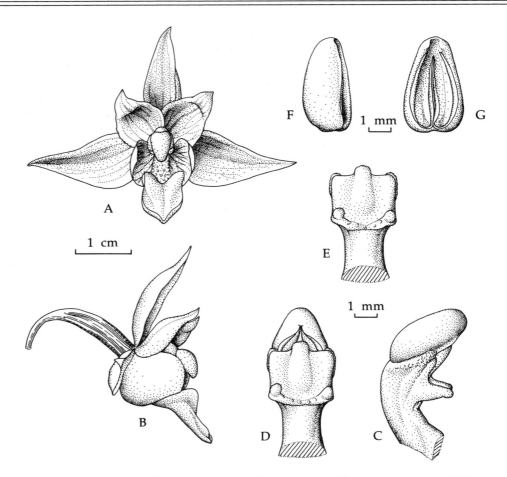

Figure 5-6. *Epipactis gigantea* (Epidendroideae: Neottieae). (A) Flower, front view. (B) Flower, side view. (C) Colunn, side view. (D) Column, ventral view, with anther in place. (E) Column, ventral view, with anther removed. (F) Anther, side view. (G) Anther, ventral view. Drawn from material preserved in liquid. (After Dressler 1981.)

intergeneric hybrids are known. Wiefelspütz (in Senghas and Sundermann 1970) considers *Epipactis gigantea* to be a link between the two genera. *Limodorum*, also, may be something of a link between these genera; superficially, at least, it resembles *Cephalanthera* more than does *Epipactis*.

The rostellum of *Epipactis* is of special interest. In autogamous species, the rostellum may be poorly developed. I had described the column as lacking a viscidium, only to be taken to task by Ed Greenwood, who sent illustrations of something very like a viscidium. The rostellar glue is covered by a membrane, as described by Darwin (1888). F. N. Rasmussen (1982) calls this a "diffuse viscidium," yet the glue is clearly delimited and quite detachable. If the pollinia were attached to the rostellar glue before the intervention of pollinators, this would be a typical (detachable) viscidium. The pollinia lack caudicles, and such pre-attachment is unlikely.

Cephalanthera is commonly said to lack a rostellum, but this assertion is questionable under any definition of rostellum. The median stigma lobe may be degraded in some autogamous forms, but it normally projects markedly from the column axis (Fig. 5-7) and has a marginal viscid glue that surely functions in pollination (F. N. Rasmussen 1982).

In some species of *Epipactis*, the lateral stigma lobes are unusually prominent, rather like those of *Corymborkis*. Intuitively, this would seem a primitive feature, but it may be a secondary adaptation that improves pollen deposition.

REFERENCES. Brantjes 1981; Dafni and Ivri 1981b; Nilsson 1978b, 1983c; Tanaka and Yokota 1982.

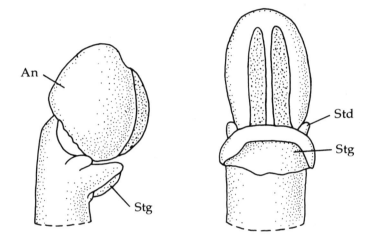

Figure 5-7. The upper column of *Cephalanthera longifolia,* showing the anther (An), stigma (Stg) and staminodia (Std). (A) ventral view. (B) lateral view. (Based on F. N. Rasmussen 1986b, Fig. 2, and other photographs).

Subtribe **Listerinae**

DESCRIPTION. Habit: terrestrial or saprophytic; small herbs with slender stems, roots fleshy but slender, without velamen. Leaves: if present usually two, subopposite, convolute, subplicate or soft herbaceous, nonarticulate. Inflorescence: terminal, simple, of several to many spiral flowers. Flowers: small, usually greenish; lip with a shallow, superficial nectary; column short, anther suberect, tilted slightly downward, subequal to the rostellum; pollinia two, soft and mealy, pollen grains in tetrads, porate or tenuate, reticulate; stigma entire, rostellum sensitive, extruding a drop of adhesive when touched.

DISTRIBUTION. Northern hemisphere.

POLLINATION. The genus *Listera* is primarily pollinated by nectar-seeking wasps and flies. Some species, such as *Listera cordata*, produce (to our senses) a foul perfume, but the fungus gnats that are the main pollinators apparently seek nectar, rather than fungi. Nilsson (1981) found *Listera ovata* to be visited by a variety of insects; ichneumons, saw-flies, and beetles were the most important pollinators.

CHROMOSOME NUMBERS. 20, 34, 36, 38, 40, 42, 46, 56; chromosomes of two size classes, some quite large.

SEED STRUCTURE. *Limodorum* type.

SPECIES. About 29.

GENERA. 2: *Listera, Neottia.*

DISCUSSION. The Listerinae seem closely related to the Limodorinae by the form of their roots, pollen structure, seed structure, and chromosome morphology. They show a derived, more condensed habit, soft herbaceous leaves, and an unusual type of rostellum. One would imagine that a membrane-covered rostellum is a necessary first step in the evolution of a sensitive rostellum, and just such a rostellum occurs in *Epipactis. Listera* and *Neottia* are probably derived from an *Epipactis*-like ancestor (see Fig. 5-5). From a cladistic viewpoint, then, the Listerinae is rather a subclade of the Limodorinae, and the division of the Neottieae into subtribes is questionable.

REFERENCES. Chen 1982; Nilsson 1981; Schick et al. 1987.

Tribe **Nervilieae**

DESCRIPTION. Habit: terrestrial, with a globose corm. Roots without velamen. Leaves: only one, subcircular or somewhat lobed, basally truncate to deeply cordate, convolute, plicate, nonarticulate. Inflorescence: terminal, simple, of one to several spiral flowers. Flowers: medium-small, resupinate or erect; column slender, anther incumbent; two pollinia, sectile, without caudicles or viscidia; exine of pollen grains coarsely reticulate to psilate; stigma entire.

DISTRIBUTION. Tropical Asia, Australasia, and Africa.

POLLINATION. Some forms are self-pollinating, and Pettersson (1989) reports pollination by wasps and small bees in Africa. The flowers of some species produce nectar, and others lack nectar and apparently achieve pollination by false advertisement.

CHROMOSOME NUMBERS. 20, 36, 40, 54.

SEED STRUCTURE. An anomalous type resembling the *Goodyera* type.

SPECIES. 65.

GENUS. *Nervilia.*

DISCUSSION. *Nervilia* differs from the Gastrodieae in habit, seed type, and flower structure. The Gastrodieae sometimes have thickened stems but not corms. The aspect of the *Nervilia* flower is reminiscent of the Triphoreae, and both were once included in *Pogonia.* The broad, plicate leaf of *Nervilia* is also rather reminiscent of that of *Monophyllorchis,* which may be an indication of phyletic relationship. Flowering is gregarious in some species (Pettersson 1989).

REFERENCES. Pettersson 1989, 1990.

Tribe **Palmorchideae**

DESCRIPTION. Habit: terrestrial, stem tough and reedlike, up to one meter tall. Roots fleshy but slender. Leaves: spiral or subdistichous, convolute, plicate, nonarticulate. Inflorescence: terminal or lateral, simple, of few to many spiral flowers. Flowers: small, resupinate; lip partly united with column along midline; column slender, anther terminal, incumbent; four pollinia, soft but coherent; stigma entire, emergent, without viscidium; exine of pollen grains psilate. Fruit: fleshy, three-chambered. (See Fig. 5-8.)

DISTRIBUTION. Tropical America.

POLLINATION. In Panama, pollination of three species by parasitic bees of the genus *Osiris* has been observed (Dressler 1983a).

CHROMOSOME NUMBER. Not known.

SEED STRUCTURE. Seeds ellipsoid, with a sclerotic seed coat, the testa cells polygonal, shallowly concave, with or without a definite cell boundary ridge, with small warts (see Fig. 2-15D, E).

SPECIES. About 12.

GENUS. *Palmorchis.*

DISCUSSION. Though widespread in tropical America, *Palmorchis* remains poorly known. Recent observations in French Guiana and Panama have given us

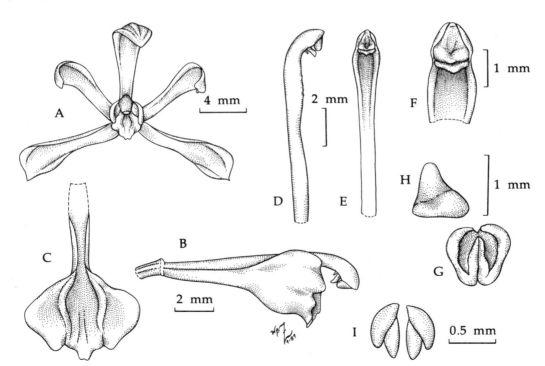

Figure 5-8. *Palmorchis silvicola* (Epidendroideae: Palmorchideae). (A) Flower, front view. (B) Lip and column, side view. (C) Lip, partially flattened. (D, E) Column side and ventral views. (F) Apex of column, ventral view. (G, H) Anther, ventral, and side views. (I) Pollinia. A through G drawn from material preserved in liquid.

better information on their biology and structure (Veyret 1981; Dressler 1983a). The plants are easily overlooked on the forest floor, especially when not in flower, as they resemble palm seedlings or broad-leaved forest grasses. Their delicate, short-lived flowers are produced gregariously, that is, the plants of a given population flower on the same day at irregular intervals. Thus, they may be overlooked by botanists, and the flowers are poorly preserved in herbarium material. The fruits are probably eaten by birds or other animals and the seeds thus dispersed.

Palmorchis is surely one of the most primitive of the Epidendroideae. The seeds are unlike those of any other orchid, and *Palmorchis* differs from the Vanilleae in the union of the lip and column (along the midline, rather than marginal) and in the thin, plicate leaves.

REFERENCES. Dressler 1983a; Veyret 1981.

Tribe **Triphoreae**

DESCRIPTION. Habit: terrestrial or saprophytic; stem slender, roots often fleshy or with nodular tuberoids. Leaves: subdistichous or solitary at midstem, or much reduced in near-saprophytes; convolute and plicate or soft herbaceous, nonarticulate. Inflorescence: terminal, simple, of one to several spiral flowers. Flowers: small, resupinate or not; column slender, anther erect or subincumbent, the thecae nearly as long as the anther, or restricted to basal portion; pollinia coherent, as tetrads, the pollen grains reticulate with a slender, open net, exine present only in the reticulum; with or without viscidium; stigma entire. (See Fig. 5-9.)

DISTRIBUTION. Tropical America and into North America.

POLLINATION. Some forms are self-pollinating; the flower structure suggests pollination by small bees in other cases. Pollination of *Triphora* by halictid bees was observed by Lownes (1920) and has been confirmed by Medley (1979), who found *Augochlora pura* to be the principal pollinator in Indiana.

CHROMOSOME NUMBER. 44 rather small chromosomes.

SEED STRUCTURE. *Eulophia* type.

SPECIES. 28.

GENERA. 3: *Monophyllorchis, Psilochilus, Triphora*.

DISCUSSION. The nearly erect anther of *Psilochilus* seems rather out of place in the Epidendroideae, and I had previously left this group in systematic limbo (Dressler 1981). The anther is somewhat incumbent in both *Triphora* and *Monophyllorchis*, and the anthers of all three genera have epidendroid thick apical beaks. *Monophyllorchis* is distinctive in its single, wide, plicate leaf. All three genera are similar in flower, seed, and pollen structure, and all three show gregarious flowering (when not self-pollinating). The pollinia of *Triphora* are very soft, but those of *Monophyllorchis* and *Psilochilus* are more coherent, at least superficially similar to those of *Palmorchis*. The rostellar glue of *Triphora trianthophora* is not quite a true viscidium, but has a thin membrane. When the membrane is broken, most of the glue is removed, as in *Epipactis* or *Dendrobium*. *Triphora trianthophora* has nectar at the base of the lip.

This is a distinctly primitive group, superficially similar to the Neottieae and, like many primitive groups, difficult to place.

REFERENCE. Medley 1979.

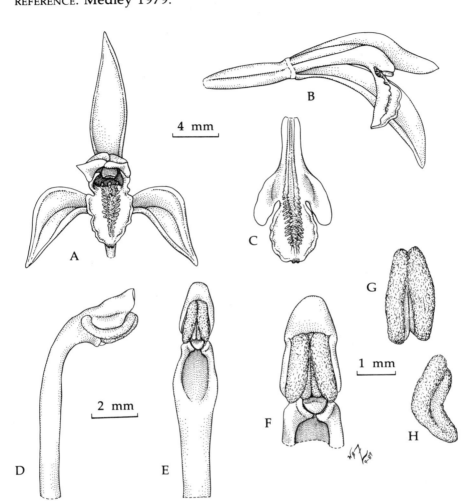

Figure 5-9. *Triphora trianthophora* (Epidendroideae: Triphoreae). (A) Flower, front view. (B) Flower, side view, with near sepal and petal removed. (C) Lip, flattened. (D, E) Column, side, and ventral views. (F) Apex of column, ventral view. (G, H) Pollinia, ventral and side views.

Tribe **Vanilleae**

DISCUSSION. The Vanilleae may be a natural group, as currently delimited, though the status of the Pogoniinae is uncertain. Most members have lens-shaped seeds with the outer testa wall thickened (*Vanilla* type), or winged seeds in which the central portion, around the embryo, shows the same seed coat features as the *Vanilla* type (*Galeola* type). The Vanillinae have some members with very wide, petiolate leaves of the *Alisma* type, and these leaves may show definite reticulate venation. This leaf type may be a primitive feature for the family, but it may equally be a specialization found only in this tribe. The Vanilleae are a relict group about equally well represented in Australasia and tropical America.

Subtribe **Galeolinae**

DESCRIPTION. Habit: saprophytic herbs or vines, stem fleshy. Leaves lacking or rudimentary. Inflorescence: terminal or axillary, simple, of few to many spiral flowers. Flowers: small to large, resupinate, usually with an abscission layer between ovary and perianth; column slender or moderately thick, anther terminal, incumbent; pollen soft and mealy or forming definite, but soft pollinia; exine of pollen grains finely punctate to granulate-psilate; stigma entire, without viscidium. Fruit: capsular or fleshy.

DISTRIBUTION. Tropical and subtropical Asia to Australia.

POLLINATION. Not known.

CHROMOSOME NUMBER. 28.

SEED STRUCTURE. *Vanilla* or *Galeola* type.

SPECIES. 27.

GENERA. 4: *Cyrtosia, Erythrorchis, Galeola, Pseudovanilla*.

DISCUSSION. All the Galeolinae are saprophytic, though the plants of *Pseudovanilla* show leaflike bracts that may have some green coloring. *Cyrtosia* has a fleshy, brightly colored, apparently indehiscent fruit with *Vanilla* type seeds. The other genera have capsules with winged seeds.

REFERENCE. Garay 1986.

Subtribe **Vanillinae**

DESCRIPTION. Habit: shrubs or vines; stem woody or fleshy; roots not thickened, velamen of the *Calanthe* type. Leaves: spiral or subdistichous, sometimes reduced to scales, convolute, fleshy or leathery and more or less net-veined, nonarticulate. Inflorescence: terminal or axillary, simple, of few to many spiral flowers. Flowers: medium to large, resupinate, usually with an abscission layer between ovary and perianth, sometimes with a three-lobed calyculus at ovary apex; lip partly united with the column by the margins, more or less trumpet-shaped; column slender, anther terminal, incumbent, generally somewhat ventral; pollen soft and mealy, as monads, polyporate, exine laevigate; stigma entire or three-lobed, usually emergent, without viscidium. Fruit one-locular or three-locular, capsular or leathery, usually opening as two unequal valves. (See Fig. 5-10.)

DISTRIBUTION. Pantropical.

POLLINATION. In tropical America, *Vanilla* is often pollinated by euglosssine bees. The pollen forms a triangular mass on the bee's scutellum (Fig. 5-11). Most Vanillinae are probably pollinated by large bees. In some cases, the same bees that pollinate the flowers may be dispersal agents for the seeds as well (Madison 1981). Vanillin is a strong attractant for the males of *Eulaema cingulata* and some other euglossine species, and it is probably this perfume that attracts the males to the

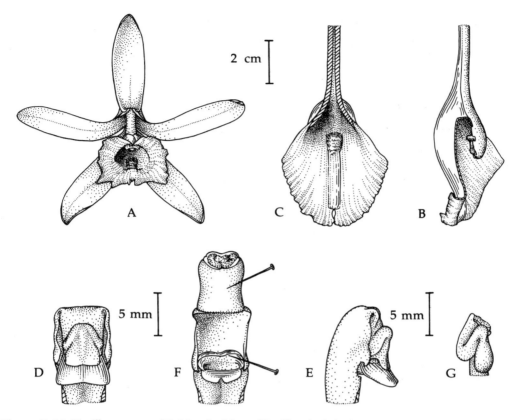

Figure 5-10. *Vanilla pompona* (Epidendroideae: Vanilleae). (A) Flower, front view. (B) Lip and Column, side view. (C) Lip, spread. (D) Apex of column, with anther in place. (E) Apex of column, side view. (F) Apex of column with both anther and midlobe of stigma tipped back. (G) Anther, side view. (After Dressler 1981.)

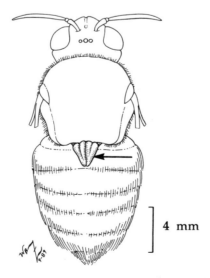

Figure 5-11. Pollen of *Vanilla* on the scutellum of *Eulaema cingulata* (arrow).

ripe fruit. The tiny seeds stick readily to forceps, and probably stick just as well to a bee's feet.

CHROMOSOME NUMBERS. 28, 30, 32.

SEED STRUCTURE. *Vanilla* or *Galeola* type.

SPECIES. 119.

GENERA. 5: *Clematepistephium, Dictyophyllaria, Epistephium, Eriaxis, Vanilla.*

DISCUSSION. Among the orchids, the Vanillinae are unusual for their often viny habit and for their sometimes leathery and net-veined leaves. Both of these may be primitive features for the Orchidaceae, though they are not found in the Apostasioideae or the primitive Cypripedioideae. The union of the lip to the column offers a derived feature that seems to unite this group.

REFERENCE. Madison 1981.

Subtribe **Lecanorchidinae**

DESCRIPTION. Habit: saprophytic, leafless, stem slender, from a scaly rhizome. Inflorescence: terminal, simple, of few to several flowers. Flowers: small or medium small, resupinate or not (?), with a prominent calyculus and an abscission layer at base of perianth, sepals and petals free, similar, lip basally united with column by margins, simple or three-lobed, the midlobe often hairy; column slender, anther incumbent, pollen soft and mealy, as monads, polyporate, exine foveolate; stigma three-lobed, without viscidium. Fruit: three-locular, capsular.

DISTRIBUTION. Tropical Asia.

POLLINATION. Not known.

CHROMOSOME NUMBER. 36.

SEED STRUCTURE. An anomalous seed type (not fitting any of the named types) with thin-walled testa cells, somewhat like the *Gastrodia* type.

SPECIES. About 20.

GENUS. *Lecanorchis.*

DISCUSSION. These small saprophytic plants appear superficially very different from the Vanillinae, but the calyculus beneath the perianth and especially the polyporate pollen grains indicate that *Lecanorchis* is either a sister group of the Vanillinae or a derivative of that group.

REFERENCES. Aoyama et al. 1987; Hashimoto 1990.

PRIMITIVE ORCHIDS OF UNCERTAIN CLASSIFICATION

Subtribe **Pogoniinae**

DESCRIPTION. Habit: terrestrial or saprophytic; may have nodular root tuberoids or produce shoots from a rhizome-like root, velamen of the *Calanthe* type. Leaves: distichous or spiral (?), scattered, basal or clustered or whorled at midstem, convolute, soft herbaceous or leathery, nonarticulate. Inflorescence: terminal, simple, of one to several spiral flowers. Flowers: medium-small to large, with an abscission layer between ovary and perianth; column usually slender, anther terminal, incumbent; pollen soft and mealy, as monads or tetrads, without viscidia; exine of pollen grains laevigate; stigma entire.

DISTRIBUTION. Tropical America, North America, and eastern Asia.

POLLINATION. As far as known, *Cleistes, Isotria,* and *Pogonia* attract bees by deceit. The cluster of hairs on the lip of *Pogonia* simulates a cluster of pollen bearing anthers, and the gullet flower of *Cleistes* is also visited by food-seeking bees. *Cleistes* offers no nectar (Gregg 1989), though the mealy pollen is sometimes gathered by bees (Gregg 1991). Only a portion of the pollen is released when the anther is tipped back, so a single flower could deposit pollen on several different visitors.

CHROMOSOME NUMBERS. 18, 20, 24; large chromosomes.

SEED STRUCTURE. *Limodorum* type or anomalous.

SPECIES. 64.

GENERA. 5: *Cleistes, Duckeella, Isotria, Pogonia, Pogoniopsis.*

DISCUSSION. F. N. Rasmussen (1982) argues that the Pogoniinae may be misplaced in the Vanilleae. Having no unusual derived features, their status and relationships remain uncertain, but they could be derived from something like the Galeolinae. On the other hand, the type of resting nucleus, the stomata without recognizable subsidiary cells, the sinuous epidermal cell walls, the seed type and the large chromosomes all suggest an alliance to the Neottieae.

REFERENCES. Gregg 1989, 1991; Mehrhoff 1983.

Thaia

Thaia was assigned to the Neottieae by Seidenfaden. The column structure is strange, but not markedly discordant in the Neottieae. Burns-Balogh et al. (1987), who place the Neottieae in the Neottioideae, suggest that *Thaia* should be referred to the Epidendroideae. W. Barthlott (personal communication) finds a seed sample (in poor condition) to be rather unlike the *Limodorum* seed type, but similar to the *Gastrodia* type.

Xerorchis

Though the presence of eight pollinia in *Xerorchis* suggests the Arethuseae or Epidendreae, the habit, persistent leaves, and seed type suggest the Pogoniinae. Little is known of this curious genus, but it could be a key to understanding the evolution of the epidendroid phylad (or phylads). Sweet (1970) interprets the leaves as floral bracts.

6

Lady's Tresses and Relatives, the Subfamily **Spiranthoideae**

The Spiranthoideae are characterized especially by the terminal viscidium, but this feature occurs independently in other groups. *Diceratostele* is treated as a primitive member of this subfamily, though it apparently has no viscidium. The Tropidieae, like *Diceratostele*, are distinctive in having slender, tough, rather "woody" stems with plicate leaves. The Cranichideae all have soft herbaceous leaves and often have fleshy roots, but not root–stem tuberoids. The Cranichideae seem to be a clearly delimited group, and the Spiranthoideae could well be monophyletic, but the evidence is inconclusive. There is little firm evidence even that the Tropidieae and the Cranichideae are close relatives. Except for *Diceratostele*, this is a group united by a derived feature, but a feature that occurs too often to place much trust in it. The spiranthoid orchids do not show close ties to any other group.

The Spiranthoideae, as a whole, are a tropical group, with some representatives in cooler areas both to the north and the south. The Tropidieae are tropical and occur in both hemispheres. Their low diversity and spotty distribution suggest a relict pattern. The Goodyerinae are mainly Old World, with a few genera in the Americas, both tropical and northern, while the rest of the Cranichideae are mainly American, except for *Manniella*, *Pachyplectron*, and a few species of *Spiranthes*.

Though Figure 3-8 suggests a sister group relationship between the Spiranthoideae and the Orchidoideae, the evidence for such a relationship is weak. Both the Spiranthoideae and some Diurideae share the *Goodyera* seed type. The seed of

111

Table 6-1. Features, their states and polarization, as used in phylogenetic diagrams in Chapter 6.[a]

 1. Resting nucleus: 1. round or rod prochromosome type
 2. Velamen: 0. lacking or *Calanthe* type; 1. *Spiranthes* type $(0 \rightarrow 1)$
 3. Roots: 0. slender but fleshy; 1. thick & fleshy $(0 \rightarrow 1)$
 4. Stem arising from: 0. base of older shoot; 1. upper part of older shoot $(0 \rightarrow 1)$
 5. Stem basally: 0. erect (from short rhizome); 1. creeping $(0 \rightarrow 1)$
 6. Leaf texture: 0. plicate; 1. soft herbaceous $(0 \rightarrow 1)$
 7. Subsidiary cells: 0. epidendroid pattern; 1. cranichid pattern $(0 \rightarrow 1)$
 8. Flower position: 0. resupinate; 1. nonresupinate $(0 \rightarrow 1)$
 9. Retrorse, nectariferous lobules at base of lip: 0. lacking; 1. present $(0 \rightarrow 1)$
10. Staminodia: 0. leaflike; 1. fingerlike; 2. lacking or obscure $(0 \rightarrow 1 \rightarrow 2 \text{ or } 1 \rightarrow 0 \rightarrow 1?)$
11. Anther apex: 1. obtuse; 2. acute
12. Anther/rostellum relationship: 0. ventral; 1. terminal $(0 \rightarrow 1)$
13. Hamulus: 0. lacking; 1. rudimentary; 2. large, with sclerenchyma core $(0 \rightarrow 1 \rightarrow 2)$
14. Pollen texture: 0. powdery; 1. sectile; 2. brittle $(0 \rightarrow 1, 0 \rightarrow 2)$
15. Intercellular spaces between testa cells: 0. absent; 1. present $(0 \rightarrow 1)$
16. Seed type: 0. *Diuris*(?); 1. *Goodyera* $(0 \rightarrow 1)$

[a] When there is a significant doubt as to polarity, a "0" state is not shown.

Diceratostele is essentially of the *Goodyera* type, but lacks the intercellular spaces so typical of the *Diuris* type and usually of the *Goodyera* type. If the *Goodyera* type seeds of the Spiranthoideae were derived from ancestors with the *Diuris* type, a close relationship between the Spiranthoideae and Orchidoideae would be implied. Equally possible, though, are independent origins from different ancestral types. The relatively primitive habit of *Diceratostele* and the Tropidieae suggests that the Spiranthoideae may be an isolated group.

Burns-Balogh and Bernhardt (1985) have interpreted the spiranthoid condition, with a terminal viscidium and with the anther base near or below the stigma base, as primitive for the monandrous orchids, but I consider it a derived condition. Terminal viscidia have evolved independently in several epidendroid genera, and in every case the anther base is near or below the stigma base (Fig. 6-1). Thus, I interpret the position of the anther base as structurally (or developmentally) linked with the terminal viscidium. I can find no independent feature correlated with the terminal viscidium. It seems clear that the primitive condition for the Orchidaceae is one with soft pollinia and with neither a rostellum nor a viscidium. An orchid flower with the anther and stigma parallel and subequal could not function without both a viscidium and coherent pollinia, so it would be an improbable starting point for the evolution of the monandrous orchids. The adaptive value of a terminal viscidium is clear, however, especially for narrow flowers into which the pollinator does not enter, inserting only its mouthparts. In such a flower, a terminal viscidium would be the most efficient in depositing pollinia on the visitor.

Some representative pollinaria of the Spiranthoideae are illustrated in Figure 6-2.

PHYLOGENY. Figure 6-3 shows the hypothesized relationships within the Spiranthoideae. The subtribes of the Cranichideae (except for Manniellinae and Pachyplectrinae) are diagrammed in Figure 6-6. Since both the Prescottiinae and Cranichidinae have nonresupinate flowers and the Spiranthinae seem very close

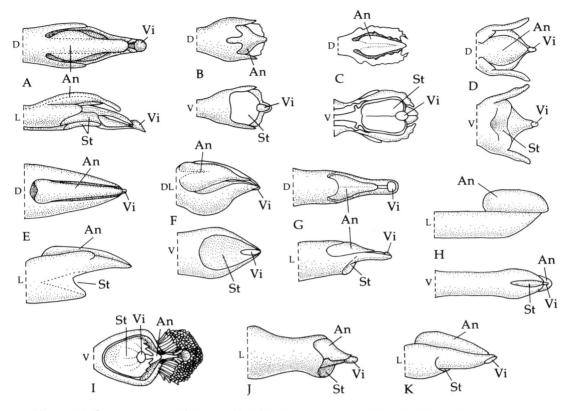

Figure 6-1. Some genera with "spiranthoid" column structure. Note that the anther base is at or below the stigma base in each case. (A) *Sarcoglottis neglecta* (Spiranthinae). (B) *Prescottia stachyodes* (Prescottiinae). (C) *Diuris punctata* (Diuridinae). (D) *Prasophyllum appendiculatum* (Prasophyllinae). (E) *Podochilus cultratus* (Podochilinae). (F) *Thelasis hongkongensis* (Thelasiinae). (G) *Meiracyllium trinasutum* (Meiracylliinae). (H) *Notylia trisepala* (Oncidiinae). (I) *Thelymitra crinita* (Thelymitrinae). (J) *Pleurothallis penduliflora* (Pleurothallidinae). (K) *Appendicula hexandra* (Podochilinae). A, B are spiranthoid in the strict sense; C, D, and I are Diurideae; E–H, J, and K are Epidendroideae. View of the column: D, dorsal; DL, dorsolateral; L, lateral; V, ventral; An, anther; St, stigma; Vi, viscidium. (Adapted from: A, Dressler 1981; B, Rodríguez 1986; C, D, I, Nicholls 1969; E, H, Dressler 1990B; F, Hu 1977; G, Dressler 1960; J, Dunsterville and Garay 1966; K, Seidenfaden 1986.)

to the Prescottiinae, treating the Spiranthinae as a "reversion" to resupination may be more parsimonious. The Cranichidinae are quite distinctive, but the proper positions of the other small Old World groups (not shown in the diagram) remain uncertain.

ALTERNATE CLASSIFICATIONS. The spiranthoids might be included in the Orchidoideae (Dressler 1986), but the evidence for a close relationship is weak. The Goodyerinae were associated with the Tropidiinae by Dressler (1981). As F. N. Rasmussen (1982) indicates, the habit is dissimilar in the Goodyerinae and the Tropidieae, leaving only the sectile pollinia as a shared derived feature. It is possible, of course, that they share a close common ancestor and that soft herbaceous leaves evolved independently in the Goodyerinae and the (other) Cranichideae.

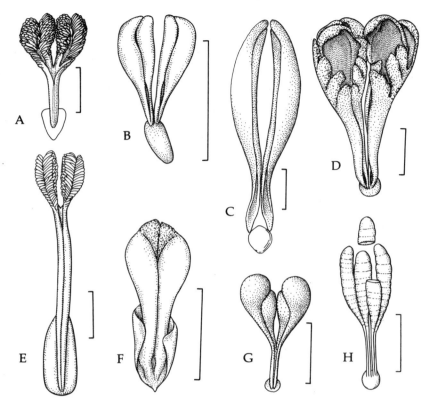

Figure 6-2. Pollinaria of Spiranthoideae. (A) *Macodes sanderiana* (Goodyerinae). (B) *Pelexia* species (Spiranthinae). (C) *Sarcoglottis neglecta* (Spiranthinae). (D) *Porphyrostachys pilifera* (Prescottiinae). (E) *Pristiglottis montanum* (Goodyerinae). (F) *Platythelys querceticola* (Goodyerinae). (G) *Ponthieva brenesii* (Cranichidinae). (H) *Solenocentrum costaricense* (Cranichidinae). Scale 1 mm. (A, C, and G after Dressler 1981.)

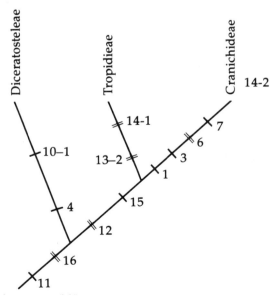

Figure 6-3. Phylogenetic diagram of Spiranthoideae. The features are given in Table 6-1.

Tribe **Diceratosteleae**

DESCRIPTION. Habit: terrestrial, with slender, reed-like stems, new stems often arising from the upper part of older stems; with stilt roots; leaves spiral, convolute, plicate, nonarticulate, scattered along the stem. Inflorescence: terminal, simple, flowers spiral. Flowers: small, resupinate; lip simple; column slender, with prominent staminodia, somewhat bent near the base of the anther; anther elongate, acute; pollinia 4, slender, without stipe or viscidium (?); stigma entire, rostellum somewhat emergent.

DISTRIBUTION. Tropical West Africa.

POLLINATION. Not known.

CHROMOSOME NUMBER. Not known.

SEED STRUCTURE. *Goodyera* type, but without intercellular spaces.

SPECIES. 1.

GENUS. *Diceratostele*.

DISCUSSION. The plants of *Diceratostele* resemble those of *Corymborkis*, but the flowers are not at all "spiranthoid" (see Fig. 6-4). It is not closely related to *Palmorchis*, as was suggested by its original description, but may be a primitive element in the Spiranthoideae. I have seen only two fully mature flowers, both of which were incomplete. The mature column is shown in Figure 6-4. The flowers of *Diceratostele* have been collected only rarely, suggesting that they might be short-lived, like those of *Tropidia*.

REFERENCE. Rasmussen and Rasmussen 1979.

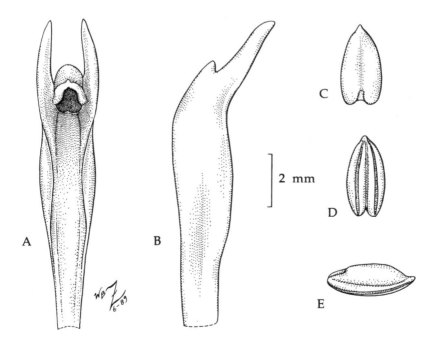

Figure 6-4. Column and anther of *Diceratostele gabonense* (Spiranthoideae: Diceratosteleae). (A) Column, ventral view. (B) Column, lateral view. (C, D, E) Anther, dorsal, ventral, and side views. (After Dressler 1990b.)

Tribe **Tropidieae**

DESCRIPTION. Habit: terrestrial, with slender, hard, rather woody, reed-like stems, to 3 m in height, stem branched or not; roots sometimes with nodular tuberoids, velamen of the *Calanthe* type. Leaves: spiral or distichous, convolute, plicate, nonarticulate, scattered along the stem. Inflorescence: terminal or lateral, simple or branched, flowers spiral. Flowers: small or medium, resupinate; lip saccate or spurred basally or narrow and not saccate; column short or elongate, with the anther dorsal, erect, and subequal to the rostellum; two pollinia, sectile, with a hamular stipe and a terminal viscidium; stigma entire. (See Fig. 6-5.)

DISTRIBUTION. Pantropical.

POLLINATION. Not known. The long, pale flowers of *Corymborkis* suggest moth pollination (F. N. Rasmussen 1977).

CHROMOSOME NUMBERS. 40, 56, 58, 60.

SEED STRUCTURE. *Goodyera* type.

SPECIES. About 43.

GENERA. 2: *Corymborkis, Tropidia*.

DISCUSSION. I have treated this problematic group as a tribe, as it does not seem closely tied to any other group. In superficial form the column is rather like that of the Cranichideae, but the structure of the hamulus is distinctive

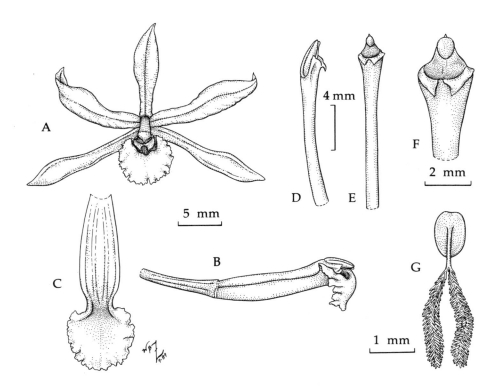

Figure 6-5. *Corymborkis veratrifolia* (Spiranthoideae: Tropidieae). (A) Flower, front view. (B) Lip and column, side view. (D, E) Column, side, and ventral views. (F) Apex of column, showing stigma and viscidium. (G) Pollinarium. A through F drawn from material preserved in liquid.

(Rasmussen 1977). The seed type is the same as that of the Cranichideae, suggesting that they may, indeed, be sister groups. Both the habit of the Tropidieae and their pantropical distribution suggest an old, relict group, but the flower structure is rather specialized. The nodular storage roots of *Tropidia* are similar to those of *Apostasia*, but the two genera are otherwise quite different. The flowers of *Tropidia* are short-lived and open successively on each branch of the inflorescence. (See Fig. 6-5.)

REFERENCE. Rasmussen 1977.

Tribe **Cranichideae**

DISCUSSION. The Cranichideae resemble the Tropidieae in the form of the column but have soft herbaceous leaves. The stigma of the Cranichideae is often bilobed, but the developmental studies of F. N. Rasmussen (1982) and Kurzweil (1988) indicate that the lateral lobes of the stigma are rudimentary or lacking. The fertile stigma is largely part of the median lobe, even when it is divided into two separate portions. As noted above, this tribe is quite distinctive. Dr. W. L. Stern (personal communication) reports that he and his coworkers have found several anatomical features that are diagnostic for the Cranichideae; these features are quite lacking in the Tropidieae, in *Diceratostele*, and in *Cryptostylis*.

REFERENCE. Thoda 1985.

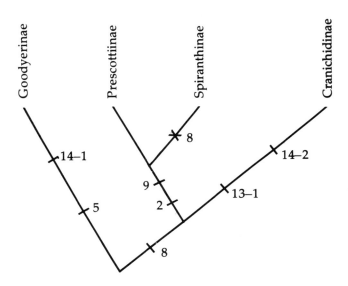

Figure 6-6. Phylogenetic diagram of the Cranichideae. The features are given in Table 6-1.

Subtribe **Goodyerinae**

DESCRIPTION. Habit: terrestrial or infrequently saprophytic or epiphytic; rhizome creeping, occasionally thicker than leafy stem, leafy stem slender; roots somewhat fleshy, as ridges only in *Cheirostylis*, roots without velamen, or velamen of the *Calanthe* type. Leaves: spiral, scattered, or clustered, convolute, soft herbaceous, nonarticulate, often marked with pale or pink spots or lines. Inflorescence: terminal, unbranched, of few to many spiral flowers. Flowers: small or small-medium, commonly white or pale green, usually resupinate; lip saccate at base, or forming a spur, often with emergent glands within the sack or spur, lip may be basally united with the column; blade often two-lobed, the claw or the blade sometimes fringed; anther dorsal, erect, subequal to rostellum; two or four pollinia, sectile, sometimes with caudicles or tegular stipe; exine of pollen grains reticulate-heterobrochate; stigma entire or bilobed, with the lateral lobes rudimentary. (See Fig. 6-7.)

DISTRIBUTION. Widespread, but mainly tropical and especially in tropical Asia.

POLLINATION. Bumblebee pollination is reported for *Goodyera*, which normally has a relatively high fruit set. Butterfly pollination is reported for *Ludisia*, and some other Asiatic genera suggest Lepidopteran pollination by their floral form and color.

CHROMOSOME NUMBERS. 20, 22, 24, 26, 28, 30, 32, 40, 42, 44.

SEED STRUCTURE. *Goodyera* type.

SPECIES. About 476.

GENERA. 35 in two alliances:

1. With a single stigmatic area: *Aspidogyne, Cystorchis, Dicerostylis, Dossinia, Erythrodes, Eurycentrum, Evrardia, Gonatostylis, Goodyera, Herpysma, Hylophila, Kreodanthus, Kuhlhasseltia, Lepidogyne, Ligeophila, Ludisia, Macodes, Moerenhoutia, Orchipedum, Papuaea, Platylepis, Platythelys, Pristiglottis, Rhamphorhynchus, Stephanothelys.*

2. With two distinct stigmatic areas: *Anoectochilus, Chamaegastrodia, Cheirostylis, Eucosia, Gymnochilus, Hetaeria, Myrmechis, Tubilabium, Vrydagzynea, Zeuxine.*

DISCUSSION. The Goodyerinae usually have a distinctive habit, with a soft, herbaceous rhizome that differs from the erect, leafy stem mainly in it's horizontal orientation and the presence of roots. *Zeuxine* has the stems densely clumped, with short rhizomes. *Cheirostylis* is unusual in that the rhizome is thick and fleshy and the roots are mere bumps or ridges on the rhizome. Many Goodyerinae have attractively marked foliage, thus giving rise to the name "jewel orchids."

Though the vegetative features of the Goodyerinae are relatively uniform, the flowers are varied and often quite specialized. We find tegular stipes, well-developed caudicles, asymmetrical, twisted columns, and much variation in the structure of the lip and column. The pollinia are sectile but vary greatly in form. In some cases, a sheath is formed about the massulae.

REFERENCES. Hurusawa and Kakadzu 1982; Sera 1990; Xu et al. 1987.

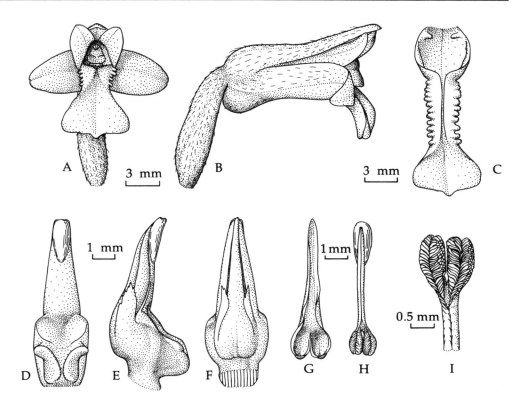

Figure 6-7. *Pristiglottis montana* (Spiranthoideae: Cranichideae). (A) Flower, front view. (B) Flower, side view. (C) Lip. (D) Column, ventral view. (E) Column, side view. (F) Column, dorsal view. (G) Anther. (H) Pollinarium. (I) Pollinia. (After Dressler 1981.)

Subtribe **Prescottiinae**

DESCRIPTION. Habit: terrestrial; leafy stem and rhizome short, roots clustered, fleshy, velamen of the *Spiranthes* type. Leaves few to several, spiral, clustered, convolute, soft herbaceous, nonarticulate, somewhat petiolate. Inflorescence: terminal or lateral (or on separate shoot), simple, of few to many spiral flowers. Flowers: small to medium, nonresupinate, lip simple, may be concave or fringed; column usually very short; anther dorsal, erect, subequal to rostellum, four pollinia, soft and mealy, with a terminal viscidium, without caudicles; exine of pollen grains reticulate-heterobrochate; stigma entire, lateral lobes rudimentary.

DISTRIBUTION. Tropical America.

POLLINATION. Eumenid wasps and a calliphorid fly have been observed pollinating *Myrosmodes cochleare* (Berry and Calvo 1991). *Porphyrostachys pilifera* shows the bird-pollination syndrome, with bright red, tubular flowers.

CHROMOSOME NUMBERS. Not known.

SEED STRUCTURE. *Goodyera* type.

SPECIES. 99.

GENERA. 7: *Aa, Altensteinia, Gomphichis, Myrosmodes, Porphyrostachys, Prescottia, Stenoptera.*

DISCUSSION. These genera have been placed in the Cranichidinae, but that group seems distinct and clearly defined, even if this one is less so. These are all rather small-flowered, except for *Porphyrostachys pilifera.* All have broad, rather laminar rostella, which contrast markedly with the pointed rostella of the Cranichidinae. Except for *Prescottia,* these are largely inhabitants of higher elevations in the Andes.

The Prescottiinae resemble the Spiranthinae in their velamen type and in the form of their pollinia, and the retrorse lobules typical of the Spiranthinae are present at least in *Prescottia.*

Subtribe **Spiranthinae** Lindley

DESCRIPTION. Habit: terrestrial or occasionally epiphytic; the leafy stem and rhizome short, roots clustered, usually fleshy, velamen of the *Spiranthes* type. Leaves: spiral, clustered, convolute, soft herbaceous, commonly petiolate, nonarticulate. Inflorescence: terminal, with several to many spiral flowers. Flowers: small to medium, resupinate; lip basally saccate or not, base of the blade commonly with two retrorse lobules or appendages; flowers commonly with a deep nectary united with the ovary, this sometimes with a prominent chin or spur; column usually erect; anther dorsal, normally erect, and subequal to the rostellum; four pollinia, soft and mealy, the viscidium terminal or rarely attached to middle of pollinia; caudicles weakly developed in a few cases; exine of pollen grains reticulate-heterobrochate; stigma entire or two-lobed, lateral lobes rudimentary. (See Fig. 6-8.)

DISTRIBUTION. Primarily American, but with a few representatives in all habitable continents; lacking in tropical and southern Africa.

POLLINATION. Even the spiral flower arrangement of *Spiranthes* is apparently an adaptation to pollination by bees and especially by bumblebees. Catling (1983) reports that *S. lucida* is pollinated mainly by halictid bees. I have observed bumblebees pollinating *Pelexia ekmanii* in southern Brazil. Euglossine bees have been collected with pollinaria of *Sarcoglottis* behind their mouthparts. The red or yellow flowers of the *Stenorrhynchos* complex appear to be adapted for hummingbird pollination, and bird pollination has been observed in *Sacoila,* though out of its normal range (Catling 1987).

CHROMOSOME NUMBERS. 24, 26, 28, 30, 32, 36, 44, 46.

SEED STRUCTURE. *Goodyera* type.

SPECIES. About 409.

GENERA. About 30–40 (41 names listed): *Aracamunia, Aulosepalum, Beloglottis, Brachystele, Buchtienia, Coccineorchis, Cotylolabium, Cybebus, Cyclopogon, Degranvillea, Deiregyne, Dichromanthus, Discyphus, Dithyridanthus, Eltroplectris, Eurystyles, Funkiella, Galeottiella, Greenwoodia, Hapalorchis, Helonema, Kionophyton, Lankesterella, Lyroglossa, Mesadenella, Mesadenus, Odontorrhynchos, Oestlundorchis, Pelexia, Pseudogoodyera, Pteroglossa, Sacoila, Sarcoglottis, Sauroglossum, Schiedeella, Skeptrostachys, Spiranthes, Stalkya, Stenorrhynchos, Stigmatosema, Thelyschista.*

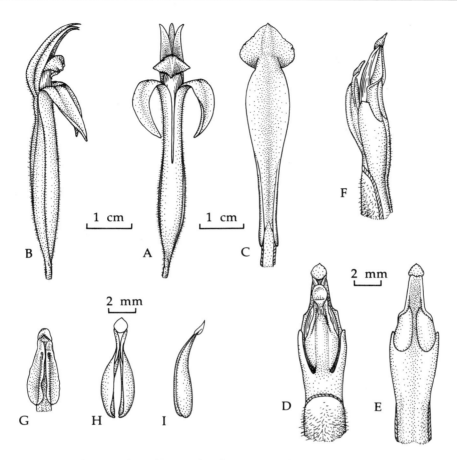

Figure 6-8. *Sarcoglottis neglecta* (Spiranthoideae: Cranichideae). (A) Flower, front view. (B) Flower, side view. (C) Lip, flattened. (D) Column, dorsal view. (E) Column, ventral view. (F) Column, side view. (G) Anther. (H) Pollinarium, dorsal view. (I) Pollinarium, side view. (After Dressler 1981.)

DISCUSSION. The Spiranthinae have received very different treatments, ranging from one extremely diverse genus to as many as 40 genera. In 1982 Garay described a number of additional genera, but offered little discussion in support of his treatment. At about the same time, Balogh published a paper recognizing about 16 genera. This paper also gave few details, though a later paper (Burns-Balogh 1986) is more detailed. I would prefer to accept a system with a moderate number of genera, but at this point, one cannot evaluate either of these classifications without redoing much of the work.

The drawing of the pollinia of *Nothostele* (Garay 1982) suggests that this genus should be referred to the Cranichidinae, as should *Pseudocranichis*, according to Burns-Balogh (1986).

Greenwood (1982) describes the viscidium structure of this group, including the wedge-type viscidium of *Cyclopogon*, *Pelexia*, and *Sarcoglottis*, which, together, form a clear clade. Greenwood describes the wedge-type viscidium as nonviscid, but, as indicated by Burns-Balogh and Robinson (1983), the wedge-type viscidium is covered by a membrane that must be ruptured to expose the glue.

REFERENCES. Balogh 1979, 1982; Burns-Balogh 1986; Burns-Balogh and Robinson 1983; Catling 1982, 1983, 1987; Garay 1982; Greenwood 1982; Larson and Larson 1987; Martínez 1985.

Subtribe **Manniellinae**

DESCRIPTION. Habit: terrestrial, the leafy stem and rhizome relatively short, roots clustered, not very fleshy. Leaves: spiral, convolute, soft herbaceous, nonarticulate, petiolate. Inflorescence: terminal, of many spiral flowers. Flowers: small, resupinate, each with a prominent cuniculus; lip simple, with two retrorse basal lobules; column bent sharply up and down again (rather like a door latch), with two prominent staminodia that clasp the anther laterally; anther dorsal, erect; two pollinia, soft and mealy, viscidium terminal; stigma entire.

DISTRIBUTION. West tropical Africa.

POLLINATION. Not known.

CHROMOSOME NUMBER. 84 (probably tetraploid).

SPECIES. 1.

GENUS. *Manniella*

DISCUSSION. This genus seems allied to the Spiranthinae and has been included in that group by Mansfeld (1937) and Garay (1982). The oddly bent column and the prominent staminodia are distinctive (illustrated by Hallé 1965).

Subtribe **Pachyplectroninae**

DESCRIPTION. Habit: terrestrial; the leafy stem and rhizome short; roots clustered or somewhat scattered, fleshy. Leaves: spiral, clustered, convolute, soft herbaceous, petiolate, nonarticulate. Inflorescence: terminal, with several to many spiral flowers. Flowers: small, resupinate; lip with a prominent basal spur; column with prominent staminodia that enfold the sides of the anther, anther erect; two pollinia, soft and mealy, with a distinct viscidium; stigma entire.

DISTRIBUTION. New Caledonia.

POLLINATION. Not known.

CHROMOSOME NUMBER. Not known.

SPECIES. 2.

GENUS. *Pachyplectron*.

DISCUSSION. *Pachyplectron* is poorly known, but appears to be spiranthoid and may be allied to *Manniella*. The red-brown leaves are very hard to find among the dead leaves of the forest floor.

Subtribe **Cranichidinae**

DESCRIPTION. Habit: terrestrial or lithophytic, rarely epiphytic; the leafy stem and rhizome short, roots clustered, rather fleshy, velamen of the *Calanthe* type. Leaves: spiral, clustered, convolute, soft herbaceous (or weakly plicate), nonarticulate, commonly petiolate. Inflorescence: terminal, with several to many spiral flowers. Flowers: small or medium, nonresupinate; lip often united with column, or spurred; column straight; anther dorsal, erect, subequal to rostellum; four pollinia, clavate, brittle, with small terminal caudicles; exine of pollen grains reticulate-heterobrochate; viscidium terminal, with a small hamular stipe; stigma entire.

DISTRIBUTION. Tropical America, with a few species in temperate North America.

POLLINATION. *Ponthieva racemosa* is reportedly visited by halictid bees (Luer 1972), but we have found that the lip produces oil, rather than nectar, so that the pollinators may be anthophorid bees that gather oil as food for their larvae.

CHROMOSOME NUMBER. 46.

SEED STRUCTURE. *Goodyera* type.

SPECIES. 152.

GENERA. 9: *Baskervilla, Cranichis, Fuertesiella, Nothostele, Ponthieva, Pseudocentrum, Pseudocranichis, Pterichis, Solenocentrum.*

DISCUSSION. Though the Cranichidinae usually have been interpreted a bit more broadly, this group is one of the clearest in the tribe. The columns are rather pointed, there is often some degree of union between lip or petals and the column, and the brittle, club-shaped pollinia with small hamular stipes are quite distinctive. In *Solenocentrum* the brittle pollinia are constricted into segments, the constrictions apparently representing breakage points (see Fig. 6-2). Thus, the pollinia function in much the same way as sectile pollinia.

In *Ponthieva*, especially, the nonresupinate flowers simulate resupinate flowers, in that the two petals, together, form a sort of "pseudolip;" the inconspicuous true lip forms an elaiophore or nectary. *Nothostele* and *Pseudocranichis* were described as members of the Spiranthinae (Garay 1982), but the drawing of the pollinia suggests that *Nothostele* is a member of the Cranichideae; Burns-Balogh (1986) considers *Pseudocranichis*, also, to be a member of the Cranichidinae.

REFERENCE. Morales 1986.

7

Orchis and Its Allies, the Subfamily **Orchidoideae**

The Orchidoideae include the plants named "*Orchis,*" part of a distinctive and quite specialized subgroup. At the same time, I include the closer relatives of that group, though most of them are less specialized. The Orchidoideae all have soft herbaceous leaves. Most genera have root–stem tuberoids, and the absence of tuberoids in a few genera may be secondary. Though the South American Chloraeinae have relatively simple flower structure, the Australian genera are quite diverse, with some relatively primitive features as well as quite specialized features. Pseudocopulation and other forms of deceit dominate their pollination spectrum. The greatest floral specialization, though, is seen in the Orchideae and Diseae, where the basal position of the viscidia seems to be immutable. *Satyrium* and members of the Coryciinae must turn their anther upside-down to have functionally "terminal" viscidia, analogous to those of the spiranthoids.

The Orchidoideae as a group may be basically southern, though the Orchideae are now widespread and are the principal orchid group of the north temperate area. Australia and/or South America may have been colonized from Antarctica in more favorable times. The African segment of the subfamily shows great specialization and considerable diversity.

The distinctive root–stem tuberoids of the Orchidoideae combine some stem structure with root structure. The plants survive the dry season or the winter as dormant tuberoids. During the growing season a new tuberoid is paired with an older one, and they are usually globose, so that the Greek word *orchis*, or testicle,

Table 7-1. Features, their states and polarization, as used in phylogenetic diagrams in chapter 7.[a]

---

1. Roots: 0. monomorphic; 1. dimorphic (root–stem tuberoids) (0-1)
2. Leaf type: 0. plicate; 1. soft herbaceous (0 → 1)
3. Subsidiary cells: 0. present; 1. absent (0 → 1)
4. Flower: 0. resupinate; 1. nonresupinate (0 → 1)
5. Dorsal sepal: 0. shallowly concave; 1. hooded (0 → 1)
6. Dorsal sepal: 0. without a spur; 1. with a spur (0 → 1)
7. Petals: 0. free from column; 1. basally fused with column (0 → 1)
8. Lip: 0. three-lobed; 1. simple (0 → 1)
9. Lip: 0. free from column; 1. basally fused with column (0 → 1)
10. Lip: 0. without a prominent appendage; 1. with an appendage (0 → 1)
11. Lip: 0. with a spur; 1. without a spur (0 → 1)
12. Column length: 0. moderate; 1. long and slender (0 → 1)
13. Anther apex: 1. obtuse; 2. acute (1 → 2?)
14. Anther attachment: 0. narrow; 1. anther base fused with column (0 → 1)
15. Anther cells basally: 0. parallel; 1. divergent (0 → 1)
16. Anther: 0. erect; 1. horizontal; 2. pendant (0 → 1 → 2)
17. Anther/rostellum relationship: 0. ventral; 1. basal (0 → 1)
18. Stigma: 0. concave; 1. convex (0 → 1)
19. Stigma: 0. sessile; 1. stalked (0 → 1)
20. Viscidium: 0. lacking or simple; 1. double (0 → 1)
21. Rostellum lateral lobes: 0. small; 1. well developed (0 → 1)
22. Pollen texture: 0. mealy; 1. sectile (0 → 1)
23. Seed types—1. *Goodyera/Diuris*; 2. *Orchis/Disa*
24. Seed types—0. *Orchis*; 1. *Disa* (0 → 1)

---

[a] When there is a significant doubt as to polarity, a "0" state is not shown.

was used for one genus and *satyrium* for another. In most Diurideae the anther tapers to a narrow point, and this is evident, too, in some of the more primitive Orchideae, though many Orchideae have the two pollen-bearing portions of the anther widely separated. Some pollinaria of the Orchidoideae are illustrated in Figure 7-1.

PHYLOGENY. Figure 7-2 shows my hypothesis about relationships within the subfamily Orchidoideae. The Orchideae and Diseae, together, seem quite clear-cut, and their origin from a common ancestor similar to the modern Diurideae is quite probable. From a strictly cladistic viewpoint, the Diurideae might be considered unresolved or paraphyletic, with the Orchideae and Diseae as a derived subclade, but their distributions suggest sister group status.

In general flower structure, some Diurideae resemble the Neottieae, and some authors place both groups in the subfamily Neottioideae. This, however, is a group based on relatively primitive features, and it is quite unlikely that it is monophyletic. There is said to be an Italian plant that may be a hybrid of *Epipactis* and *Gymnadenia* (mentioned in Dressler 1981). Unfortunately, nothing has been published on this intriguing plant, so it is difficult to judge its significance. In any case, the Neottieae differ from the Orchideae in several important features, and there are few derived features to suggest a close relationship (unless the absence of distinct subsidiary cells is uniquely derived). Both cytology (Yokota 1987) and floral development (Kurzweil 1987b, 1988) indicate that the Neottieae and Orchideae are only distantly related. The Orchidoideae may be closely related to the Spiranthoideae, but the evidence for such a relationship is weak.

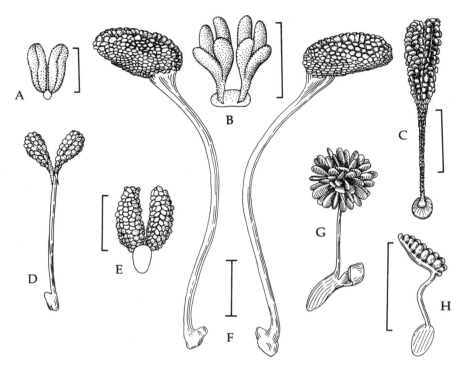

Figure 7-1. Pollinaria of Orchidoideae. (A) *Townsonia viridis.* (B) *Eriochilus cucullatus.* (C) *Disa venosa.* (D) *Prasophyllum striatum.* (E) *Piperia elongata.* (F) *Habenaria avicula.* (G) *Disperis fanniniae.* (H) *Disperis pusilla.* Scale 1 mm. (A, B, D after Nicholls 1969; C, after Vogel 1959; E, after Ackerman 1977; H, after Verdcourt 1968.)

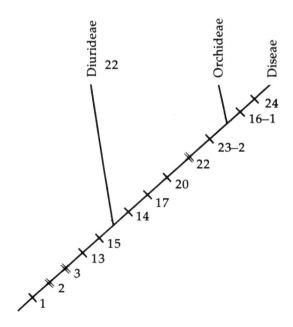

Figure 7-2. Phylogenetic diagram of Orchidoideae. The features are given in Table 7-1.

Tribe **Diurideae**

DISCUSSION. Australian workers generally recognize two complexes within this group (see Table 7-2). The Diuridinae, Thelymitrinae, and Prasophyllinae make up the *Diuris* complex, with prominent, vascularized staminodia that are at least partially free from the column. In the *Caladenia* complex, on the other hand, the staminodia may be prominent but are fully united with the column. Presumably they are not vascularized, but I find no clear information. Each of these groups is specialized in different ways. Intuitively, one would consider *Diuris* to be primitive, without a true column. It is quite possible, though, that this condition has come about through the evolution of a very short column from an ancestor more like *Caladenia* or *Chiloglottis* (see Fig. 7-3). Thus, polarity is not clear in this tribe.

Though some genera of the *Caladenia* complex share the *Diuris* seed type with the Diuridinae, others show the *Goodyera* type, and some genera have both types in different species. In this complex, these two types apparently correlate with ecological factors, rather than with phyletic groups. According to M. Clements (personal communication), species that produce their seeds before an unfavorable season have the *Diuris* seed type, while the *Goodyera* seed type is found in other species of the same genera whose seeds may germinate soon after being shed.

ALTERNATE CLASSIFICATION. A case can be made for limiting the Diurideae to the subtribes of the *Diuris* group (including Rhizanthellinae and probably Cryptostylidinae), and treating the remaining subtribes as the Geoblasteae Barbosa Rodrigues (tribal name published in *Vellozia* 1:132. 1891). Szlachetko (1991) treats the Diurideae as a distinct subfamily, the Thelymitroideae, and considers something near *Diuris* to be ancestral to the Thelymitrinae and Caladeniinae.

REFERENCES. Ackerman and Williams 1981; Jones 1981; Peakall and James 1989; Szlachetko 1991.

Table 7-2. A comparison of the *Diuris* and *Caladenia* groups.

Feature	*Diuris* group	*Caladenia* group
Root–stem tuberoids	present	present (except Chloraeinae)
Leaves generally	narrow	wide
Column	short	long
Seed type	*Diuris*	*Diuris* or *Goodyera*
Staminodia	prominent, vascularized	smaller, not vascularized?
Viscidium	present	present or absent

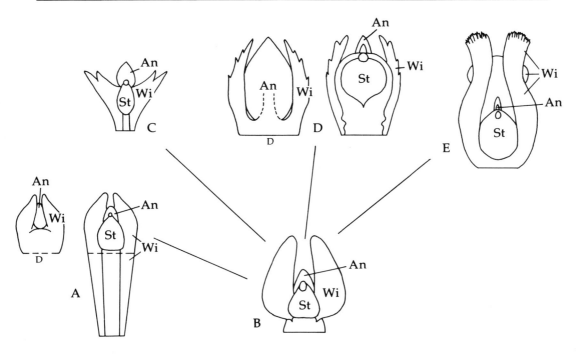

Figure 7-3. The hypothetical derivation of the *Diuris* group from a *Caladenia*-like ancestor by reducing the length of the column. (1) A column like that of *Chiloglottis* (but with a viscidium). (2) A column like that of *Epiblema*. (3) The column of *Genoplesium*. (4) The column of *Diuris*. (5) The column of *Thelymitra*. An, anther; D, dorsal view; St, stigma; Wi, staminode or staminodial wing.

Subtribe **Chloraeinae**

DESCRIPTION. Habit: terrestrial or rarely epiphytic, roots fleshy, or with root–stem tuberoids (*Codonorchis*), velamen of the *Calanthe* type; stems slender. Leaves: spiral, clustered or scattered on stem, whorled in middle of stem in *Codonorchis*, convolute, soft herbaceous, nonarticulate. Inflorescence: terminal, simple, of one to many spiral flowers. Flowers: medium, often with osmophores on lateral sepals; lip usually adorned with warts or calluses; may have paired nectaries between lip and column; column slender, usually arched; anther dorsal, erect, projecting beyond the rostellum; two pollinia, of tetrads or monads, soft and mealy, without viscidium; exine of pollen grains foveolate to reticulate; stigma entire. (See Fig. 7-4.)

DISTRIBUTION. Southern South America north to Peru, and New Caledonia.

POLLINATION. The pollination of *Chloraea* by *Colletes*, a rather primitive bee, has been reported (Gumprecht 1975).

CHROMOSOME NUMBER. 16.

SEED STRUCTURE. *Goodyera* type.

SPECIES. About 80.

GENERA. 6: *Bipinnula, Chloraea, Codonorchis, Gavilea, Geoblasta, Megastylis*.

ALTERNATE CLASSIFICATION. There is no sharp distinction between the

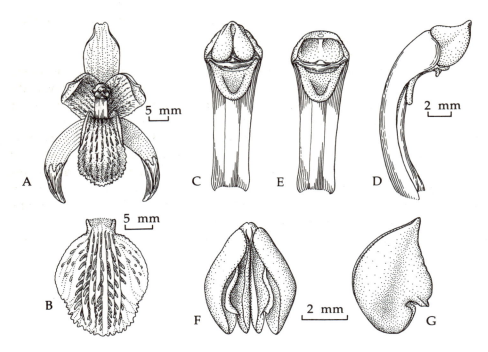

Figure 7-4. *Chloraea lamellata* (Orchidoideae: Diurideae). (A) Flower, front view. (B) Lip, flattened. (C) Column, ventral view. (D) Column, side view. (E) Column, with anther removed. (F) Anther, ventral view, (G) Anther, side view. Drawn from material preserved in liquid. (After Dressler 1981.)

Chloraeinae and the Caladeniinae, and it may be that the two (or parts of each) should be combined. Chloraeinae is the earlier name.

DISCUSSION. The Chloraeinae lack root–stem tuberoids except in *Codonorchis*. I assume that the tuberoids have been lost in most Chloraeinae; an alternative might be to place *Codonorchis* in the Caladeniinae or in its own subtribe. The Chloraeinae appear relatively primitive in lacking a viscidium, and show few of the bizarre specializations that occur in some of their Australian relatives. The two groups appear to be closely related, however, and may not be clearly separable. According to Bernhardt (manuscript) the bases of the column wings fuse to form a tube between the column and the lip in some species of *Gavilea*.

REFERENCES. Ackerman and Williams 1981; Correa 1956; Izaguirre de Artucio 1973.

Subtribe **Caladeniinae**

DESCRIPTION. Habit: terrestrial, with root–stem tuberoids, velamen of the *Calanthe* type. Leaves: 1 or 2, basal, convolute, soft herbaceous, nonarticulate. Inflorescence: terminal, simple, of one to several spiral flowers. Flowers: resupinate, small to medium; lip usually adorned with fringes, hairs, or calluses, hinged at base or not; column arched, often broadly winged above; anther conic, dorsal, erect, projecting beyond the rostellum; four or eight pollinia, of monads or tetrads, soft and mealy; exine of pollen grains foveolate to reticulate; stigma entire, with or without a distinct viscidium.

DISTRIBUTION. Australasia.

POLLINATION. Some species of *Caladenia* are pollinated by male thynnid wasps (pseudocopulation) and evidently produce a fragrance that mimics the pheromone of the female wasp. The calli of the lip may mimic the form of a wingless female wasp, as in the Drakaeinae, but to our eyes, at least, the mimicry is usually less impressive (except in *C. barbarossa*). Pseudocopulation may have evolved independently in these two groups. Peakall (1984) and Peakall et al. (1987) report the pollination of *Leporella fimbriata* by winged male ants, also a case of pseudocopulation. *Eriochilus* and some species of *Caladenia* are pollinated by other bees, wasps or syrphid flies, apparently attracted by false advertisement.

CHROMOSOME NUMBERS. 38, 42, 44, 46, 48.

SEED STRUCTURE. The *Diuris* type, the *Goodyera* type, and intermediates

Table 7-3. Fungal symbionts of the Diurideae.[a]

	Tulasnella	Sebacina	Ceratobasidium
Caladeniinae (*sensu stricto*)			
A			
Adenochilus		+	
Caladenia		+	
Elythranthera		+	
Eriochilus		+	
Glossodia		+	
Leporella		+	
B			
Burnettia	+		
Lyperanthus	+		
Drakaeinae			
Arthrochilus	+		
Caleana	+		
Chiloglottis	+		
Drakaea	+		
Paracaleana	+		
Spiculaea	+		
Acianthinae			
Acianthus	+		
Corybas	+		
Cyrtostylis		+	
Pterostylidinae			
Pterostylis			+
Diuridinae			
Diuris	+		
Orthoceras	+		
Calochilus	+		
Thelymitra	+		
Rhizanthellinae			
Rhizanthella			+
Prasophyllinae			
Microtis		+	
Genoplesium			+
Prasophyllum			+

[a] Based on Clements 1988 and personal communication; see also Warcup 1981.

between *Diuris/Disa* or *Diuris/Goodyera*.

SPECIES. 121.

GENERA. 10:

1. With definite column wings, usually prominent: *Adenochilus, Aporostylis, Caladenia, Elythranthera, Eriochilus, Glossodia, Leporella*.

2. Without evident column wings: *Burnettia, Lyperanthus, Rimacola*.

DISCUSSION. With the segregation of the Drakaeinae, the Caladeniinae becomes a more homogeneous group. As far as known, all except *Lyperanthus* have mycorrhizal symbionts of the genus *Sebacina* (Table 7-3). *Lyperanthus*, however, is associated with *Tulasnella*, but it surely would not fit in the Drakaeinae. *Burnettia, Lyperanthus* and *Rimacola* are similar in column structure and might form a natural group. The rostellum in the Caladeniinae may be associated with the ventral surface of the anther or with the base, and viscidia occur in some genera.

REFERENCE. Stoutamire 1983.

Subtribe **Drakaeinae**

DESCRIPTION. Habit: terrestrial, with root–stem tuberoids. Leaves: 1 to several, basal, convolute, soft herbaceous, nonarticulate. Inflorescence: terminal, simple, of one to several spiral flowers. Flowers: resupinate or nonresupinate, small to medium; lip usually adorned with warts or hairs, borne on a hinged stalk; column arched, broadly winged above; anther conic, dorsal, erect, projecting beyond the rostellum; four pollinia, of monads or tetrads, soft and mealy; exine of pollen grains laevigate to foveolate; stigma entire, with a distinct viscidium in *Spiculaea*.

DISTRIBUTION. Australasia.

POLLINATION. In all of this group part of the lip mimics a female thynnid wasp, and the flowers are pollinated by the male wasps (pseudocopulation). The "pseudowasp" is borne on a hinged stalk; in *Chiloglottis* the stalk is short, but in the other genera the stalk is longer and may be hinged either at the base or the middle (the base apparently representing a column foot). In *Drakaea* and *Spiculaea*, when the male wasp seizes the decoy and attempts to fly off with it, it can only swing against the column, where pollinia may be received or deposited. The movement is thus passive in these genera. In *Caleana* the movement is active according to both Cady (1965) and Firth (1965).

CHROMOSOME NUMBERS. 40, 44.

SEED STRUCTURE. The *Diuris* type, or *Diuris/Disa* intermediates.

SPECIES. 21.

GENERA. 5: *Arthrochilus, Caleana, Chiloglottis, Drakaea, Spiculaea*.

DISCUSSION. Reinstating the Drakaeinae in the sense of Schlechter, including *Chiloglottis*, appears to form a monophyletic group, while the Caladeniinae become somewhat less diverse. It is possible, though, that this is a subclade of the Caladeniinae.

REFERENCES. Clemesha 1968; Stoutamire 1985.

Subtribe **Pterostylidinae**

DESCRIPTION. Habit: terrestrial with root–stem tuberoids. Leaves: several, spiral, clustered or scattered, broad, convolute, soft herbaceous, nonarticulate. Inflorescence: terminal, simple, of one to several spiral flowers. Flowers: small to medium; resupinate; the dorsal sepal commonly hoodlike, and the laterals often forming slender tails; lip narrow, hinged, with a basal appendage, sensitive; column arched, slender below, winged above, the wings with projections both distally and basally; anther dorsal, more of less erect; four pollinia, soft and mealy, of monads, without viscidium; exine of pollen grains reticulate; stigma entire.

DISTRIBUTION. Australasia.

POLLINATION. The greenhoods are fly traps with sensitive lips that trap small gnats (Diptera) against the column when touched. Bates (1977) finds that the flowers of *Pterostylis boormanii* attract only male fungus gnats. All members of the *Pterostylis rufa* group may be pollinated by pseudocopulation (Beardsell and Bernhardt 1983).

CHROMOSOME NUMBERS. 42, 50.

SEED STRUCTURE. *Goodyera* type.

SPECIES. 100.

GENUS. *Pterostylis*.

DISCUSSION. The greenhoods are quite distinct; to which other subtribe they are most closely related is difficult to determine.

REFERENCES. Bates 1977; Stoutamire 1982.

Subtribe **Acianthinae**

DESCRIPTION. Habit: terrestrial, with tuberoids or tuberoidlike structures. Leaves: basal or at midstem, solitary, broad and cordate or lobed, convolute, soft herbaceous or weakly plicate, nonarticulate. Inflorescence: terminal, simple, of one to several spiral flowers. Flowers: small or medium, resupinate; column slender and arched or short and relatively thick, without wings; anther terminal, pollinia of monads or tetrads, four, in two pairs, with a viscidium; exine of pollen grains coarsely reticulate, rugulate or clavate; stigma entire.

DISTRIBUTION. Australasia and tropical Asia.

POLLINATION. Though the flowers of *Acianthus* and *Corybas* seem very different, both are reported to be pollinated by fungus gnats. The flowers of *Corybas* may be fungus mimics.

CHROMOSOME NUMBERS. 38, 44, 54.

SEED STRUCTURE. *Goodyera* or *Diuris* type.

SPECIES. About 147.

GENERA. 5: *Acianthus, Corybas, Cyrtostylis, Stigmatodactylus, Townsonia*.

DISCUSSION. Though *Acianthus* and *Corybas* are vegetatively similar and apparently pollinated by similar insects, their flowers are quite different. They may not be close allies, but the variation in both groups needs careful study. *Cyrtostylis* proves to be distinctive in its seed (*Diuris* type), and its fungal symbiont is *Sebacina*, rather than *Tulasnella*. It is quite possible that this subtribe is artificial.

Burns-Balogh and Funk (1986) treat the Acianthinae as members of the tribe Triphoreae, a viewpoint supported by Hesse et al. (1989). Quite aside from what appear to be root–stem tuberoids, the Acianthinae agree with the Diurideae in their seed type and stomata without subsidiary cells. Unlike the anther of the Triphoreae, the anther of *Acianthus* is very short, quite without a fleshy beak, and the locules diverge markedly toward the base of the anther. Further, the anther is strongly united with the column apex basally and dorsally, paralleling the Orchideae in this feature. Hesse et al. (1989) argue that the staminode margins are united between the anther and the stigma in both *Acianthus* and *Triphora*. According to Nicholls (1969), the structure in *Acianthus* is a bifid rostellum with a separate viscidium formed by each lobe. In *Triphora*, also, I suspect that the lobes clasping the rostellar glue are rostellar or stigmatic. The internal pollen structure may be similar in *Acianthus* and *Triphora*, but we need more work on the internal structure of the pollen grains of the Neottieae and other Diurideae before these resemblances can be evaluated.

REFERENCES. Dransfield et al. 1986; Jones and Clements 1987; Van Royen 1983.

Subtribe **Cryptostylidinae**

DESCRIPTION. Habit: terrestrial or saprophytic; leafy stem and rhizome short, roots clustered, fleshy, velamen of the *Calanthe* type. Leaves few to several, spiral, clustered, convolute, soft herbaceous, nonarticulate, somewhat petiolate. Inflorescence: terminal, simple, of few to many spiral flowers. Flowers: small to medium, nonresupinate, lip simple, column very short, rather conical; anther dorsal, erect, subequal to rostellum; four pollinia, soft and mealy, with a terminal viscidium, without caudicles; stigma entire.

DISTRIBUTION. Australasia, Pacific Islands, and tropical Asia.

POLLINATION. The pollination of *Cryptostylis*, studied by Edith Coleman, is a classic case of pseudocopulation. The flowers attract male ichneumon wasps, and the male wasps react to the signals normally found in the female of their species. In Australia, at least, all species of *Cryptostylis* are pollinated by the same wasp, *Lissopimpla excelsa* (syn. *L. semipunctata*), and the orchid species are intersterile (Stoutamire 1975).

CHROMOSOME NUMBERS. 42, 56.

SEED STRUCTURE. *Goodyera* type.

SPECIES. About 16.

GENERA. 2: *Coilochilus, Cryptostylis*.

DISCUSSION. *Cryptostylis* is best known for its pollination system, studied in Australia, but the genus is widespread in the Pacific and in tropical Asia. Structurally, the most obvious feature of these two genera is the very short column. *Cryptostylis* was placed in the Diurideae by Schlechter (1926), but Dressler and Dodson (1960) assigned it to the Cranichideae because of its column shape and the lack of root–stem tuberoids. Now I am convinced that Schlechter's original placement was correct.

The epidermal pattern of *Cryptostylis* is unlike that of the Cranichideae but very like that of the Diurideae (N. H. Williams, unpublished). When Freuden-stein first told me that the endothecial thickenings of *Cryptostylis* were of type I, like the majority of the Diurideae, I was prepared to place this subtribe among the misfits in Chapter 11, if not in the Diurideae. Further study showed type I thickenings to occur in some Cranichideae (Freudenstein 1991a), so I remained in doubt through the penultimate revision of this manuscript. Now W. L. Stern (per-sonal communication) and his coworkers find a distinctive starch storage organelle (amyloplast) and three other anatomical features that are diagnostic for the Cranichideae; none of these features is found in *Cryptostylis*. The main feature favoring a cranichid relationship for *Cryptostylis* is the column shape, but this fea-ture occurs too often in different groups to be convincing. The epidermal pattern, the anatomical details to be reported by Dr. Stern, the type I endothecial thickenings, and the geographic distribution all favor placing the Cryp-tostylidinae in the Diurideae. The lack of root–stem tuberoids, the small staminodia, and the seed type are unusual in the Diuridinae, but all of these occur in other Diurideae.

REFERENCE. Wallace 1978.

Subtribe **Diuridinae**

DESCRIPTION. Habit: terrestrial, with root–stem tuberoids, roots without velamen. Leaves few to several, largely basal, convolute, soft herbaceous, usually narrow, nonarticulate, subcylindrical in *Epiblema*. Inflorescence: terminal, simple, of several to many spiral flowers. Flowers: medium; column short, with promi-nent, vascularized staminodia that may equal or surpass the fertile anther, these united with the median filament basally; anther conic, dorsal, erect, subequal to the stigma, or projecting beyond the rostellum; four pollinia, of monads or tetrads, soft and mealy, with a subterminal viscidium; exine of pollen grains reticulate; stigma entire. (See Fig. 7-5.)

DISTRIBUTION. Australasia.

POLLINATION. Recent observations indicate that *Diuris* flowers mimic the flowers of sympatric legumes (Fabaceae). Small bees accustomed to gathering nectar or pollen from legume flowers occasionally visit the similar flowers of *Diuris*, but the *Diuris* offer no reward (Beardsell et al. 1986). This may explain the curiously "unorchid" look of *Diuris* flowers.

CHROMOSOME NUMBERS. 38, 56.

SEED STRUCTURE. *Diuris* type.

SPECIES. 41.

GENERA. 3: *Epiblema, Diuris, Orthoceras*.

DISCUSSION. The prominent staminodia of this group have definite vascular strands and would seem to be a primitive feature. *Diuris*, especially, appears to have primitive column structure. Though the large and nearly free staminodia might be considered primitive, the well-developed viscidium is surely a derived feature, and not one to be expected in a primitive genus. The viscidium is subter-minal with respect to the anther, and Burns-Balogh (1986) assigns *Diuris* and *Orthoceras* to the Spiranthoideae on the basis of the anther/stigma relationship.

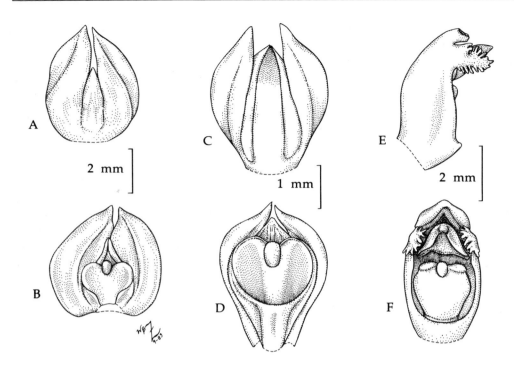

Figure 7-5. Column structure of some genera in the *Diuris* group. (A, B) *Epiblema grandiflorum*, dorsal and ventral views. (C, D) *Diuris laxiflora*, dorsal and ventral views. (E, F) *Thelymitra rubra*, side and ventral views. Drawn from material preserved in liquid.

The column, however, is very different from any of the spiranthoid group, and the root–stem tuberoids and seed structure, especially, place *Diuris* in the "Australian" complex. P. Bernhardt (personal communication) argues that there is no staminodial rim ventral to the column in *Diuris*, considering the appearance of such a rim in some drawings to be an artifact caused by cutting the lip from the column. F. N. Rasmussen (1982), however, shows a definite staminodial rim in developing buds of *Diuris punctata*. In any case, the ventral margins of the staminodia are generally lobed or undulate near their bases and often fleshy.

   Epiblema has been considered a close ally of *Thelymitra*, but its column structure is basically like that of *Diuris*. The blue color and ornate flower may reflect a similar pollination system, rather than a close relationship to *Thelymitra*.

   REFERENCE. Beardsell et al. 1986.

Subtribe **Thelymitrinae**

   DESCRIPTION. Habit: terrestrial, with root–stem tuberoids. Leaves 1 or 2, largely basal, convolute, soft herbaceous, usually narrow, nonarticulate, sometimes cylindrical or subcylindrical. Inflorescence: terminal, simple, of several to many spiral flowers. Flowers: medium; column short, with prominent, vascularized staminodia united with the median filament and forming a hood over or around the style, the hood often ornate; anther conic, dorsal, erect, or somewhat bent forward, subequal to the stigma, or projecting beyond the rostellum; four pollinia, of monads or tetrads, soft and mealy, with a viscidium, viscidium subbasal to subterminal; exine of pollen grains reticulate; stigma entire. (See Fig. 7-5.)

DISTRIBUTION. Australasia and into Malaysia.

POLLINATION. The flowers of *Thelymitra* are pollinated by pollen-gathering female bees, though the flowers offer no usable pollen. The ornate, commonly yellow hood over the anther and stigma simulates a cluster of pollen-bearing anthers, so that bees are occasionally deceived into seeking pollen on the flowers. *Thelymitra nuda* appears to mimic flowers such as *Dichopogon* and *Thysanotus* (Liliaceae), whose pollen is normally gathered by vibration. The halictid bees that visit the *Thelymitra* vibrate (quite futilely) on the false anthers of the orchid (Bernhardt and Burns-Balogh 1986a). Dafni and Calder (1987) report similar mimicry of pollen flowers (but without vibration) for *T. antennifera*. The unusual color and nearly radial symmetry of the perianth of *Thelymitra* have long been noted. Selection has apparently favored the aspect of an ordinary, radially symmetrical flower with several anthers. The pollinia of *Thelymitra* may be deposited behind the mouthparts, beneath the thorax, or either above or beneath the abdomen (Burns-Balogh and Bernhardt 1988). *Calochilus* is pollinated through pseudocopulation by male wasps, though to human eyes the flowers do not resemble wasps. The pollinia are placed on top of the head.

CHROMOSOME NUMBERS. 24, 26, 32, 56.

SEED STRUCTURE. *Diuris* type.

SPECIES. 75.

GENERA. 2: *Calochilus, Thelymitra.*

DISCUSSION. The staminodia of this group are united with the median filament to form an ornate hood, or mitra, and the base of the hood surrounds the base of the column, possibly involving staminodia of the ventral side of the flower. The column of *Gavilea* (Chloraeinae) resembles that of the Thelymitrinae at least superficially, but their structure may not be the same. In *Thelymitra*, the ornate hood surpasses the anther, but in *Calochilus* the hood reaches only the base of the anther. *Calochilus* has auricle like structures much like those of *Lyperanthus*, apparently on the lateral staminodia. Burns-Balogh and Bernhardt (1988) interpret the auriclelike structures of *Calochilus* as reduced staminodial arms, and consider *Calochilus* to be derived from *Thelymitra*. Unfortunately, we have little information on the ontogeny of the column of either genus.

REFERENCES. Bernhardt and Burns-Balogh 1986a; Burns-Balogh and Bernhardt 1988; Dafni and Calder 1987; Jones & Gray 1974.

Subtribe **Rhizanthellinae**

DESCRIPTION. Habit: saprophytic, subterranean or barely reaching the soil surface; leafless, stem fleshy. Inflorescence: dense, headlike, of many spiral flowers, surrounded by large bracts; Flowers: small, fleshy, erect; lip hinged; column small, with small arm-like appendages; anther terminal, erect; pollen soft and mealy; exine of pollen grains reticulate; stigma entire; fruit fleshy.

DISTRIBUTION. Australia.

POLLINATION. In *Rhizanthella gardneri* the flowers may be pollinated by fungus gnats that enter cracks in the soil to reach the flowers (George 1981). Dixon (1985) illustrates termites bearing pollinaria and indicates wasps as probable pollinators.

CHROMOSOME NUMBER. Not known.

SEED STRUCTURE. Intermediate between *Diuris* and *Disa* types.

SPECIES. 2.

GENUS. *Rhizanthella*.

DISCUSSION. We now know enough about these curious plants that their place in or near the Diurideae seems quite secure. The fleshy fruits are probably eaten by small marsupials, and the hard coated seeds pass through the digestive tract undamaged and are thus dispersed. *Rhizanthella gardneri* requires a particular mycorrhizal fungus, a *Rhizoctonia* species, and this fungus forms a mycorrhiza with *Melaleuca uncinata* (Warcup 1985). Thus, the orchid is associated with *Melaleuca* in nature, and may even be an indirect parasite on the shrub. Clements and Cribb (1984) transfer *Cryptanthemis slateri* to *Rhizanthella*. Dixon (1985) suggests that we know too little about *Cryptanthemis* to be sure of its relationships.

Clements and Cribb (1984) compare *Rhizanthella* with *Genoplesium*, with which a number of features are shared. Most illustrations of *Rhizanthella* do not show a viscidium, but the drawing in George (1979), apparently prepared from fresh material, does show what may be a viscidium. There is apparently no hamulus, but it is quite likely that the Rhizanthellinae and Prasophyllinae are sister groups.

REFERENCES. Clements and Cribb 1984; Dixon 1985; George 1979, 1981; Warcup 1985.

Subtribe **Prasophyllinae**

DESCRIPTION. Habit: terrestrial, with root–stem tuberoids. Leaves: solitary, cylindrical, nonarticulate. Inflorescence: terminal, simple, of few to many spiral flowers. Flowers: small to medium, resupinate or not; column short, with two prominent staminodia partially united with the column; anther dorsal, either shorter or longer than the rostellum; two or four pollinia, granular or sectile, with a viscidium and a short to very long hamular stipe; exine of pollen grains reticulate to rugulate-baculate; stigma entire.

DISTRIBUTION. Australasia and into tropical Asia.

POLLINATION. The Prasophyllinae are among the few Diurideae that produce nectar. The flowers of *Prasophyllum* are pollinated especially by wasps, small bees, and syrphid flies; the pollinaria are generally attached to the head or mouthparts (Beardsell and Bernhardt 1983; Bernhardt and Burns-Balogh 1986b). The flowers of *Genoplesium* are pollinated by small flies (drosophilid and chironomid), and the pollinaria are attached to the thorax. The tiny flowers of *Microtis* are pollinated by a variety of insects, including flies, ants, beetles and wasps.

CHROMOSOME NUMBER. 44.

SEED STRUCTURE. *Diuris* type.

SPECIES. 99.

GENERA. 3: *Genoplesium, Microtis, Prasophyllum.*

DISCUSSION. The genera of the Prasophyllinae have relatively short columns with prominent staminodia, but the staminodia are partially united with the column and the lower margins are not developed as in the Diuridinae. The cylindric leaves, possibly a uniquely derived feature, suggest a relationship to *Epiblema.* The Prasophyllinae all have a hamulus, cylindric leaves, and sectile pollinia. In spite of these shared features, Burns-Balogh (1984) has assigned *Prasophyllum* to the Spiranthoideae on the basis of the anther position. This is one-character taxonomy carried to an extreme. Note, however, how nearly terminal the viscidium may be even in *Microtis* (F. N. Rasmussen 1982, Fig. 26). If the drawings of Garnet (1940) are to be trusted, *Genoplesium* also may have a terminal viscidium. *Genoplesium baueri* is merely a saprophytic member of what has been called *Prasophyllum* section *Micranthum,* and the species of that group have been transferred to *Genoplesium* (Jones and Clements 1989). Though there is good reason to treat the traditional sections of *Prasophyllum* as distinct genera, the differences between them scarcely justify placing them in different subfamilies or even subtribes. *Microtis,* however, is unlike either genus in several features, and Bates (1984b) suspects that it may be a misfit in the Prasophyllinae. *Microtis* has mycorrhizal symbionts of the genus *Sebacina,* rather than *Ceratobasidium;* its flowers are resupinate, the exine of the pollen grains is reticulate, rather than clavate; the lip is sessile, rather than hinged, and its staminodia are not, or rarely, bilobed. The pollinia of *Microtis* may be either sectile or mealy, but the loss of the sectile character may be associated with tiny flowers.

REFERENCES. Bates 1981, 1984a,b; Bernhardt and Burns-Balogh 1986b; Burns-Balogh 1984; Garnet 1940; Jones and Clements 1989; Peakall and Beattie 1989.

Tribe **Orchideae**

The Orchideae and Diseae, together, are surely monophyletic and quite advanced relative to the other basically terrestrial groups. Some authors have divided the monandrous orchids into two groups, these and everything else, but that is a bit extreme. A number of features have been taken to distinguish the Orchideae:

> Root–stem tuberoids
> Auricles present at base of anther
> Anther base fused with column
> Anther cells divergent basally
> Rostellar beak or strap between anther cells
> Viscidia attached to base(s) of pollinia
> Pollinia sectile
> Caudicles prominent
> Caudicles including anther wall tissue

Some of these features are shared with the Diurideae, and others are derived from features similar to those of the Diurideae. The Orchideae are surely either a sister group or a derivative of the Diurideae. To place them in separate subfamilies obscures their close relationship. The relationships between the Orchideae and the Diseae are not altogether clear, and Kurzweil's continuing studies are likely to reorganize both groups. For now, I leave both groups relatively unchanged, and discuss some of Kurzweil's suggestions where appropriate.

The auricles, small lateral appendages near the anther base, have been considered a characteristic of the Orchideae. Kurzweil's developmental study (1987b) indicates that the auricles of the Orchideae are appendages of the anther base or of the filament (sometimes combined with staminodial structures), while the auricle-like structures of *Calochilus*, *Lyperanthus* and some Spiranthoideae appear to be formed on the staminodia.

This tribe and the Diseae have been distinguished as the "Basitonae," because the viscidia are associated with the base of the anther, or "basitonic." In most other orchid groups, the viscidia (or the rostellum) are associated with the apex of the anther ("acrotonic"). In the Diurideae, one finds more variation, and the rostellum is often somewhere between these extremes. There are often two distinct viscidia in the Orchideae, but even when there is only one, a rostellar beak extends beyond the viscidium, between the anther locules. In the ancestor of the Orchideae, the anther cells must have been divergent, with a rostellar beak between them. The viscidia evolved not at the very tip of the rostellum, but more basally, leaving a distal beak between the anther cells. The primitive condition for this group was probably two separate viscidia, or at least this condition is very common.

The interlocular caudicles are said to include some anther wall tissue (Vogel 1959), and it may be for this reason that the connection between the pollinia and the viscidia seems fixed in the basal position. In *Satyrium* (Diseae), we find a functionally terminal viscidium, analogous to that in the spiranthoid orchids, but in *Satyrium* this condition is achieved by turning the anther upside-down.

J. Renz (in Senghas and Sundermann 1980) outlines the origin of the Orchideae from a tropical Asiatic center of origin, but the Orchideae seems tied to southern hemisphere groups. Linder and Kurzweil (1990) offer a consensus tree of subtribes that shows the Diseae as a derivitive of the Orchideae, rather than a sister group, and shows *Huttonaea* as essentially intermediate between these tribes (the sister group of the other Diseae). H. Kurzweil (personal communication) suggests, though, that the Diseae are a monophyletic group, though the Orchideae may be either paraphyletic or polyphyletic. The above-mentioned consensus tree also suggests that the crustose seeds of the *Disa* complex are secondary, as suggested in Chapter 2, and implies that the relationships between the Orchideae and the Diurideae should be sought in the Orchidinae and the Chloraeinae/Caladeniinae. That is, any close resemblances between the *Disa* and *Diuris* complexes are probably parallelisms.

Some Orchideae are well represented in Europe and have been studied much more than most other orchid groups. Nevertheless, a good overall study of the group is lacking, and how best to classify the group is not at all clear (see Linder and Williamson 1986, for example).

PHYLOGENY. I have adapted the consensus tree of Linder and Kurzweil (1990) with slight modification. I have placed both the Orchidinae and the Habenariinae as part of a trichotomy at the base of the diagram (Fig. 7-6). The separation between the Orchidinae and the Habenariinae is not wholly convincing, and the Habenariinae may be a subclade of the Orchidinae. Similarly, some Orchidinae have no spur, and the Diseae also may be a subclade of the Orchidinae. In other words, the Orchidinae may be paraphyletic.

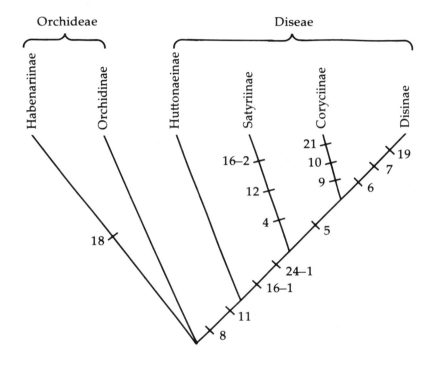

Figure 7-6. Phylogenetic diagram of Orchideae and Diseae.

Subtribe **Orchidinae**

DESCRIPTION. Habit: terrestrial, rarely saprophytic, usually with root–stem tuberoids, roots without velamen; stem slender. Leaves: spiral, scattered or basal and clustered, convolute, soft herbaceous, nonarticulate. Inflorescence: terminal, simple, of few to many spiral flowers. Flowers: small to medium-large, resupinate; lip usually with a basal spur; column short or moderately long; the anther erect, basally firmly united with the column; two or four pollinia, sectile, with two basal, interlocular caudicles attached to two or one basal viscidia; exine of pollen grains variable; stigma entire, concave.

DISTRIBUTION. Primarily northern hemisphere, ranging into Africa and tropical Asia.

POLLINATION. Two species of *Orchis* attract pollinators by offering nectar, but most lack nectar. Most seem not to mimic other flowers, though Dafni and Ivri (1981a) consider *Orchis israelitica* to be a mimic of *Bellevalia flexuosa* (Liliaceae). One may well wonder how such a bizarre system as the pseudocopulation of *Ophrys* could arise. Bino et al. (1982) find that *Orchis galilea* produces a musky odor and attracts males of *Halictus marginatus*. The relationship appears to be a sexual attraction, and it would seem that pseudocopulation may function if the flower has the right odor. Selection can bring about structural and textural mimicry later, thus improving the system.

Galearis spectabilis is pollinated by bumble-bee queens (Dieringer 1982). White or greenish flowers of *Platanthera* are usually pollinated by moths, though the very short-spurred *P. chorisiana* is normally pollinated by a small beetle. Inoue (1983) found the moth species visiting *Platanthera* to be correlated with flower structure and phenology. The short-spurred *P. stricta* is pollinated by various short-tongued insects (Patt et al. 1989). The brightly colored North American species are pollinated by butterflies (Folsom 1984).

Dafni et al. (1981) find that male solitary bees commonly sleep in the flowers of *Serapias vomeracea*, often shifting from flower to flower and effecting pollination. They term this relationship a "mimicry," though, in fact, the flower offers a usable resting place. Selection seems to favor a flower that is just a bit too shallow for comfort, so that bees often move from one flower to another, rather than sleeping in the first flower they find.

CHROMOSOME NUMBERS. 20, 30, 32, 36, 38, 40, 42, 44, 46, 48.

SEED STRUCTURE. *Orchis* type, or *Dactylorhiza* variant.

SPECIES. 371.

GENERA. 34 in four tentative alliances:

1. Lacking thickened tuberoids: *Amerorchis, Aorchis, Chondradenia, Galearis.*

2. Tuberoids palmate or attenuate: *Brachycorythis, Chusua, Coeloglossum, Dactylorhiza, Gymnadenia, Nigritella, Platanthera, Pseudodiphryllum, Pseudorchis.*

3. Tuberoids spheroid: *Aceras, Amitostigma, Anacamptis, Barlia, Chamorchis, Comperia, Hemipilia, Himantoglossum, Neobolusia, Neotinea, Neottianthe, Ophrys, Orchis, Piperia, Schizochilus, Serapias, Steveniella, Symphyosepalum, Traunsteinera.*

4. Hairy plants with flat, basal leaves, petals and lip often fimbriate, African: *Bartholina, Holothrix.*

DISCUSSION. Though the European members of this group have been much studied, we greatly need a balanced review of the entire group. For the European groups, the division on the basis of tuberoid form seems to approximate relationships better than any one floral detail, but needs to be checked carefully. The European genera appear rather finely split as compared to most other groups. The Orchidinae are traditionally characterized by having a concave stigma, but Kurzweil and Weber (1991) find the stigma to be usually convex or padlike, though often in a depression. The distinction between the Orchidinae and the Habenariinae needs to be reevaluated. At the same time, Kurzweil and Weber show that *Holothrix* and *Bartholina*, though doubtfully distinct from each other, form a very distinctive group without auricles.

Cauwet-Marc and Balayer (1984b) find *Neotinea* to be very close to *Orchis* and

Traunsteinera to be very close to *Dactylorhiza*. A study of enzyme electrophoresis confirms the close relationship between *Gymnadenia* and *Dactylorhiza*, and indicates that *Orchis* is paraphyletic, that is, that *Gymnadenia* and *Dactylorhiza* form a subclade of *Orchis*. Clearly we need more careful work on the European Orchideae, and such work will probably reduce the number of genera.

Chen (1982) considers *Aceratorchis* to be a very primitive member of the Orchideae, because the lip is quite like the petals. Others have interpreted this as a peloric variant of some other genus. It is apparently normally cross-pollinated, but its radial symmetry may be secondary, as in *Thelymitra*.

REFERENCES. Bino et al. 1982; Catling and Catling 1989; Cauwet-Marc and Balayer 1984b, 1986; Dafni 1983, 1987; Dafni and Ivri 1979; Dafni et al. 1981; Del Prete 1984; Folsom 1984; Fritz 1990; Greilhuber and Ehrendorfer 1975; Inoue 1983; Kurzweil and Weber 1991; Nilsson 1978a, 1980, 1983a,b, 1984; Patt et al. 1989; Paulus and Gack (in Senghas and Sundermann 1986); Sheviak and Bowles 1986; Tang et al. 1982; Tohda 1983; Vöth 1984, 1987.

Subtribe **Habenariinae**

DESCRIPTION. Habit: terrestrial, rarely epiphytic, with spheroid or oblong root–stem tuberoids, roots without velamen; stems slender. Leaves: spiral, scattered, or basal and clustered, convolute, soft herbaceous, nonarticulate. Inflorescence: terminal, simple, of one to many spiral flowers. Flowers: small to medium-large, usually resupinate; lip usually with a basal spur; column short; anther erect, basally firmly united with the column; two or four pollinia, sectile, with two basal, interlocular caudicles attached to two basal viscidia; the viscidia often borne on long rostellar stalks; exine of pollen grains laevigate to baculate-pilate; stigma convex, entire or two-lobed, the lobes often stalked. (See Fig. 7-7.)

DISTRIBUTION. Africa and pantropical, ranging into Eurasia and North America.

POLLINATION. Moth pollination has been reported for *Habenaria*, and butterfly pollination for *Bonatea*. Flower form would suggest lepidopteran pollination for most of this group. *Herminium* has a very short spur and is visited by various insects, and especially by parasitic wasps. Nilsson (1979b) considers *Herminium* to be derived from longer spurred ancestors and its wasp-pollination secondary.

CHROMOSOME NUMBERS. 28, 30, 32, 34, 36, 38, 40, 42, 44, 46, 48.

SEED STRUCTURE. *Orchis* type, or *Habenaria* variant, with some anomalous sorts not assigned to any type.

SPECIES. 932.

GENERA. 23 in two tentative alliances:

1. Stigmas sessile or very short-stalked, entire or partly divided: *Androcorys*(?), *Benthamia*, *Diphylax*, *Gennaria*, *Herminium*, *Oligophyton*, *Pecteilis*, *Peristylus*, *Smithorchis*, *Tylostigma*.

2. Stigmas 2, distinctly stalked: *Arnottia*, *Bonatea*, *Centrostigma*, *Cynorkis*, *Diplomeris*, *Habenaria*, *Megalorchis*, *Physoceras*, *Platycoryne*, *Roeperocharis*, *Stenoglottis*, *Thulinia* (?), *Tsaiorchis*.

DISCUSSION. The Habenariinae are notable for long-stalked stigmas, long

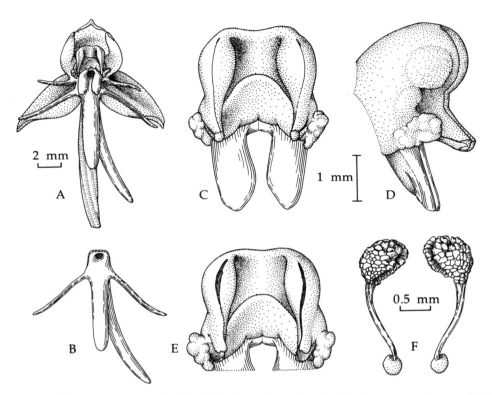

Figure 7-7. *Habenaria entomantha* (Orchidoideae: Orchideae). (A) Flower, front view. (B) Lip and spur. (C) Column, front view. (D) Column, side view. (E) Column, after removal of pollinaria. (F) Pollinaria. (After Dressler 1981.)

rostellar arms, and long caudicles. Most have prominent spurs. Though Renz (1980) postulates a tropical Asiatic origin, the group is most diverse in Africa and could well be African in origin. Kurzweil and Weber (1992) suggest that several African genera might be reduced to *Habenaria* and reemphasize the difficulty of delimiting the Habenariinae and Orchidinae.

REFERENCES. Hesse and Burns-Balogh 1984; Kurzweil and Weber 1992; Nilsson 1979b; Nilsson and Jonsson 1985; Renz 1980; Stewart 1989.

Tribe **Diseae**

The subtribes treated as the Diseae are primarily African, though *Disperis* and *Satyrium* each have representatives in Asia. The anther is usually much more strongly bent backward than in the Orchidinae, and the lip is simple, either without a spur or (in the Satyriinae) with paired spurs. As Vogel (1959) has indicated, these genera are designed to place their pollinia on the undersides of the pollinators, but the Orchidinae place the pollinia on the dorsal surface or on the head (see Fig. 7-8). This complex shows the *Disa* seed type essentially through-

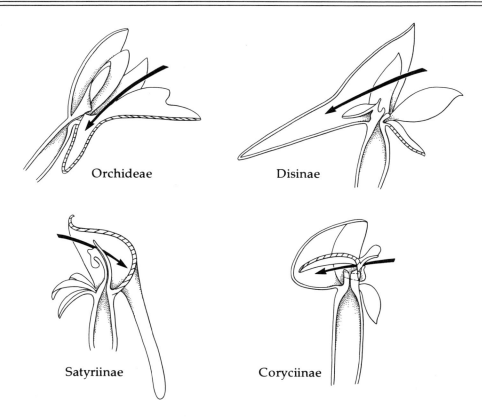

Figure 7-8. Diagrams showing the relationships between floral parts and pollinators in the Orchideae and Diseae. The section of the lip is cross-hatched, and the *arrow* indicates the path of the pollinator (or its mouthparts). The flowers of the Orchideae place the pollinaria on the head or dorsal surface of the pollinator, while the flowers of the Diseae place their pollinaria on the ventral surface. (After Linder 1986.)

out, though *Huttonaea* is said to have seed of the *Orchis* type (B. Ziegler, unpublished). The column structure of the Diseae is remarkably diverse and complex (Fig. 7-9). Linder (1986) offers an interesting analysis of this group, and Kurzweil's recent studies suggest that the group needs some revision.

Subtribe **Huttonaeinae**

DESCRIPTION. Habit: terrestrial, with globose root–stem tuberoids; stems slender. Leaves: spiral, few, convolute, soft herbaceous, nonarticulate. Inflorescence: terminal, simple, of several spiral flowers. Flowers: medium, resupinate, sepals and petals fimbriate; lip without a spur; column short; anther erect, the locules together apically and diverging basally; two pollinia, sectile, with two basal, interlocular caudicles and two viscidia; stigma entire.

DISTRIBUTION. Southern Africa.

POLLINATION. Pollinated by oil-gathering bees of the genus *Rediviva*, according to Steiner (1987).

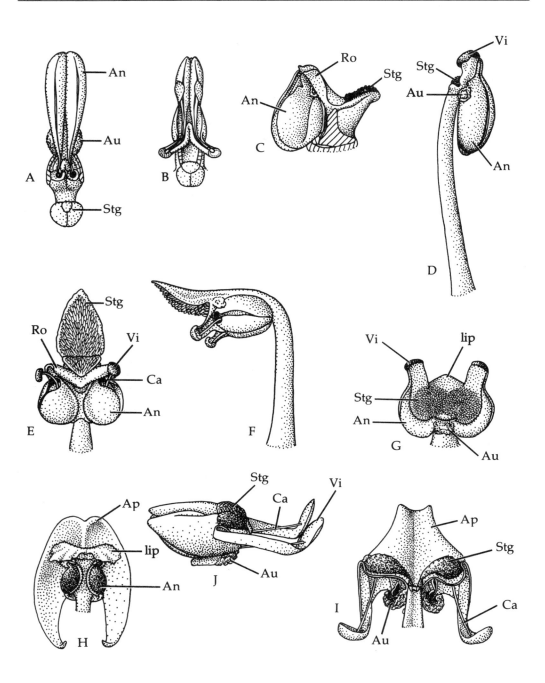

Figure 7-9. Columns of Diseae. (A) *Disa bivalvata*. (B) *Disa filicornis*. (C) *Disa draconis*, side view. (D) *Satyridium rostratum*, side view. (E, F) *Satyrium saxicolum*, front and side views. (G) *Ceratandra globosa*. (H) *Corycium orobanchoides*. (I, J) *Disperis paludosa*, front and side views. An, anther; Ap, appendage of the lip; Au, auricle; Ca, caudicle; Ro, rostellum; Stg, stigma; Vi, viscidium. (After Vogel 1959.)

CHROMOSOME NUMBER. Not known.

SEED STRUCTURE. *Habenaria* variant of *Orchis* type.

SPECIES. 5.

GENUS. *Huttonaea*.

DISCUSSION. *Huttonaea* is very distinctive in the form of the column and apparently has no close allies. Linder (1986) suggested that *Huttonaea* might be allied to the Coryciinae, and Kurzweil (1989a) finds the column structure to agree well with the Diseae. The seed type is apparently similar to that of the Orchideae. Linder and Kurzweil (1990) suggest, however, that the *Disa* complex is a subclade of the Orchideae, in the narrow sense, implying that the crustose seed coat of other Diseae may be a secondary feature.

REFERENCES. Kurzweil 1989a; Steiner 1987.

Subtribe **Disinae**

DESCRIPTION. Habit: terrestrial, with root–stem tuberoids, roots without velamen; stems slender. Leaves: spiral, scattered, or basal and clustered, convolute, soft herbaceous, nonarticulate. Inflorescence: terminal, simple, of few to many spiral flowers. Flowers: small to medium-large, resupinate, dorsal sepal saccate or deeply spurred; lip usually very small or narrow, may be stalked or fringed; column very short, anther erect or usually inclined dorsally, basally completely united with column; two pollinia, sectile, with basal caudicles and one or two viscidia; exine of pollen grains varied, often hamulate; stigma entire, very near base of perianth. (See Figs. 7–9, 7–10.)

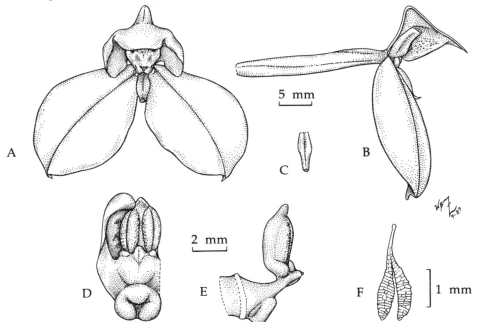

Figure 7-10. *Disa tripetaloides* (Orchidoideae: Diseae). (A) Flower, front view. (B) Flower, side view, with part of dorsal sepal cut away. (C) Lip. (D) Column, ventral view, with one petal cut away. (E) Column, side view. (F) Pollinarium.

DISTRIBUTION. Africa, and especially southern Africa.

POLLINATION. The few pollination records for *Disa* indicate pollination by butterflies and flies, including the flower-visiting bee flies (Bombylidae), for different species. Vogel's detailed study (1959) indicates that the Disinae have radiated to a number of pollination systems.

CHROMOSOME NUMBERS. 36, 38.

SEED STRUCTURE. *Disa* type.

SPECIES. 143.

GENERA. 5: *Brownleea, Disa, Herschelia, Monadenia, Schizodium.*

DISCUSSION. The listing of genera follows Linder (1986). Kurzweil (1990) finds primordia apparently representing the lateral staminodia that form keels connecting the petals to the column. Linder and Kurzweil (1990) present a phylogenetic analysis of the Disinae based on floral morphology. Some groups are clearly delimited, but Linder and Kurzweil find that floral morphological data, by themselves, are not enough to place the taxonomy of the Disinae on a sound footing.

REFERENCES. Kurzweil 1990; Linder 1981a–f, 1986; Linder and Kurzweil 1990; Wimber 1987.

Subtribe **Satyriinae**

DESCRIPTION. Habit: terrestrial, with root–stem tuberoids. Leaves: spiral, scattered or basal and more or less clustered, convolute, soft herbaceous, nonarticulate. Inflorescence: terminal, simple, of few to many spiral flowers. Flowers: small to medium, nonresupinate; lip with two saccate nectaries or spurs; column elongate, arched, the anther basally firmly united with the column, bent back so that the base is uppermost; two pollinia, sectile, with two basal interlocular caudicles and two viscidia; exine of pollen grains baculate-pilate; stigma somewhat two-lobed, somewhat overtopping the anther.

DISTRIBUTION. Primarily African; *Satyrium* extends into Asia.

POLLINATION. The few available records indicate fly pollination, and *Satyrium pumilum* shows the syndrome of carrion fly pollination. Some species, however, have much longer spurs and may be pollinated by Lepidoptera or bees.

SEED STRUCTURE. *Disa* type.

CHROMOSOME NUMBERS. 36, 42.

SPECIES. About 103.

GENERA. 3: *Pachites, Satyridium, Satyrium.*

DISCUSSION. The Satyriinae are closely allied to the Disinae, but have nonresupinate flowers, double nectaries or spurs on the lip, and a slender column with the anther quite reversed.

Chen (1979) finds that plants of *Satyrium ciliatum* may have staminate, pistillate, or bisexual flowers. Whether or not other species of *Satyrium* show a similar pattern is not known.

REFERENCES. Chen 1979; Hall 1982; Rao and Sood 1979.

Subtribe **Coryciinae**

DESCRIPTION. Habit: terrestrial or saprophytic, with root–stem tuberoids, roots without velamen, stem slender. Leaves: spiral, subbasal, or much reduced, convolute, soft herbaceous, nonarticulate. Inflorescence: terminal, simple, of one to many spiral flowers. Flowers: small to medium, resupinate or not; lateral sepals often each with a small spur-like sack; petals often united to dorsal sepal; lip united with column, usually very small, may bear an "appendage" that is larger than the blade itself; column relatively short, anther basally firmly united with column, bent backward so that the apex is toward the base of the column, the anther thecae widely separated, each projecting forward with a viscidium at apex of the anther/rostellar arm; two pollinia, sectile, each with a prominent interlocular caudicle and a viscidium; exine of pollen grains hamulate or rugose; stigma two-lobed, dorsal, well below the viscidia. (See Fig. 7-11.)

DISTRIBUTION. Primarily African, with *Disperis* extending into tropical Asia.

POLLINATION. About one fourth of the *Disperis* species and all of the other Coryciinae, produce oil rather than nectar, and they are pollinated by bees of the genus *Rediviva* (Melittidae) (Steiner 1989). This suggests that *Disperis* may be paraphyletic, a possibility that seems compatible with the phylogenetic analysis by Kurzweil et al. (1991) (see Fig. 7-12).

CHROMOSOME NUMBER. Not known.

SEED STRUCTURE. *Disa* type.

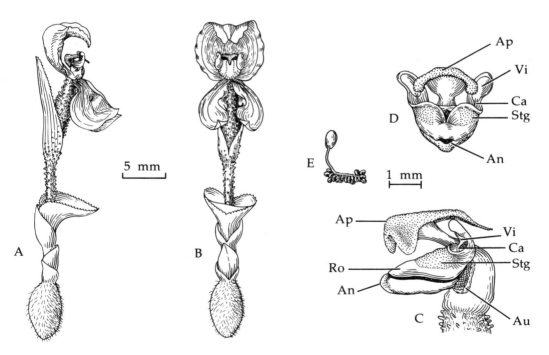

Figure 7-11. *Disperis pusilla* (Orchidoideae: Diseae). (A) Habit, showing flower in side view. (B) Habit, front view. (C) Lip and column, side view. (D) Lip and column, front view. (E) Pollinarium. An, anther. Ap, appendage of lip; Au, auricle; Ca, caudicles; Ro, rostellum; Stg, stigma; Vi, viscidium. (After Verdcourt 1968.)

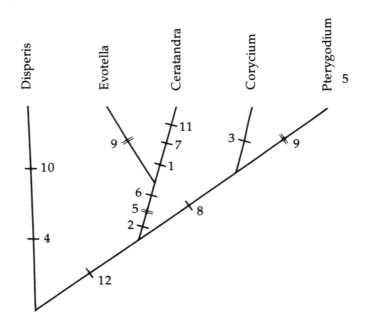

Figure 7-12. Phylogenetic diagram of Coryciinae. Derived features: (1) underground organs thickened roots; (2) cauline leaves all of same size; (3) flower shape globose; (4) lateral sepals saccate to spurred; (5) lip blade with a distinct claw; (6) lip anchor-shaped; (7) lip with a prominent callus; (8) lip appendage entire or shallowly bifid; (9) lip appendage elongate or undivided; (10) rostellum lobes covering the anther cells; (11) anther pendant; (12) anther cells divergent at least at the apex. (Adapted from Linder 1986 and Kurzweil et al. 1991.)

SPECIES. 107.

GENERA. 5: *Ceratandra, Corycium, Disperis, Evotella, Pterygodium.*

DISCUSSION. The Coryciinae show very complicated flower structure; the lip and column are somewhat united; the lip itself is quite small, but the lip appendage becomes a major floral structure. Some species appear to be partially saprophytic. The sacklike structures on the lateral sepals of *Disperis* are not nectaries, but serve as covers for the rostellar arms during development (Steiner 1989).

Kurzweil (1991) treats the column structure of this group in detail, and finds that the Coryciinae, *Huttonaea*, and *Brownleea* all lack lateral stigma lobes. This condition is analogous to that in the Cranichideae, and Kurzweil suggests that both *Brownleea* and *Huttonaea* may form part of the Coryciinae clade. Kurzweil et al. (1991) discuss the phylogeny of the *Pterygodium–Corycium* complex in detail, and revise the genera, reducing *Anochilus* and *Evota* to synonymy but creating the genus *Evotella* for a species that had been a misfit in both *Pterygodium* and *Ceratandra.*

REFERENCES. Kurzeil 1991; Kurzweil et al. 1991; Steiner 1989; Verdcourt 1986.

8

The Advanced **Epidendroideae,** General Discussion and Phylogeny

The Epidendroideae is by far the largest of the orchid subfamilies, with many more genera and species than all of the other subfamilies together. There are uncertainties about the more primitive tribes, yet their removal would scarcely affect either the size or the diversity of the Epidendroideae. This subfamily demonstrates some of the problems that may be found in classifying quite natural groups. One may list a number of derived features that characterize the more advanced epidendroid orchids:

> Epiphytism
> Pseudobulbs or corms
> Distichous leaves
> Caducous leaves
> Fleshy leaves
> Conduplicate leaves (duplicate vernation)
> Lateral inflorescences
> Hard pollinia
> Caducous anther

Some of these features are found in most epidendroid orchids, and others occur only in the most highly evolved groups. None of these features is to be found in all epidendroid orchids, and no two of them show exactly the same distribution.

The best feature to characterize the subfamily as a whole is the incumbent anther, which is clearly recognizable in nearly all of the primitive and moderately advanced epidendroid orchids. In the early stages of development, the epidendroid anther is erect and parallel with the axis of the column, as in other orchid groups. As the column grows in the bud, the anther bends downward at the column apex, until it is nearly at right angles to the axis of the column, or in many cases pointing back toward the base of the column (Fig. 2-13). One finds the anther to be tipped a bit downward in several orchid genera, such as *Epipactis, Neottia,* or *Gavilea* (F. N. Rasmussen 1982). Even this slight bending makes it more probable that the pollinator will touch the anther and the pollen masses on leaving the flower. In several cases, the anther is tipped downward and the anther cells are markedly convex. This may be the primitive pattern from which the fully incumbent anther has evolved by reduction of the terminal portion of the anther (see Fig. 3-3). There is often a flat, terminal beak on the anther in the Epidendroideae.

The terminal beak, the incumbent anther, and the parallel position of the median stigma lobe lead to a truly elegant mechanism (Fig. 8-1). A pollinator may enter the flower without disturbing the anther or touching the stigma, but as it leaves the flower, it almost inevitably touches the median stigma lobe and then the beak of the anther. On touching the stigma lobe, the insect receives some of the sticky stigmatic fluid on its back; as it continues, it tilts the anther back, exposing

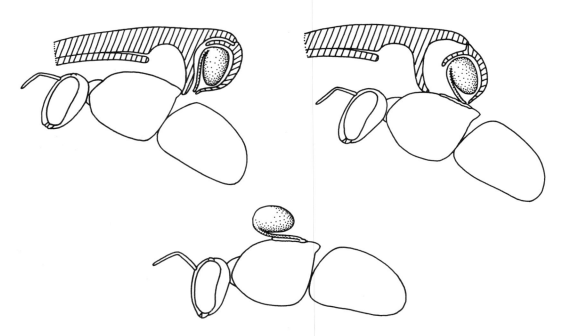

Figure 8-1. The function of the incumbent anther of the Epidendroideae in pollination. The anther swivels on its point of attachment when an insect brushes against the anther beak, depositing the pollinia on the glue just removed from the rostellum. (After Dressler 1981.)

the pollen masses, which then touch the stigmatic glue already on the insect's back. If the insect visits a second flower, at least a portion of the pollen mass will surely remain in the stigmatic cavity as the insect backs out of the flower.

Burns-Balogh and Bernhardt (1985) point out an important corollary of this bending of the anther and stigma lobe. In orchids with soft pollen and shallow or convex stigmas, a portion of a pollen mass is generally sufficient to fertilize a stigma, and a single pollen mass may pollinate a number of flowers. In the Epidendroideae, the scoop-like structure of the stigma permits the removal of whole pollen masses from the pollinator. This is the general pattern in the subfamily, and this feature probably facilitated the evolution of hard pollinia. In the other subfamilies, the ovary has many ovules, but the number is minor, compared to that of most epidendroid orchids. In most Epidendroideae, then, each pollination delivers enough pollen grains to fertilize an enormous number of ovules.

The mechanism described above works very well without a viscidium or with soft pollen masses, but most epidendroid orchids have evolved hard pollinia, and many of them have a distinct viscidium, a clearly defined part of the stigma (or rostellum) that is removed with the pollinia as a unit. In these more advanced epidendroid orchids, the incumbent anther is no longer necessary for the transfer of pollinia, and we find a great diversity of anther form and position. In the most advanced, or vandoid, groups, the incumbent position of the anther is not nearly so clear, and I once thought that this could give a clear distinction between the epidendroid and vandoid groups (Dressler 1981). Now, it is clear that the anther of the vandoid orchids bends downward, but that the bending of the anther in these advanced Epidendroideae occurs very early in development, so that it is not easily seen (Kurzweil 1987a).

Some authors have preferred to limit the Epidendroideae to those orchids with hard pollinia, but there is continuous gradation from the loose, sticky pollen grains of *Vanilla* to the hard pollinia of *Oncidium* or *Vanda*. This problem is well illustrated by what we have called *Crybe rosea*, long considered a member of the Neottioideae (under one name or another). "*Crybe*" is normally self-pollinating, and the pollinated flowers certainly have soft (germinating) pollen masses. If self-pollination does not occur, the pollen masses are relatively firm, and this plant should probably be treated as a species of *Bletia* (*B. purpurata*). The pollinia of the Bletiinae are nicely intermediate between "soft" and "hard." Even if one were to draw an arbitrary line between soft and hard to separate the Epidendroideae from a very broad Neottioideae, the Epidendroideae so delimited would probably be polyphyletic, and the Neottioideae would be paraphyletic, if not polyphyletic as well.

The incumbent anther gives a much clearer delimitation of the Epidendroideae than any other feature known to me, but there are exceptions. The highly advanced Epidendroideae that have lost or obscured the incumbent anther present no problem, as they are clearly epidendroid on a number of features. One finds, however, more or less erect anthers in a few primitive Epidendroideae, notably *Triphora*, *Psilochilus*, the Neottieae, and some Gastrodieae. The Gastrodieae may represent secondary "reversions" in self-pollinating populations. In *Triphora*, the anther appears erect, because of the very fleshy apex, but the anther cells are basal and actually somewhat incumbent. In the related genus *Monophyllorchis*, the anther is more nearly incumbent, though here, too, the very fleshy beak makes it appear erect. In *Psilochilus*, the anther is erect, but *Psilochilus* has a distinct viscidium (when not self-pollinating), and may represent a sort of "reversion," as in *Isochilus*.

The Epidendroideae are mainly a tropical group, and their present distribution gives no clear clues as to their origin. The cymbidioid phylad is likely to be Old World in origin, but the pattern is more complex in the epidendroid phylad. If we accept the Sobraliinae as a basal group of the Epidendreae, that tribe appears to be American in origin. If this is the case, a probable subclade (or sister group?) of the Epidendreae invaded the Old World at an early date, and the dendrobioid subclade is the major group of Old World orchids. At the same time, the Maxillarieae are the major advanced group of New World orchids, and they are a subgroup of the seemingly Old World cymbidioid phylad.

The "Vandoid" Character Complex

One of the most persistently frustrating problems in orchid classification has been the status of the vandoid orchids, that is, those with a viscidium and a well developed stipe. If one compares the "fully vandoid" condition with the more ordinary "epidendroid" condition, an impressive number of features seem to distinguish the vandoid orchids:

> Lateral inflorescence
> Anther partitions reduced
> Bending of anther very early in ontogeny
> Pollinia two or four
> Pollinia superposed (if four)
> Well developed viscidium
> Tegular stipe

On close inspection, one finds that the correlation between these features is imperfect. Some species of *Cymbidium* or *Maxillaria* have little or no stipe. *Polystachya* seems quite vandoid, except that the pollinia may become superposed only after the anther is removed. To treat the Vandoideae as a distinct subfamily, I placed *Calypso* in the Epidendroideae (Dressler 1981), for its anther is obviously incumbent, and the ontogenetic bending is easily seen. The stipes of some Bulbophyllinae are structurally different from those of the Vandeae, but de Vogel (1986) has described *Geesinkorchis*, a member of the "epidendroid" Coelogyneae, with a well-developed and apparently tegular stipe. Finally, Kurzweil (1988) has shown that developmental differences cannot be taken to distinguish the vandoid and epidendroid orchids. Thus, it is almost a relief that Møller and Rasmussen (1984) have given such convincing evidence that the Vandeae are not closely allied to the other "vandoids."

The evidence that the vandoid level is polyphyletic seems quite convincing (see Fig. 8-2; Dressler 1989b). This implies a great deal of parallelism in several

Group	Vandeae	Polystachyinae	Cymbidieae, Maxillarieae
Silica bodies	 ⬡		 ◁---◁
Pollinia	 ⊜ ⊜	 ⌐◑ ◐⌐	 ⊜⊜
Primitive stem type			
Inflorescence	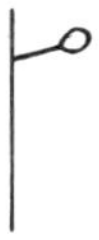		
Chromosome nos.	**38–42**	40	10–60
Related groups	Eriinae, Dendrobiinae	Glomerinae, Adrorhizinae	Malaxideae

Figure 8-2. A graphic comparison of the principal clades of vandoid orchids. (After Dressler 1989b.)

features. Once an epidendroid orchid has evolved a well-developed viscidium, it appears that selection favors the development of a stipe, superposed pollinia (already present in the Coelogyneae), and bending of the anther very early in ontogeny. This pattern strongly supports the idea that similar plants (and especially closely related plants) are likely to evolve in a parallel fashion.

Phylogeny of the Advanced Epidendroideae

The Epidendroideae show a great deal of parallelism, and the natural phylads within this large group are difficult to determine. For several years I tried to distinguish a "reed-stem" phylad and a "cormous" phylad, but some genera refuse to fit these otherwise clear groups. By emphasizing the relatively few features that appear to show little or no parallelism, it is possible to divide the advanced Epidendroideae into two major phylads, the cymbidioid phylad and the epidendroid phylad (see Fig. 8-3). The epidendroid phylad is distinguished by the presence of eight pollinia in its primitive members, and includes most of the plants with a reed-stem habit of growth. By this interpretation, the problem with the reed-stem phylad was simply that its most primitive members are cormous. Most of the cymbidioid phylad are vandoid, with superposed pollinia, viscidia, and usually stipes. The pollinia of the Malaxideae lack caudicles (as do those of the Dendrobieae), and it is possible that the Malaxideae should be treated as a third

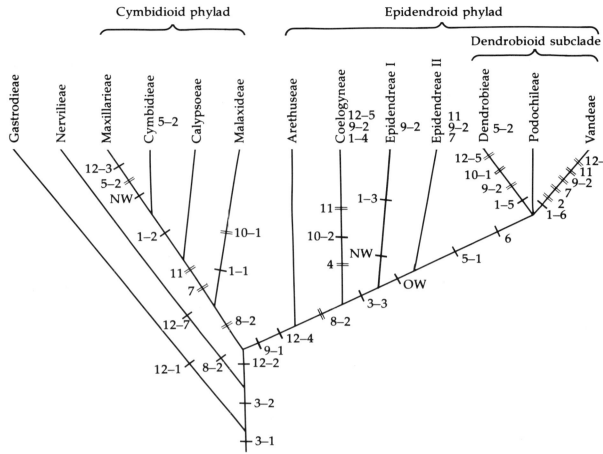

Figure 8-3. Phylogenetic diagram of the advanced Epidendroideae. The features are given in Table 8-1.

Table 8-1. Features, their states and polarization, as used in the phylogenetic diagram in Figure 8-3.[a]

1. Velamen: 0. lacking or *Calanthe* type; 1. *Malaxis* type; 2. *Cymbidium* type; 3. *Epidendrum* type; 4. *Coelogyne* type; 5. *Dendrobium/Bulbophyllum* types, 6. *Vanda* type
2. Growth habit: 0. sympodial; 1. monopodial $(0 \rightarrow 1)$
3. Stem: 0. slender; 1. base fleshy; 2. corm; 3. secondarily reed–stem $(0 \rightarrow 1 \rightarrow 2 \rightarrow 3)$
4. Corm or pseudobulb of: 0. several internodes; 1. a single internode $(0 \rightarrow 1)$
5. Inflorescence: 0. terminal; 1. upper lateral; 2. basal $(0 \rightarrow 1, 0 \rightarrow 2)$
6. Silica cells: 0. conical; 1. spherical $(0 \rightarrow 1)$
7. Tegular stipe: 0. lacking; 1. present $(0 \rightarrow 1)$
8. Pollinia texture: 0, soft; 1. sectile; 2. firm or hard $(0 \rightarrow 1, 0 \rightarrow 2, 1 \rightarrow 2?)$
9. Pollinia number: 0. four; 1. eight; 2. reduced from eight to four or two $(0 \rightarrow 1 \rightarrow 2)$
10. Caudicles: 0. present; 1. secondarily lacking, (naked pollinia); 2. massive $(0 \rightarrow 1 \rightarrow 2, 1 \rightarrow 3)$
11. Pollinia: 0. parallel; 1. superposed $(0 \rightarrow 1)$
12. Seed type: 1. *Gastrodia*; 2. *Eulophia*; 3. *Maxillaria*; 4. *Bletia*; 5. *Dendrobium*; 6. *Vanda*; 7. other distinctive type $(2 \rightarrow 3, 2 \rightarrow 4,$ other polarities unclear)

Distribution: NW = New World, OW = Old World.

[a]When there is a significant doubt as to polarity, a "0" state is not shown.

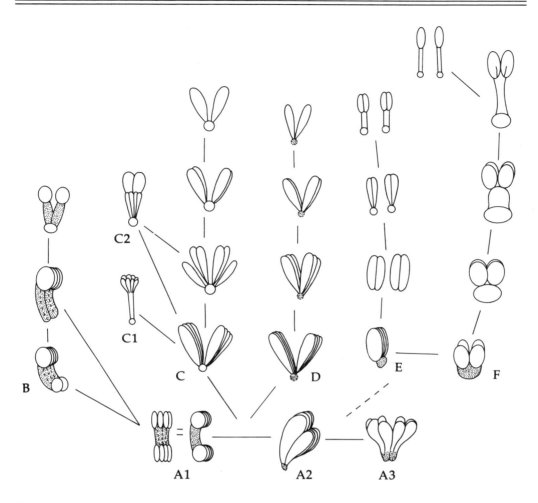

Figure 8-4. Diagram showing the major patterns of pollinia evolution in the Epidendroideae. (A) Eight pollinia, as in the Arethuseae or primitive Epidendreae. These may be discoid (A1), club-shaped (A2), or obovoid (A3). (B) Reduction in number in discoid pollinia, with superficially "vandoid" pollinaria, as in *Epidanthus*. (C) Reduction with club-shaped pollinia and viscidia, as in the Podochilinae. (C1) Prominent caudicles, as in the Thelasiinae. (C2) Reduced pollinia simulating a stipe, as in the Podochilinae. (D) Reduction in number in club-shaped pollinia without viscidia, as in some Pleurothallidinae. (E) The loss of caudicles, and the evolution of viscidia and hamular stipes, as in the Dendrobieae. A similar pattern (without stipes) occurs in the Malaxideae, and was probably not derived from an ancestor with eight pollinia. (F) The evolution of the "vandoid" pattern, either from an ancestor with eight pollinia (Epidendroid phylad) or from one with four (Cymbidioid phylad).

phylad, independent of the cymbidioid phylad. The general pattern of evolution of epidendroid pollinaria is shown in Figure 8-4.

The reed-stem members of the epidendroid phylad all have their rhizomes distinctly thickened, and I have suggested (Dressler 1990c) that this habit may be derived either from a cormous habit or from a saprophytic seedling stage with thickened stems (see Fig. 3-1).

More data are clearly needed, and we should consider alternative patterns of phylogeny. I had, at first, treated the Coelogyneae as part of the "reed-stem" complex, possibly because of similar seed structure, but none of the Coelogyneae has a truly reed-stem habit and some are clearly cormous. Placing the Coelogyneae next to the Arethuseae makes a clearer diagram and suggests the possibility that the Arethuseae plus Coelogyneae might be treated as a distinct phylad. In such a case, one could hypothesize that the cymbidioid phylad is closest to the ancestral type because of pollinia number, that the Arethuseae is more primitive because of soft pollinia, or that the truly reed-stem group represents the primitive condition, because of habit and soft pollinia. One can easily imagine schemes other than those diagrammed in Figure 8-5, but other schemes seem less probable. The seemingly primitive *Xerorchis*, with eight pollinia, suggests that the reed-stem complex might represent a separate phylad not derived from cormous ancestors, though both seed types and pollinia number seem to indicate a close alliance between the Arethuseae and the reed-stem complex. The pattern shown in Figure 8-3 may well change when more information is available.

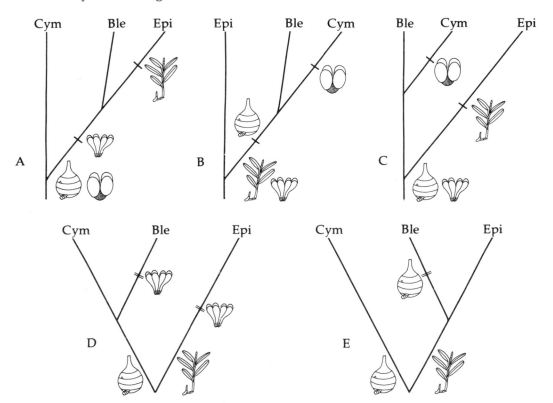

Figure 8-5. Possible phylogenies of the advanced epidendroid groups. (A) The cymbidioid pattern is primitive; eight pollinia evolved only once. (B) The reed-stem pattern is primitive; number of pollinia was reduced in the common ancestor of the cymbidioid phylad. (C) The bletioid pattern is primitive, the cymbidioid and epidendroid phylads each evolving from a bletioid ancestor. If the cormous and reed-stem phylads diverged very early, the bletioid pattern could have evolved by the independent evolution of eight pollinia in a cormous group (D), or the independent evolution of corms from a reed-stem phylad (C). Ble, Bletiinae/Coelogyneae complex; Cym, cymbidioid phylad; Epi, epidendroid phylad (except Arethuseae and Coelogyneae).

9

The Cormous
or Cymbidioid Phylad

The Malaxideae, Calypsoeae, Cymbidieae, and Maxillarieae appear to form a natural group, with distinct seed types in the Maxillarieae and some Cymbidieae. The *Cymbidium* seed type is probably derived from the *Eulophia* seed type, mainly through the predominance of longitudinal, rather than reticulate, thickenings and the distinctive, emergent cell corners.

The relationships within the Maxillarieae are rather clear, and the primitive Maxillarieae resemble the Calypsoeae and Cymbidieae in habit and general flower structure. The seed types are distinct, but the *Maxillaria* seed type also may be derived from the *Eulophia* type. Both the velamen type and the viable hybrids between *Cymbidium* and Maxillarieae (Tanaka et al. 1987), support a close relationship between these tribes. Numerous intermediates connect the *Maxillaria* seed type and the *Vanda* type, surely independently evolved in this tribe. Similarly, the seeds of some Stanhopeinae clearly approach the *Maxillaria* type, and the more primitive Zygopetalinae and Stanhopeinae resemble each other in habit.

The Calypsoeae, Cymbidieae, and Malaxideae occur in both the Americas and the Old World, but the Maxillarieae are strictly American. Representative pollinaria of the cymbidioid phylad are illustrated in Figure 9-1.

Tribe **Malaxideae**

DESCRIPTION. Habit: terrestrial or epiphytic, with pseudobulbs or corms of one or several internodes, or stems slender. Velamen diverse, usually of the *Malaxis* type. Leaves: spiral or distichous, convolute or duplicate, plicate, conduplicate or

159

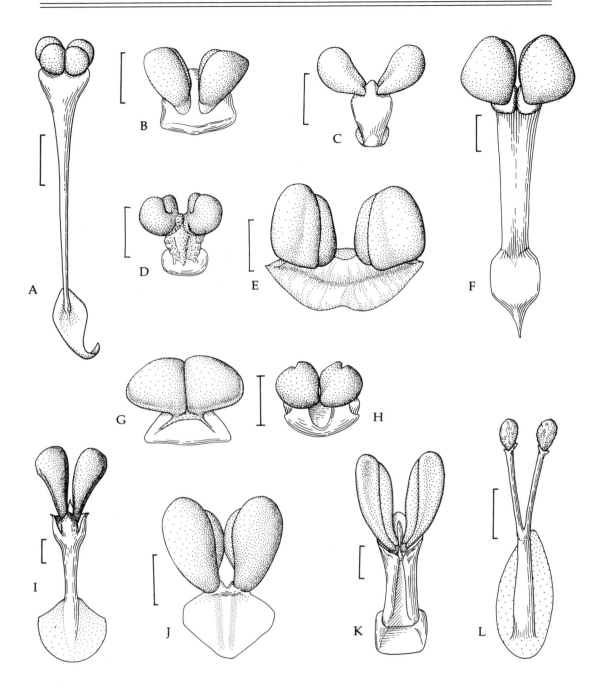

Figure 9-1. Pollinaria of the Cymbidioid Phylad. (A) *Ornithocephalus powellii*. (B) *Calypso bulbosa*. (C) *Miltonia regnellii*. (D) *Govenia liliacea*. (E) *Scuticaria steelii*. (F) *Anguloa dubia*. (G) *Eulophia petersii*. (H) *Grammatophyllum scriptum*. (I) *Houlletia lowiana*. (J) *Warrea costaricensis*. (K) *Catasetum trulla*. (L) *Fernandezia sanguinea*. Scale 1 mm. (After Dressler 1981.) See also Figures 9-9, 9-10.

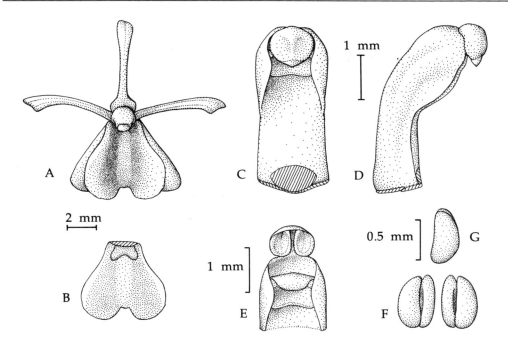

Figure 9-2. *Liparis nervosa* (Epidendroideae: Malaxideae). (A) Flower, front view. (B) Lip, flattened. (C) Column, ventral view. (D) Column, side view. (E) Apex of column, with anther tipped back. (F) Pollinia. (G) Pollinia, side view. (After Dressler 1981.)

laterally flattened and fleshy, articulate or not. Inflorescence: terminal, simple, of few to many flowers, spiral or distichous. Flowers: small to medium-small, resupinate or not; column short or elongate; anther terminal or subdorsal, incumbent or more or less erect, two-celled; four pollinia, hard, oblong or often somewhat clavate; stigma entire, often with two small viscidia. (See Fig. 9-2.)

DISTRIBUTION. World-wide (except New Zealand).

POLLINATION. Almost unknown; we have reports of fly pollination and hemipteran pollination in *Liparis*, and pollination by small flies is indicated for *Malaxis paludosa* (Reeves and Reeves 1984).

CHROMOSOME NUMBERS. 20, 26, 28, 30, 32, 34, 36, 38, 42, 44, 46.

SEED STRUCTURE. *Eulophia* type, *Vanda* type, or *Dactylorhiza* variant of *Orchis* type.

SPECIES. 960.

GENERA. 6: *Hippeophyllum, Liparis, Malaxis, Oberonia, Orestias, Risleya.*

DISCUSSION. The Malaxideae have been characterized by "naked" pollinia, though many species have tiny viscidia. These tiny, often clavate pollinia appear to have no caudicles, and, unlike the Dendrobieae, there is no clear indication that they were derived from ancestors with distinct caudicles. This tribe may be a natural group, but it is diverse in nearly every feature. One finds, for example, three seed types in *Liparis*. Other genera, as far as sampled, show either the *Eulophia* type or the *Vanda* type.

REFERENCE. Schick et al. 1987; Sehgal and Sehgal 1989.

Tribe **Calypsoeae**

DESCRIPTION. Habit: terrestrial, or saprophytic, with corms of several inter-nodes. Velamen of the *Calanthe* type. Leaves: convolute, plicate, nonarticulate or articulate, sometimes petiolate. Inflorescence: lateral, simple, of few to many spiral flowers. Flowers: small to medium, resupinate, lip with or without a spur; column may be broad or have a prominent foot; anther ventral or terminal and opercu-late, with poorly developed partitions; four pollinia, superposed, with a distinct viscidium and a stipe; stigma entire. (See Fig. 9-3.)

DISTRIBUTION. North temperate and subtropical.

POLLINATION. *Tipularia* is pollinated by noctuid moths, and the asymmetric column places the pollinarium on the insect's eye. *Aplectrum* is pollinated by halictid bees (Hogan 1983). *Corallorhiza* is reportedly pollinated by syrphid flies. *Calypso* offers no reward, but deceives bumblebees by the clump of anther-like hairs on the lip.

CHROMOSOME NUMBERS. 24, 28, 32, 36, 40, 42, 46, 48, 50, 52, 54.

SEED STRUCTURE. *Eulophia* type.

SPECIES. 35.

GENERA. 9: *Aplectrum, Calypso, Corallorhiza, Cremastra, Dactylostalix, Ephippian-thus, Oreorchis, Tipularia, Yoania.*

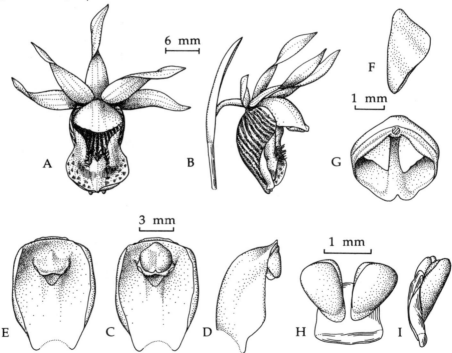

Figure 9-3. *Calypso bulbosa* (Epidendroideae: Calypsoeae). (A) Flower, front view. (B) Flower, side view. (C) Column, ventral view. (D) Column, side view. (E) Column, with anther and pollinarium removed. (F) Anther, side view. (G) Anther, ventral view. (H) Pollinarium, top view. (I) Pollinarium, side view. A through G drawn from material preserved in liquid. (After Dressler 1981.)

DISCUSSION. The Calypsoeae are characterized by four superposed pollinia, stipe, and viscidium. As far as known, all show the *Eulophia*-type seed. *Calypso* has an incumbent anther and, at the same time, a clearly "vandoid" pollinary apparatus, with both viscidium and stipe. Now that the supposed distinctions between "epidendroid" and "vandoid" have crumbled away, *Calypso* is no longer an embarrassment, but merely an early stage in the evolution of one of the vandoid clades.

Recent work by Freudenstein (1991b; personal communication) indicates that this is a paraphyletic grade that may be basal to the other groups in the cymbidioid clade. *Aplectrum*, *Corallorrhiza*, *Cremastra*, and *Oreorchis* have a typical hamular stipe, similar to that of some Bulbophyllinae. *Dactylostalix*, *Ephippianthus*, and *Tipularia* have a distinctive type of hamular stipe, in which the apex of the rostellum forms the viscidium. *Calypso* and *Govenia* both have tegular stipes.

ALTERNATIVE CLASSIFICATION. The work of Freudenstein (see above) suggests that this grade might better be divided into several different subtribes; most or all of these groups appear to be primitive elements of the Cymbidieae. Still, it is quite likely that both the Malaxideae and the Maxillarieae also have been derived from this grade. The cladogram in Chase and Hills (1992) suggests that we are not even close to having a phylogenetic classification of the Calypsoeae and Cymbidieae.

REFERENCES. Ackerman 1981; Boyden 1982; Currah et al. 1988; Hogan 1983; Lund 1987b.

Tribe **Cymbidieae**

DISCUSSION. The advanced members of this tribe are characterized by a distinctive seed type, but *Cymbidium* subgenus *Jensoa* has seeds of the *Eulophia* type (Du Puy and Cribb 1988). Also, the flowers may have either two or four pollinia, even within *Cymbidium* (Seidenfaden 1983). The leaves may be either plicate or conduplicate, and there are often several or many leaves spaced along the pseudobulbs.

REFERENCE. Seidenfaden 1983.

Subtribe **Goveniinae**

DESCRIPTION. Habit: terrestrial, with a globose or irregular (rhizome-like) corm. Velamen of the *Cymbidium* type. Leaves: one or two, plicate, soft, articulate. Inflorescence: lateral, simple, of few to many flowers. Flowers: small to medium, resupinate, lip hinged, simple; column with a slight foot; anther terminal, operculate, with reduced partitions; four pollinia, with viscidium and stipe; stigma entire.

DISTRIBUTION. Tropical America.

POLLINATION. Not known, but probably by bees.

CHROMOSOME NUMBER. Not known.

SEED STRUCTURE. *Eulophia* type.

SPECIES. 20.

GENUS. *Govenia*.

DISCUSSION. *Govenia* seems a bit of a misfit in the Corallorhizinae (= Calypsoeae) but is not discordant in the Cymbidieae. It could be a sister group to the Maxillarieae, but we have only the distribution and the velamen type to suggest such a relationship.

Subtribe **Bromheadiinae**

DESCRIPTION. Habit: terrestrial or epiphytic; stems slender. Velamen of the *Cymbidium* type. Leaves distichous, conduplicate or fleshy, articulate. Inflorescence: terminal, simple, of few to many spiral flowers. Flowers: small to large, resupinate; column slender, slightly arched; anther terminal, operculate, with reduced partitions; two pollinia, notched, with broad or lunate stipe continuous with viscidium, produced behind two rostellar flaps; stigma entire.

DISTRIBUTION. Tropical Asia.

POLLINATION. Not known, though probably by bees; the lunate viscidium of *B. truncata* resembles that of *Galeandra* or *Maxillaria* and is probably deposited on a bee's scutellum.

CHROMOSOME NUMBER. Not known.

SEED STRUCTURE. *Eulophia* type.

SPECIES. 12.

GENUS. *Bromheadia*.

DISCUSSION. The stipe is formed behind a curious double flap (Fig. 9-4). Though the viscidium/stipe is quite large, there is no sharp division into two structures. In this, as in habit, it is quite unlike the Eulophiinae. The flowering of *Bromheadia* is gregarious; that is, the flowers are short-lived and produced on the same days throughout a population. *Bromheadia* might well merit tribal status, as it is very distinctive and tied to other Cymbidieae only by the velamen type. In both habit and gregarious flowering, this genus resembles the epidendroid phylad rather than the other members of the cymbidioid phylad.

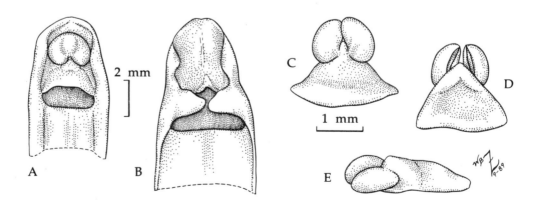

Figure 9-4. *Bromheadia finlaysoniana* (Epidendroideae: Cymbidieae). (A) Apex of young column, with anther in place. (B) Apex of mature column, with anther and pollinarium removed. (C, D, E) Pollinarium, dorsal, ventral and side views. A and B drawn from material preserved in liquid.

Subtribe **Eulophiinae**

DESCRIPTION. Habit: terrestrial, rarely saprophytic, with corm or pseudobulb of one or several internodes. Velamen of the *Cymbidium* type. Leaves: spiral or solitary, convolute or duplicate, plicate or conduplicate, articulate. Inflorescence: lateral, simple, of few to many spiral flowers. Flowers: small to large, resupinate, lip may be saccate or deeply spurred; column may have a prominent foot or wings; anther terminal, operculate, with reduced partitions; two pollinia, notched or cleft, with viscidium and stipe; stigma entire.

DISTRIBUTION. Pantropical, mainly Old World.

POLLINATION. Pollination by carpenter bees has been observed in *Eulophia*. A female megachilid bee has been observed pollinating *Dipodium* (Bernhardt and Burns-Balogh 1983).

CHROMOSOME NUMBERS. 32, 34, 36, 38, 40, 42, 44, 46, 48, 52, 54, 56, 58, 60.

SEED STRUCTURE. *Eulophia* type.

SPECIES. 264.

GENERA. 6: *Cyanaeorchis, Dipodium, Eulophia, Geodorum, Oeceoclades, Pteroglossaspis.*

DISCUSSION. The members of this group are generally cormous and have two perforate or slightly cleft pollinia. All have the *Eulophia* seed type. *Dipodium* has a slender, often climbing, stem with plicate leaves and lateral inflorescences. The pollinarium has two stipes, superficially like those of *Grammatophyllum*. *Oeceoclades* often has conduplicate or fleshy leaves, and in some species the base of the leaf and the abscission layer are widely separated.

REFERENCE. Bernhardt and Burns-Balogh 1983; Poggio et al. 1986.

Subtribe **Thecostelinae**

DESCRIPTION. Habit: epiphytic, with pseudobulbs of one internode. Velamen of the *Cymbidium* type. Leaves: terminal on pseudobulb, duplicate, articulate. Inflorescence: lateral, of several to many spiral flowers. Flowers: medium-small, resupinate, lip hinged to column foot; column arched, winged; column foot hollow (a nectary?); anther terminal, operculate, with reduced partitions; two or four pollinia, grooved if two, with viscidium and a broad stipe; stigma entire.

DISTRIBUTION. Tropical Asia.

POLLINATION. Not known.

CHROMOSOME NUMBER. Not known.

SEED STRUCTURE. W. Barthlott (personal communication) reports on the seed of *Thecostele alata*:

> The seeds are even more bizarre than the flowers. Each seed contains about 4 to 12 embryos, which fill up the seed coat like a sack of potatoes. The isodiametric testa cells and the absence of *Randleisten* makes the seed look rather little advanced. Seed coat ... more or less the *Limodorum* type. In some respects the seed coat reminds [one] of the eulophioid and cymbidioid ... genera.

SPECIES. 4.

GENERA. 2: *Thecopus, Thecostele.*

DISCUSSION. *Thecostele* has been treated as a close ally of *Acriopsis*, but I can find no sign of close relationship. Each has a tubular structure in the flower; in *Acriopsis* the tube is formed by the lip and column, while *Thecostele* has a hollow column foot. The other floral details are very different in the two genera, as are the plants. *Acriopsis* has several leaves scattered along an elongate pseudobulb (like a miniature *Catasetum*), but *Thecostele* has grooved pseudobulbs of a single internode with a single terminal leaf. In its vegetative features, *Thecostele* resembles the Eulophiinae more than the Cyrtopodiinae; it might be allied to the Collabiinae.

Seidenfaden (1983) has described the genus *Thecopus* for *Thecostele maingayi* and *T. secunda*, and these species share all of the discordant features of *Thecostele*. The leaves of some *Thecostele alata* have glands on the upper surface. The foliar glands of a plant cultivated in Panama attracted ants, and the glands may attract "ant guards" in Malaysia.

Subtribe **Cyrtopodiinae**

DESCRIPTION. Habit: epiphytic, terrestrial, or occasionally saprophytic, with corm or pseudobulb of several to many internodes, a few species with pseudobulbs of one internode. Velamen of the *Cymbidium* type. Leaves: spiral or distichous, convolute or duplicate, plicate or conduplicate, articulate, often scattered along pseudobulb. Inflorescence: lateral or occasionally terminal, simple or branched, of few to many spiral flowers. Flowers: small to large, resupinate, lip may be saccate or deeply spurred, or hinged; column usually with a prominent foot, may have wings; anther terminal, operculate, with reduced partitions, two or four pollinia, with viscidium and usually with stipe; stigma entire. (See Fig. 9-5.)

DISTRIBUTION. Pantropical but mainly Old World.

POLLINATION. Pollination of *Cymbidium* by honey bees, wasps, stingless bees, carpenter bees, and bumblebees has been reported. Kjellsson et al. (1985) report that *Cymbidium insigne* is a mimic of *Rhododendron*. Nilsson et al. (1986) find that *Cymbidiella flabellata* is pollinated by sphecid wasps, and indicate that *C. flabellata* is a generalized food-flower mimic, though, curiously, it attracts primarily large wasps. *Cyrtopodium* is reported to be pollinated by *Euglossa*, and the pollinia of *Galeandra* have been found on *Euglossa* from several localities.

CHROMOSOME NUMBERS. 38, 40, 42, 46, 52, 54, 56.

SEED STRUCTURE. *Cymbidium* type, or *Eulophia* type in *Cymbidium* subgenus *Jensoa*.

SPECIES. 139.

GENERA. 12: *Acrolophia, Ansellia, Cymbidiella, Cymbidium, Cyrtopodium, Eulophiella, Galeandra, Grammangis, Grammatophyllum, Graphorkis, Grobya, Porphyroglottis*.

DISCUSSION. Even with the Eulophiinae removed, the Cyrtopodiinae remains a rather diverse group. The chromosome numbers range from 38 to 56 and flower structure is diverse. Most of the intergeneric hybrids involving *Cymbidium* are with the genera having 40 chromosomes, but crosses have been made with *Eulophiella*, with 56 chromosomes (Aoyama et al. 1986). Artificial hybrids are also

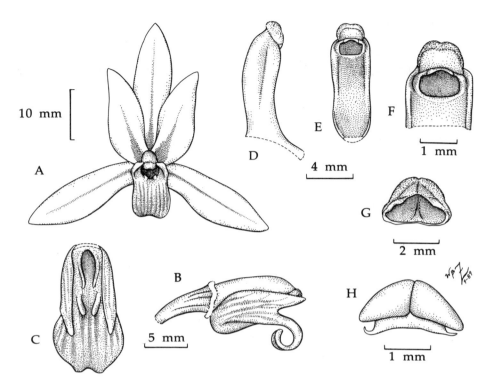

Figure 9-5. *Cymbidium aloifolium* (Epidendroideae: Cymbidieae). (A) Flower, front view. (B) Lip and column, side view. (C) Lip, not flattened. (D, E) Column, side and ventral views. (F) Apex of column, ventral view. (G) Anther. (H) Pollinarium.

recorded between *Cymbidiella* and *Eulophiella*, and between *Galeandra* and *Catasetum*. Some authors treat the Cymbidiinae and Cyrtopodiinae as separate subtribes, as was done by Schlechter, but how to delimit these groups is not clear. Both *Ansellia* and *Grammatophyllum* appear to be close allies of *Cymbidium*, but whether *Cymbidiella* or *Grammangis* are so closely allied is unclear. The relatively simple flower structure of *Cymbidium* may be near the primitive state for this complex, so this feature does not necessarily delimit a natural group.

REFERENCES. Aoyama 1989; Du Puy 1987; Du Puy and Cribb 1988; Kjellson et al. 1985; Nilsson et al. 1986; Vij and Shekhar 1987.

Subtribe **Acriopsidinae**

DESCRIPTION. Habit: epiphytic, with pseudobulbs of several internodes. Velamen of the *Cymbidium* type. Leaves: distichous, duplicate, articulate (above the base). Inflorescence: lateral, simple or branched, of several to many spiral flowers. Flowers: small, resupinate, lip and column highly united, column with armlike wings and hooded clinandrium; anther suberect, with reduced partitions; two pollinia, deeply notched, laterally flattened, bladelike, with a slender stipe and viscidium; stigma entire, narrowly elliptical.

DISTRIBUTION. Tropical Asia.

POLLINATION. Not known.

CHROMOSOME NUMBER. 40.

SEED STRUCTURE. *Cymbidium* type.

SPECIES. 6.

GENUS. *Acriopsis.*

DISCUSSION. Though its flower structure is bizarre, both habit and seed structure place this genus near the Cyrtopodiinae. *Acriopsis* seems clearly a derivative of the Cyrtopodiinae and is cladistically a subgroup of that subtribe. Some authors describe the pollinia as four, though the pairs are firmly united along one side in all that I have seen. Traditionally, *Acriopsis* has been classified with *Thecostele*, but these genera differ in virtually all details.

REFERENCE. Minderhoud and de Vogel 1986.

Subtribe **Catasetinae**

DESCRIPTION. Habit: epiphytic, with pseudobulbs of several internodes. Velamen of the *Cymbidium* type. Leaves: distichous, scattered on pseudobulbs, convolute, plicate, articulate. Inflorescence: lateral, simple, of few to many spiral flowers. Flowers: medium-small to rather large, resupinate or not, often unisexual; column usually with an elastic pollinarium-throwing device; anther rather ventral, with reduced partitions; two pollinia, with a viscidium and an elastic stipe; stigma entire.

DISTRIBUTION. Tropical America.

POLLINATION. All Catasetinae are pollinated by male euglossine bees.

CHROMOSOME NUMBERS. 54, 56, 64, 68.

SEED STRUCTURE. *Cymbidium* type.

SPECIES. 194.

GENERA. 5: *Catasetum, Clowesia, Cycnoches, Dressleria, Mormodes.*

DISCUSSION. The Catasetinae are a distinctive American group with unisexual flowers in most species. Their relationship with the Cyrtopodiinae has recently been confirmed by an artificial hybrid between *Catasetum* and *Galeandra.* In cladistic terms, the Catasetinae are surely a subgroup of the Cyrtopodiinae. In the genera with unisexual or functionally unisexual flowers, the pollinaria are expelled violently when the proper trigger is moved, so that the pollinaria strike the pollinators with some force and usually in the proper spot, where they will be ready to pollinate a pistillate flower if it is visited by the same insect.

Mormodes is often described as bisexual, but many species have distinct floral dimorphism, with "pistilloid" and "staminoid" flowers. There is no information available on the degree to which these types are functionally bisexual or unisexual.

Romero and Nelson (1986) suggest that the violent placement of the pollinaria may be advantageous to the plant and also contribute to the strong dimorphism of the flowers. That is, any bee that has been slapped by a male flower may be cautious about visiting another flower of the same sort. The advantage to a plant with male flowers is that the bee is less likely to visit another male flower and

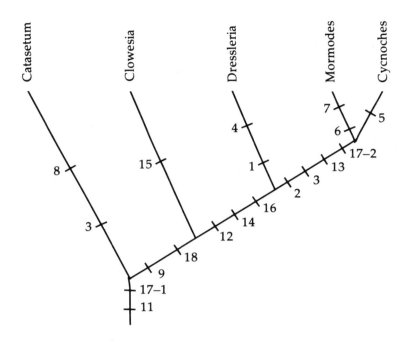

Figure 9-6. Phylogenetic diagram of the Catasetinae, adapted from Romero (1990) and Chase and Pippen (1990). Features as in Table 9-1.

Table 9-1. Features, their states and polarization, as used in Figure 9-6.[a]

1. Leaves: 0. deciduous; 1. persistent (0 → 1)
2. Inflorescence: 0. basal; 1. from mid- to upper pseudobulb (0 → 1)
3. Flowers: 0. protandrous; 1. unisexual (0 → 1)
4. Lip: 0. free; 1. united with column (0 → 1)
5. Column length: 0. short; 1. long and slender (0 → 1)
6. Column symmetry: 0. symmetrical; 1. twisted (0 → 1)
7. Column movement: 0. absent; 1. present (0 → 1)
8. Antennae: 0. absent; 1. present (0 → 1)
9. Clinandrial filament: 0. absent; 1. present (0 → 1)
10. Pollinarium hinged: 0. no; 1. yes (0 → 1)
11. Stipe elastic and stretched taut: 0. no; 1. yes (0 → 1)
12. Stipe after discharge: 0. straight; 1. curled (0 → 1)
13. Stipe axial folding: 0. rostellum side up; 1. rostellum side down (0 → 1)
14. Stipe lateral folding: 0. toward rostellum; 1. away from rostellum (0 → 1)
15. Stipe jointed: 0. no; 1. yes (0 → 1)
16. Stipe lower end: 0. straight; 1. doubly bent (0 → 1)
17. Discharged pollinarium: 0. attached; 1. partially detached; 2. completely free (0 → 1 → 2)
18. Cell surface of seed: 0. longitudinally ridged; 1. irregularly thickened (0 → 1)

[a] When there is a significant doubt as to polarity, a "0" state is not shown.

receive a second pollinarium that might prevent the first from functioning. The female flowers are usually very different in form, and should not frighten away potential pollinators. At female flowers of *Catasetum ochraceum*, Romero and

Nelson found the percentage of bees carrying pollinaria to be about eight times that observed at male flowers.

Chase and Pippen (1990) find that in *Catasetum*, the periclinal cell walls of the seed coat have longitudinal thickenings, like those of most Cyrtopodiinae; the other Catasetinae, however, have irregular thickenings. Chase and Pippen suggest that unisexual flowers, as in *Catasetum*, may be the primitive condition, with bisexual flowers secondarily derived in other Catasetinae (but see below). On the basis of chloroplast DNA, Chase and Hills (1992) find the Catasetinae closely allied to *Cyrtopodium*.

Romero (1990) offers an excellent cladistic study of the Catasetinae (Fig. 9-6). He finds the bisexual flowers to be effectively protandrous, as the stigma cannot function until the viscidium has been removed. This feature, in conjunction with the somewhat sensitive rostellum, has probably led to the independent evolution of unisexual flowers in *Catasetum* and in the *Cycnoches-Mormodes* clade. Romero finds his analysis compatible with the Chase and Pippen hypothesis of secondary bisexuality, but the hypothesis that unisexual flowers have twice evolved from protandrous ancestors is as parsimonious and possibly more credible. Romero confirms the relationship suggested by Dodson (1975) between *Dressleria* and *Cycnoches*, but does not support Dodson's suggestion of a close relationship between *Clowesia* and the *Catasetum discolor* group. In fact, the resemblances between *Clowesia* and *Catasetum discolor* may represent features found in the common ancestor of the Catasetinae.

REFERENCES. Chase and Hills 1992: Chase and Pippen 1990; Gregg 1983; Romero 1990; Romero and Nelson 1986.

Tribe **Maxillarieae**

DISCUSSION. This group includes most of the tropical American orchids with stipe and viscidium. It is a diverse group, but also seems to be a natural one. In overall appearance, the most primitive Zygopetalinae resemble the Calypsoeae, so much so that I placed most of that tribe in the Maxillarieae in 1981. Though the seed structure of the Maxillarieae is distinctive, the Maxillarieae and the Cymbidieae seem clearly linked by the *Cymbidium* type velamen and by the intertribal hybrids reported by Tanaka et al. (1987). The connections between the *Maxillaria* seed type and the other seed types in the Maxillarieae seem quite clear.

ALTERNATIVE CLASSIFICATION. The Oncidiinae, Ornithocephalinae, and Telipogoninae might be placed in a separate tribe, the Oncidieae; and the Stanhopeinae might be treated as a separate tribe, for which the correct name would be Gongoreae.

Subtribe **Cryptarrheninae**

DESCRIPTION. Habit: epiphytic, with pseudobulbs or short, unthickened stems. Leaves: distichous, duplicate, articulate. Inflorescence: lateral, simple, of many spiral flowers. Flowers: small, resupinate; lip clawed, with a thick callus on claw, four-lobed; column short, with a hooded clinandrium over the anther; anther terminal, incumbent but with an erect beak, two-celled; four pollinia, superposed, with two hyaline cylindrical caudicles that may be basally joined, and a distinct viscidium; stigma entire. (See Fig. 9-7.)

DISTRIBUTION. Tropical America.

POLLINATION. Not known.

CHROMOSOME NUMBER. Not known.

SEED STRUCTURE. *Maxillaria* type.

SPECIES. 7.

GENUS. *Cryptarrhena*.

DISCUSSION. When the epidendroid and vandoid orchids were considered two distinct phylads, *Cryptarrhena* was hard to place. It is vandoid in most of its features, but it lacks a stipe. We now know that its seeds are of the *Maxillaria* type, and its position is less puzzling. It was probably derived from some primitive member of the Maxillarieae. Rather than evolving a stipe, as have most Maxillarieae, this genus evolved long, stipelike caudicles.

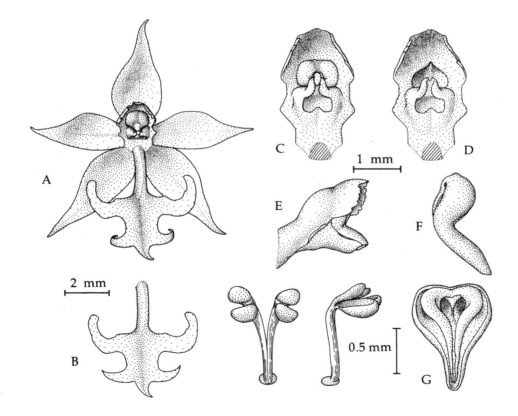

Figure 9-7. *Cryptarrhena guatemalensis* (Epidendroideae: Maxillarieae). (A) Flower, front view. (B) Lip, flattened. (C) Column, ventral view. (D) Column, with anther and pollinarium removed. (E) Column, side view. (F) Anther, side view. (G) Anther, ventral view. (H) Pollinarium, top view. (I) Pollinarium, side view. (After Dressler 1981.)

Subtribe **Zygopetalinae**

DESCRIPTION. Habit: terrestrial or epiphytic, with pseudobulbs of several internodes or of one internode, or stems slender, short or elongate. Velamen of the *Cymbidium* type. Leaves: convolute or duplicate, plicate or conduplicate, articulate. Inflorescence: lateral, of one to many spiral flowers, often from a young shoot. Flowers: small to large, resupinate or not; column short or elongate, often winged or flattened, usually with a distinct foot; anther terminal or rather ventral, operculate, with reduced partitions; four pollinia, superposed, with a prominent viscidium and usually with a stipe; stigma entire.

DISTRIBUTION. Tropical America.

POLLINATION. All records indicate that the members of the Zygopetalinae are pollinated by euglossine bees, and in most cases by perfume-gathering males. Most genera place their pollinaria on the bee's head, thorax, or scutellum, but *Chaubardiella* places its pollinaria on the trochanters of the bees' legs, and *Kefersteinia* places its pollinaria on the bases of antennae. Some species of *Cochleanthes* appear to be deceit flowers rather than perfume flowers, attracting naive, nectar-seeking bees (Ackerman 1983).

CHROMOSOME NUMBERS. 46, 48, 52.

SEED STRUCTURE. *Maxillaria* type, or *Chondrorhyncha* variant.

SPECIES. 331.

GENERA. 30: Tentatively divided into three alliances and four distinctive genera:

1. *Warrea* alliance; pseudobulbs of several internodes, leaves plicate, inflorescence racemose—*Otostylis, Warrea, Warreella, Warreopsis.*

2. *Zygopetalum* alliance; pseudobulbs of one internode, inflorescence racemose, of few to many flowers, callus generally high and ridged—*Aganisia, Batemania, Galeottia, Koellensteinia, Neogardneria, Pabstia, Paradisianthus, Promenaea, Zygopetalum, Zygosepalum.*

3. *Chondrorhyncha* alliance (Huntleyinae); pseudobulbs small or lacking, leaves conduplicate, inflorescence one-flowered—*Benzingia, Bollea, Chaubardia, Chaubardiella, Chondrorhyncha, Cochleanthes, Dodsonia, Hoehneella, Huntleya, Kefersteinia, Pescatorea, Stenia.*

4. *Cheiradenia*—pseudobulb of several internodes, leaves conduplicate, flowers fascicled.

5. *Dichaea* (Dichaeinae)—plants monopodial, inflorescence of one flower.

6. *Scuticaria*—leaves cylindrical, inflorescence of one or few flowers.

7. *Vargasiella*—plants monopodial (?), inflorescence racemose.

DISCUSSION. One might divide this group into several subtribes, but the features of pseudobulb, leaf, inflorescence, and flower do not correlate well enough to permit clear subgroups. I separate the Lycastinae at the suggestion of Mark Chase, whose work on molecular systematics indicates this to be a distinct group. I am not satisfied with the Zygopetalinae as it stands, but it does appear to be a natural group. It is characterized by the *Maxillaria* seed type, and the complex with reduced pseudobulbs, nearly conduplicate leaves, and one-flowered inflorescences shows the *Chondrorhyncha* variant. The *Chondrorhyncha* variant is also found in *Promenaea*, which does not fit the usual characterization of the Huntleyinae, as it has well developed pseudobulbs and may have more than one flower on an inflorescence. Still, placing *Promenaea* in the *Chondrorhyncha* alliance (or the Huntleyinae) might give us a monophyletic subgroup, though it is closely allied to the rest of the Zygopetalinae.

The *Warrea* alliance surely represents the most primitive element in the subtribe, with pseudobulbs of several internodes, plicate leaves, racemes, and relatively simple pollinaria.

Scuticaria is usually included in the Maxillariinae, but the often two-flowered inflorescence is quite out of place there, and the flower is more reminiscent of the *Zygopetalum* complex.

Recent and more complete specimens of *Vargasiella* have definite stipe and viscidium (G. Romero, personal communication), so its position in the Maxillarieae now seems clear. The subtribe Vargasiellinae was not validly published, but might well be validated for this isolated genus.

Subtribe **Lycastinae**

DESCRIPTION. Habit: terrestrial or epiphytic, with pseudobulbs of one internode. Velamen of the *Cymbidium* type. Leaves: convolute or duplicate, plicate or conduplicate, articulate. Inflorescence: lateral, of one to many spiral flowers. Flowers: medium to large, resupinate; column elongate, often winged or flattened, usually with a distinct foot; anther terminal, operculate, with reduced partitions; four pollinia, superposed, with a prominent viscidium and a well-developed stipe; stigma entire. (See Fig. 9-8.)

DISTRIBUTION. Tropical America

POLLINATION. By euglossine bees in *Anguloa*, *Lycaste*, and some species of *Bifrenaria*. Pollinaria of *Lycaste* have been found on female bees, so some may practice false advertisement. *Xylobium* does not show the euglossine syndrome, and is probably pollinated by some other insect, possibly stingless bees.

CHROMOSOME NUMBERS. 38, 40, 44, 48, 50.

SEED STRUCTURE. *Maxillaria* type.

SPECIES. 127.

GENERA. 8: in three tentative alliances:

1. *Bifrenaria* alliance (Bifrenariinae); leaves plicate to conduplicate, inflorescence of one to many flowers; stipe short, often divided into two arms—*Bifrenaria*, *Horvatia*, *Rudolfiella*, *Teuscheria*, *Xylobium*.

2. *Lycaste* alliance (Lycastinae in the narrow sense); leaves plicate; inflorescence of one flower, stipe very long—*Anguloa*, *Lycaste*.

3. *Neomoorea*—pseudobulb with many grooves, leaves plicate; inflorescence racemose; stipe very long.

DISCUSSION. *Xylobium* may be out of place in the *Bifrenaria* complex, though I find no clear features to separate it. *Neomoorea* agrees closely with *Lycaste* in the column and pollinarium, though the aspect of the inflorescence and flower is quite different.

REFERENCE. Aoyama and Karasawa 1988.

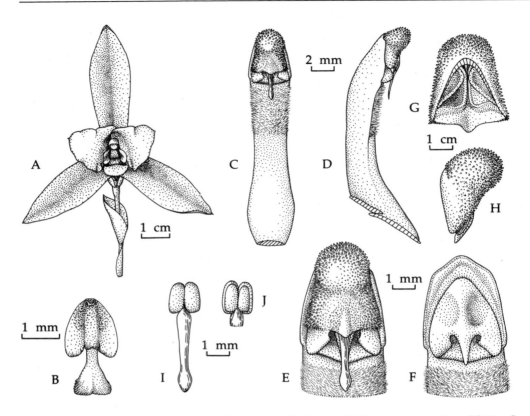

Figure 9-8. *Lycaste tricolor* (Epidendroideae: Maxillarieae). (A) Flower, front view. (B) Lip, flattened. (C) Column, ventral view. (D) Column, side view. (E) Apex of column, with anther in place. (F) Apex of column, with anther and pollinarium removed. (G) Anther, ventral view. (H) Anther, side view. (I) Pollinarium. (J) Pollinia, ventral view. (After Dressler 1981.)

Subtribe **Maxillariinae**

DESCRIPTION. Habit: epiphytic or terrestrial, with pseudobulbs of one internode, or stem slender, short or elongate. Velamen of the *Cymbidium* type. Leaves: distichous or secondarily spiral, duplicate or fleshy, articulate. Inflorescence: lateral, of single flowers. Flowers: small to large, lip usually hinged, sometimes forming a saccate nectary with the column foot, with a distinct spur in some taxa; column slender or short, anther terminal, operculate, with reduced partitions, four pollinia, superposed, with viscidium and more or less well-developed stipe; stigma entire.

DISTRIBUTION. Tropical America.

POLLINATION. We have found *Maxillaria* pollinia on stingless bees (*Trigona*), and some species are reported to attract bees or wasps by offering wax on the lip (used as nest-building material?). We have also observed *Melipona* gathering pseudopollen from the lip of a *Maxillaria* in Ecuador.

CHROMOSOME NUMBERS. 40, 42.

SEED STRUCTURE. *Maxillaria* type.

SPECIES. **472.**

GENERA. **8:** *Anthosiphon, Chrysocycnis, Cryptocentrum, Cyrtidiorchis, Maxillaria, Mormolyca, Pityphyllum, Trigonidium.*

DISCUSSION. The *Maxillaria* complex seems clearly distinct from the Zygopetalinae, but the generic lines within the complex are not very convincing.

Subtribe **Stanhopeinae**

DESCRIPTION. Habit: epiphytic, with pseudobulbs usually of one internode. Velamen of the *Cymbidium* type. Leaves: terminal, convolute, plicate, articulate, usually petiolate. Inflorescence: lateral, of one to several spiral flowers. Flowers: small to very large, resupinate or not, often pendant; column winged or not; anther terminal or ventral, operculate, with reduced partitions; two pollinia, cleft, with viscidium and usually a promiment stipe; stigma entire. (See Fig. 9-9.)

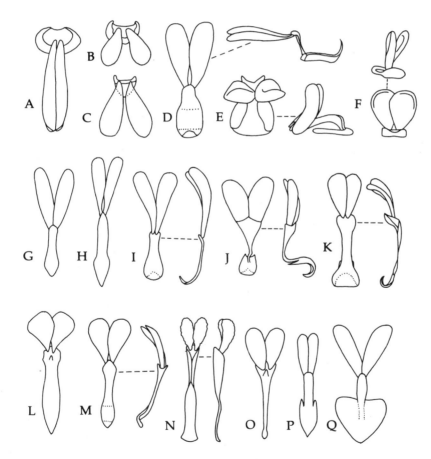

Figure 9-9. Pollinaria of the Stanhopeinae. (A) *Lycomormium.* (B) *Peristeria.* (C) *Coeliopsis.* (D) *Acineta.* (E) *Coryanthes.* (F) *Lueddemannia.* (G) *Gongora.* (H) *Polycycnis.* (I) *Cirrhaea.* (J) *Sievekingia.* (K) *Trevoria.* (L) *Houlletia* section *Houlletia.* (M) *Lacaena.* (N) *Paphinia.* (O) *Schlimia.* (P) *Kegeliella.* (Q) *Stanhopea.* Both dorsal and side views shown in D, F, I, J, K, M, and N. (After Dressler 1981.)

DISTRIBUTION. Tropical America.

POLLINATION. The Stanhopeinae are always pollinated by male euglossine bees, and the more highly evolved genera, such as *Stanhopea, Polycycnis, Cirrhaea,* and *Coryanthes,* include some of the most bizarre and complicated pollination mechanisms to be found in the plant kingdom.

CHROMOSOME NUMBERS. 38, 40, 42.

SEED TYPE. *Stanhopea* type, or in a few genera, transitions between *Maxillaria* and *Stanhopea* types.

SPECIES. 248.

GENERA. 22: *Acineta, Braemia, Cirrhaea, Coeliopsis, Coryanthes, Embreea, Gongora, Horichia, Houlletia, Kegeliella, Lacaena, Lueddemannia, Lycomormium, Paphinia, Peristeria, Polycycnis, Schlimia, Sievekingia, Soterosanthus, Stanhopea, Trevoria, Vasqueziella.*

DISCUSSION. The relationships of this interesting subtribe are now quite clear. Not only do the more primitive genera, *Coeliopsis, Lacaena, Lycomormium,* and *Peristeria,* resemble the primitive Zygopetalinae in general appearance, they also have seeds intermediate between the *Maxillaria* type and the distinctive *Stanhopea* type found in the other genera.

REFERENCES. Jenny 1979–1991, 1985, 1987–.

Subtribe **Telipogoninae**

DESCRIPTION. Habit: epiphytic, with pseudobulbs of one internode, or slender stems. Velamen of the *Cymbidium* type. Leaves: distichous, duplicate, articulate, occasionally leafless. Inflorescence: terminal or lateral, of few to many flowers, spiral or distichous. Flowers: tiny to medium-large, resupinate or erect; column short, usually bristly; anther dorsal, erect, with reduced partitions; four unequal pollinia, with a long stipe and a hooked viscidium; stigma entire.

DISTRIBUTION. Tropical America.

POLLINATION. The Telipogoninae offer no reward to their pollinators but mimic bristly female flies (Tachinidae). The hooklike viscidia attach the pollinaria to the legs of the deceived males.

CHROMOSOME NUMBER. Not known.

SEED STRUCTURE. *Vanda* type.

SPECIES. 126.

GENERA. 4: *Hofmeisterella, Stellilabium, Telipogon, Trichoceros.*

DISCUSSION. The Telipogoninae have four pollinia, but their seed structure is like that of the Oncidiinae, thus suggesting that they may be a link between the Oncidiinae and the Maxillarinae. Dr. C. H. Dodson suggests the inclusion of *Hofmeisterella* in this group. He finds the seedling morphology of this group to be similar to that of some Vandeae, evidently a parallelism. The diminutive plants of *Stellilabium* appear to approach the "shootless" condition of some Vandeae. Mature plants are not infrequently found without leaves, but in this case, the flattened inflorescence axis surely contributes significantly to the plant's photosynthesis.

REFERENCE. Dodson and Escobar 1987.

1

2

3

4

5

6

Plate 1. Apostasioideae, Cypripedioideae: 1. *Neuwiedia veratrifolia* (Apostasioideae), Malaysia (Courtesy P. Vaughn). 2. *Apostasia wallichii* (Apostasioideae), Malaysia. 3. *Selenipedium chica* (Cypripedioideae), Panama. 4. *Cypripedium irapeanum* (Cypripedioideae), Mexico. 5. *Paphiopedilum sukakhulii* (Cypripedioideae), Thailand. 6. *Phragmipedium schlimii* (Cypripedioideae), Colombia.

Plate 2. Spiranthoideae; Orchidoideae Diurideae: 7. *Tropidia polystachya* (Tropidieae), Florida, United States (Courtesy C. A. Luer). 8. *Ludisia discolor* (Goodyerinae), tropical Asia. 9. *Altensteinia fimbriata* (Prescottiinae), Ecuador (Courtesy N. H. Williams). 10. *Pelexia* sp. (Spiranthinae), Panama. 11. *Solenocentrum costaricense* (Cranichidinae), Panama. 12. *Cryptostylis arachnites* (Cryptostylidinae), Malaysia.

Plate 3. Orchidoideae Diurideae: 13. *Chloraea gavilu* (Chloraeinae), Chile (Courtesy C. A. Luer).
14. *Caladenia catenata* (Caladeniinae), Australia. 15. *Pterostylis nana* (Pterostylidinae), Australia.
16. *Corybas mucronatus* (Acianthinae), Malaysia. 17. *Diuris longifolia* (Diuridinae), Australia. 18.
Calochilus robertsonii (Diuridinae), Australia (Courtesy J. W. Green).

Plate 4. Orchidoideae Orchideae, Diseae: 19. *Galearis spectabilis* (Orchidinae), North America (Courtesy C. A. Luer). 20. *Platanthera ciliaris* (Orchidinae), North America (Courtesy J. P. Folsom). 21. *Habenaria entomantha* (Habenariinae), Panama. 22. *Brownleea coerulea* (Disinae), southern Africa (Courtesy E. A. Schelpe). 23. *Satyrium ocellatum* (Satyriinae), southern Africa (Courtesy J. Stewart). 24. *Disperis capensis* (Coryciinae), southern Africa (Courtesy J. Stewart).

Plate 5. Epidendroideae primitive groups: 25. *Epipactis gigantea* (Limodorinae), western United States (Courtesy C. A. Luer). 26. *Listera cordata* (Listerinae), Colorado, United States (Courtesy C. A. Luer). 27. *Palmorchis nitida* (Palmorchideae), Panama. 28. *Psilochilus carinatus* (Triphoreae), Panama. 29. *Vanilla barbellata* (Vanillinae), Florida, United States (Courtesy C. A. Luer). 30. *Cleistes rosea* (Pogoniinae), Panama.

Plate 6. Gastrodieae, Malaxideae, Calypsoeae: 31. *Gastrodia sesamoides* (Gastrodiinae), Australia (Courtesy P. Lavarack). 32. *Wullschlaegelia calcarata* (Wullschlaegeliinae), Panama. 33. *Liparis lacerata* (Malaxideae), Malaysia. 34. *Oberonia* sp. (Malaxideae), Papua New Guinea. 35. *Calypso bulbosa* (Calypsoeae), western United States (Courtesy C. A. Luer). 36. *Corallorhiza striata* (Calypsoeae), Michigan, United States (Courtesy C. A. Luer).

37

38

39

40

41

42

Plate 7. Cymbidieae: 37. *Bromheadia finlaysoniana* (Bromheadiinae), Malaysia. 38. *Geodorum* sp. (Eulophiinae), Thailand. 39. *Thecostele alata* (Thecostelinae), Malaysia. 40. *Cymbidium lancifolium* (Cyrtopodiinae), Asia. 41. *Acriopsis javanica* (Acriopsidinae), Malaysia. 42. *Catasetum barbatum* (Catasetinae), South America.

Plate 8. Maxillarieae Cryptarrheninae, Lycastinae, Stanhopeinae, Zygopetalinae: 43. *Cryptarrhena guatemalensis* (Cryptarrheninae), Panama. 44. *Zygopetalum maxillare* (Zygopetalinae), Brazil. 45. *Lycaste deppei* (Lycastinae), Mexico. 46. *Maxillaria fulgens* (Maxillariinae), Panama. 47. *Polycycnis aurita* (Stanhopeinae), Colombia (Courtesy R. Escobar R.). 48. *Stanhopea gibbosa* (Stanhopeinae), Costa Rica.

Plate 9. Maxillarieae Ornithocephalinae, Telipogoninae, Oncidiinae: 49. *Ornithocephalus inflexus* (Ornithocephalinae), Panama. 50. *Telipogon elcimeyae* (Telipogoninae), Costa Rica. 51. *Oncidium hastilabium* (Oncidiinae), Colombia. 52. *Sigmatostalix unguiculata* (Oncidiinae), Costa Rica. 53. *Ionopsis satyrioides* (Oncidiinae), Colombia. 54. *Fernandezia* sp. (Oncidiinae), Colombia.

Plate 10. Arethuseae; Coelogyneae; Epidendreae Sobraliinae: 55. *Arethusa bulbosa* (Arethusinae), northern United States (Courtesy C. A. Luer). 56. *Calanthe* sp. (Bletiinae), New Guinea. 57. *Hexalectris spicata* (Bletiinae), Mexico (Courtesy C. A. Luer). 58. *Coelogyne corymbosa* (Coelogyninae), India. 59. *Sobralia bouchei* (Sobraliinae), Panama. 60. *Elleanthus* sp. (Sobraliinae), Panama.

Plate 11. Epidendreae Arpophyllinae, Meiracylliinae, Coeliinae, Laeliinae: 61. *Arpophyllum* sp. (Arpophyllinae), Guatemala (Courtesy F L Stevenson). 62. *Meiracyllium trinasutum* (Meiracylliinae), Mexico (Courtesy H. Page). 63. *Coelia triptera* (Coeliinae), Mexico. 64. *Schomburgkia splendida* (Laeliinae), Colombia. 65. *Encyclia venosa* (Laeliinae), Mexico. 66. *Epidendrum hunterianum* (Laeliinae), Panama.

Plate 12. Epidendreae Pleurothallidinae, Glomerinae, Polystachyinae: 67. *Pleurothallis ignivomis* (Pleurothallidinae), Ecuador. 68. *Platystele ovalifolia* (Pleurothallidinae), Panama. 69. *Lepanthes pantomima* (Pleurothallidinae), Panama. 70. *Glomera obtusa* (Glomerinae), Papua New Guinea. 71. *Aglossorhyncha jabiensis* (Glomerinae), Papua New Guinea. 72. *Polystachya virginea* (Polystachyinae), central Africa.

Plate 13. Podochileae: 73. *Eria hyacinthoides* (Eriinae), Malaysia. 74. *Eria imitans* (Eriinae), Papua New Guinea. 75. *Eria* (*Trichotosia*) *aporina* (Eriinae), Malaysia. 76. *Mediocalcar abbreviatum* (Eriinae), Papua New Guinea. 77. *Podochilus falcatus* (Podochilinae), Ceylon. 78. *Octarrhena condensata* (Thelasiinae), Malaysia.

80

79

81

82

83

84

Plate 14. Dendrobieae: 79. *Dendrobium cuthbertsonii* (Dendrobiinae), Papua New Guinea. 80. *Dendrobium draconis* (Dendrobiinae), Thailand. 81. *Dendrobium torresae* (Dendrobiinae), Australia. 82. *Bulbophyllum subcubium* (Bulbophyllinae), Papua New Guinea. 83. *Hapalochilus callipes* (Bulbophyllinae), Papua New Guinea. 84. *Pedilochilus flavum* (Bulbophyllinae), Papua New Guinea.

Plate 15. Vandeae: 85. *Pteroceras chrysanthum* (Aeridinae), Papua New Guinea. 86. *Dimorphorchis lowii* (Aeridinae), Borneo. 87. *Neofinetia falcata* (Aeridinae), Japan. 88. *Chiloschista lunifera* (Aeridinae), Thailand. 89. *Rangaeris muscicola* (Aerangidinae), Africa (Courtesy E. A. Schelpe). 90. *Microcoelia guyoniana* (Aerangidinae), Africa (Courtesy E. A. Schelpe).

Plate 16. Problem genera and structure: 91. *Dilochia cantleyi* (Arundinae), Malaysia. 92. *Claderia viridiflora* (Cymbidieae?), Malaysia. 93. *Eriopsis biloba* (Maxillarieae?), Panama. 94. *Chloraea crispa* (Chloraeinae), nectaries (Courtesy C. A. Luer). 95. *Chloraea gavilu* (Chloraeinae), column (Courtesy C. A. Luer). 96. *Palmorchis nitida* (Palmorchideae), fruit in section.

Subtribe **Ornithocephalinae**

DESCRIPTION. Habit: dwarf epiphytes, with or without pseudobulbs, stem short, sympodial or monopodial. Velamen not of a defined type. Leaves: distichous or secondarily spiral, duplicate, may be laterally flattened, articulate or not. Inflorescence: lateral, simple, of few to many flowers, flowers spiral, secund, or (?) distichous. Flowers: small, lip with a prominent oil gland, or deeply saccate to somewhat spurred; column slender; anther terminal, operculate, beaked, with reduced partitions; four pollinia, superposed or obovoid, with a viscidium and a long stipe; stigma often basal on the column, entire; rostellum beaked, short to very long.

DISTRIBUTION. Tropical America.

POLLINATION. Oil-gathering bees of the genus *Paratetrapedia* have been observed pollinating *Ornithocephalus*. Apparently all of this group have oil glands, and thus attract those small anthophorid bees that gather oil as food for their larvae.

CHROMOSOME NUMBER. Not known.

SEED STRUCTURE. *Gomesa* variant of *Vanda* type.

SPECIES. 76.

GENERA. 14: *Caluera, Centroglossa, Chytroglossa, Dipteranthus, Dunstervillea, Eloyella, Hintonella, Ornithocephalus, Phymatidium, Platyrhiza, Rauhiella, Sphyrastylis, Thysanoglossa, Zygostates*.

DISCUSSION. The Ornithocephalinae may be related to the Telipogoninae. In any case, they show the same combination of four pollinia and seeds like the Oncidiinae. Some species of *Ornithocephalus* have hooked viscidia like those of the Telipogoninae.

Subtribe **Oncidiinae**

DESCRIPTION. Habit: epiphytic or terrestrial, usually with pseudobulbs of a single internode, sometimes with short or elongate, slender stems, sometimes monopodial. Velamen of the *Cymbidium* type. Leaves: distichous, duplicate, articulate or not, may be cylindric or laterally flattened. Inflorescence: lateral, simple or branched, of one to many flowers, flowers spiral or distichous. Flowers: tiny to very large, resupinate, lip may be spurred, or may have basal nectariferous appendages extending into a sepaline spur, often with large calli; column winged or not, anther terminal and operculate or erect and dorsal, with reduced partitions; two pollinia, with viscidium and stipe, the stipe and viscidium usually appearing different in color or texture; stigma entire or two-lobed. (See Fig. 9-10.)

DISTRIBUTION. Tropical America.

POLLINATION. Wasp pollination is reported in *Ada, Brassia*, and *Leochilus*. Euglossine pollination is reported in *Aspasia, Lockhartia, Macroclinium, Notylia, Rodriguezia, Trichocentrum*, and *Trichopilia*. Pollination of both *Gomesa* and *Miltoniopsis* by other bees has been observed. Some species of *Odontoglossum* are pollinated by bumblebees, and many species of *Oncidium* appear to be pollinated by anthophorid bees, especially *Centris*. Pollination by *Centris* includes a variety of relationships. Female *Centris* gather oil from the lips of some species. In other

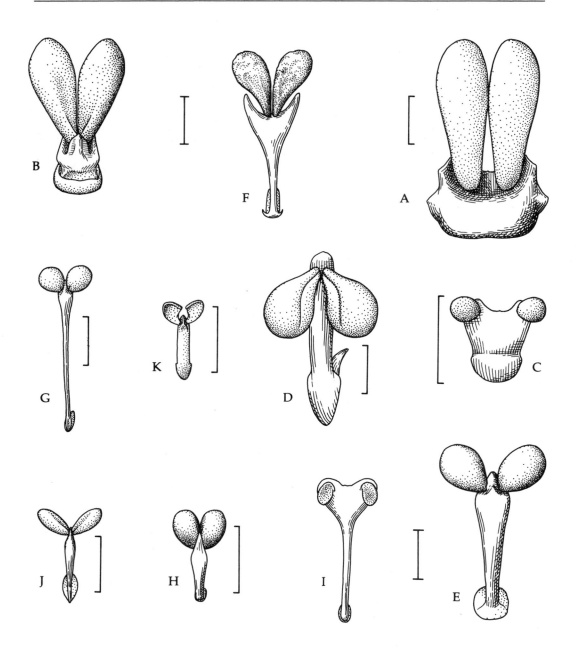

Figure 9-10. Representative pollinaria of the subtribe Oncidiinae. (A) *Oncidium cavendishianum.* (B) *Brassia arcuigera.* (C) *Systeloglossum costaricense.* (D) *Lemboglossum maculatum.* (E) *Comparettia macroplectron.* (F) *Trichopilia turialbae.* (G) *Oncidium cheirophorum.* (H) *Oncidium ansiferum.* (I) *Macroclinium bicolor.* (J) *Sigmatostalix picta.* (K) *Hybochilus inconspicuus.* Scale 1 mm. (After Dressler 1981.)

cases, the *Oncidium* flower simulates oil-bearing flowers (other *Oncidium* species or Malpighiaceous vines) and attracts female *Centris* without offering any real reward. Male *Centris* often show territorial behavior near masses of flowers, and some *Oncidium* species are apparently pollinated by these males, who react to the flowers as though they were intruding insects when the inflorescence is moved by a breeze. *Sigmatostalix* is pollinated by small oil-gathering bees. A few cases of hummingbird pollination have been observed, and a number of species or genera show features that suggest bird pollination.

CHROMOSOME NUMBERS. 10, 14, 24, 28, 30, 36, 38, 40, 42, 44, 48, 50, 56, 60.

SEED STRUCTURE. *Vanda-Maxillaria* transition, *Vanda* type, with *Gomesa, Oncidium variegatum,* and *Thrixspermum* variants.

SPECIES. 1232.

GENERA. About 77: *Ada, Amparoa, Antillanorchis, Aspasia, Binotia, Brachtia, Brassia, Buesiella, Capanemia, Caucaea, Cischweinfia, Cochlioda, Comparettia, Cuitlauzinia, Cypholoron, Diadenium, Dignathe, Erycina, Fernandezia, Gomesa, Helcia, Hybochilus, Ionopsis, Konantzia, Lemboglossum, Leochilus, Leucohyle, Lockhartia, Macradenia, Macroclinium, Mesoglossum, Mesospinidium, Mexicoa, Miltonia, Miltoniopsis, Neodryas, Neokoehleria, Notylia, Odontoglossum, Oliveriana, Oncidium, Osmoglossum, Otoglossum, Pachyphyllum, Palumbina, Papperitzia, Plectrophora, Polyotidium, Psychopsiella, Psychopsis, Psygmorchis, Pterostemma, Quekettia, Raycadenco, Rodriguezia, Rodrigueziella, Rodruguieziopsis, Rossioglossum, Rusbyella, Sanderella, Saundersia, Scelochiloides, Scelochilus, Sigmatostalix, Solenidiopsis, Solenidium, Stictophyllum, Suarezia, Sutrina, Symphyglossum, Systeloglossum, Ticoglossum, Tolumnia, Trichocentrum, Trichopilia, Trizeuxis, Warmingia.*

DISCUSSION. The Oncidiinae represent the most highly derived of the New World orchids, paralleling the Vandeae in seed structure and floral complexity, while showing great diversity in chromosome number and vegetative features. About half of the species are classified in two genera, *Oncidium* and *Odontoglossum.* Traditionally, these genera have been separated by the angle between lip and column, though this feature varies greatly and does not separate two clear groups. There is no doubt that the generic classification is artificial, but, to date, no clear resolution is available. In recent years, *Odontoglossum* has been divided into more homogeneous groups, though in some cases, the division may have been carried a bit far. There has been no clear resolution of the *Miltonia/Oncidium* complex. M. W. Chase (personal communication) feels that it may be most practical to abandon both *Miltonia* and *Odontoglossum* for the present and use a very broad concept of *Oncidium* until a more natural classification is available.

Chase (1986a) divides the Oncidiinae into two main complexes: group A, with conduplicate, dorsoventral leaves as both seedlings and adults, and with chromosome numbers of mostly 56 or 60; and group B, with laterally flattened leaves in the seedlings, lower chromosome numbers, and often with sepaline nectaries. Chase suggests that $2n = 60$ may be the primitive number, and that reduction in chromosome number is especially typical of group B, which includes many short-lived twig epiphytes. In recent years, some authors have separated very small, "splinter" genera in the Oncidiinae. Chase (1986a) argues, very reasonably, that this does not add to our understanding of the subtribe. We need, rather, further work to determine the major phylads within the Oncidiinae. When this has been done, we may be able to delineate a reasonable number of natural and meaningful genera.

Preliminary results are now available from Chase's work on the chloroplast DNA of the Oncidiinae (Chase and Palmer 1988, 1992). These results confirm the relationship between the Oncidiinae and the Maxillariinae, the close relationship between the Ornithocephalinae and Oncidiinae, and the position of *Lockhartia* in the Oncidiinae. Also, molecular systematics shows clearly that the traditional concepts of *Oncidium, Odontoglossum*, and *Miltonia* are artificial. In their latest paper Chase and Palmer (1992) recognize 16 distinct clades within the Oncidiinae, ten of which include members of *Oncidium* and/or *Odontoglossum* in the traditional sense. Of special interest are the following:

1. The Altissimum clade includes species that have been treated as *Oncidium, Odontoglossum, Miltonia*, and *Miltonioides*. It includes the type species of *Oncidium*, and about 160 species of *Oncidium* in the traditional sense.

2. The Crispum clade is primarily a Brazilian group that includes *Gomesa* and about 80 species that have been treated as *Oncidium*. The names *Baptistonia, Coppensia*, and *Waluewa*, all of which have been considered synonyms of *Oncidium*, appear to be based on members of this group.

3. The Macranthum clade is a high Andean group of species that have been treated as members of *Oncidium* and *Odontoglossum*, and includes the *Cyrtochilum* group of "*Oncidium*" (about 70 species).

4. The *Rodriguezia* clade corresponds to Chase's group B, with about 35 species of "*Oncidium*" of the Variegatum group. This West Indian group may be treated as the genus *Tolumnia*. By separating *Tolumnia* from *Oncidium*, the *Rodriguezia* clade becomes a natural group with very little revision. These derived twig epiphytes sort out by floral features much more clearly than do other clades within the Oncidiinae.

5. The *Lophiaris* clade includes the Mule-ear and Rat-tail Oncidiums, *Oncidium ampliatum, Oncidium* section *Pulvinatum, Psychopsis*, and *Trichocentrum*. Most or all of this group could be treated as a single genus. It is clear now that *Trichocentrum* is a subgroup of the "Mule-ear" group, and the name *Trichocentrum* is older than *Lophiaris*, so it would have priority. There is reduction in chromosome number in this clade, though it does not go as far as in the twig-epiphytes of the *Rodriguezia* clade.

Chase and Palmer (1992) suggest that the Oncidiinae early divided into several distinct clades, and that something similar to the classic *Oncidium* flower is the primitive condition for the subtribe, so that this sort of flower occurs in all of the major clades. Chase and Palmer support a chromosome number of $n = 28$ or 30 as the ancestral condition for the subtribe.

We need more data from chloroplast DNA (and other methods) to determine the pattern of relationships in the Oncidiinae, but a new classification is clearly needed. This statement does not reflect any conflict between traditional methods and molecular systematics. It was already quite clear that the traditional classification of the Oncidiinae was artificial. As Chase and Palmer stress, no one feature gives a clear picture of relationships within the Oncidiinae, but a combination of anatomy, chromosome number, seed structure, vegetative features, floral features, and crossability already showed the necessity of revision and pointed toward some of the changes indicated by molecular systematics.

I have placed *Pachyphyllum* and *Fernandezia* in the Oncidiinae. A recently described Ecuadorian genus, *Raycadenco*, combines features of *Lockhartia*, *Fernandezia*, and typical *Oncidium*. *Lockhartia* and *Fernandezia* share very similar pollinaria, and, together with *Raycadenco* and *Pachyphyllum*, may form a quite natural subclade of the Oncidiinae.

REFERENCES. Bock 1986, 1988; Bockemühl 1983–1988; Braem 1986a, 1988; Brieger and Lückel 1983–1984; Chase 1986a–c, 1987a–c; Chase and Palmer 1988, 1992; Halbinger 1982; Lückel and Braem 1982; Senghas 1987; Sinoto 1962.

10

The Reed-stem
or Epidendroid Phylad

The epidendroid phylad is approximately the "reed-stem" phylad as I visualized it in 1986, except that it includes the largely cormous Arethuseae. If my analysis is correct, the primary derived feature for this phylad is not the reed-stem habit but the possession of eight pollinia. This may have evolved only once in the Epidendroideae, and this phylad shows reduction series to six, four, and two in most subtribes. Except for the primitive *Eulophia* seed type and the highly derived *Vanda* type, the seed types of the epidendroid phylad are largely distinct from those of the cymbidioid phylad. The *Vanda* seed type has certainly evolved more than once. Representative pollinaria of the epidendroid phylad are illustrated in Figure 10-1.

Phylogeny

For convenience, I separate the New World and Old World elements of the Epidendreae in Figure 8-3. The American Epidendreae are a relatively clear group, but their Old World counterparts are both less clear and less well studied. Their velamen types are different, and their seed types are largely distinct. The Old World complex is probably a subclade of the New World Epidendreae, but the relationships are not clear. The Dendrobioid subclade seems clearly related to the Old World Epidendreae. The Glomerinae and the Podochileae each have similar seed types, and some of these seed types approach the *Vanda* type, so that the independent evolution of this seed type in the Vandeae and in the Old World Epidendreae is reasonable. The Eriinae/Glomerinae complex needs more

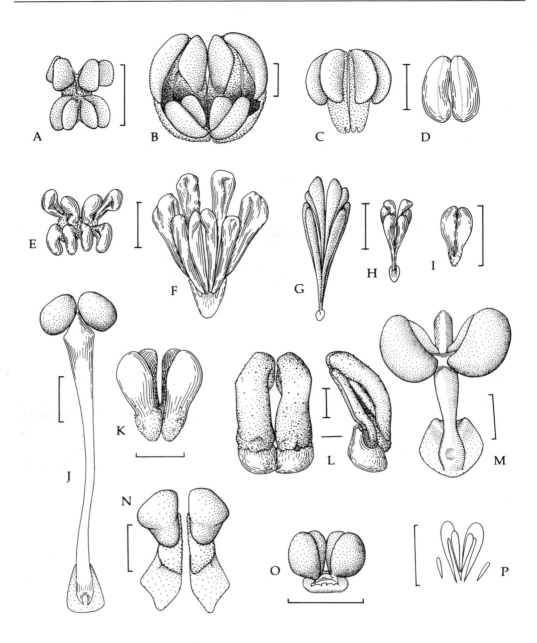

Figure 10-1. Pollinaria of the Epidendroid phylad. (A) *Eria andersonii*. (B) *Chysis maculata*. (C) *Epidendrum ciliare*. (D) *Dendrobium fimbriatum*. (E) *Elleanthus capitatus*. (F) *Calanthe brevicornu*. (G) *Meiracyllium wendlandii*. (H) *Appendicula cornuta*. (I) *Masdevallia zahlbruckneri*. (J) *Micropera philippinensis*. (K) *Coelogyne ochracea*. (L) *Sobralia powellii*. (M) *Trichoglottis fasciata*. (N) *Dendrophylax fawcettii*. (O) *Polystachya bella*. (P) *Brachionidium kuhniarum*. Scale 1 mm. (A–O after Dressler 1981.)

detailed study, and until this complex has been resolved, the classification of the Old World Epidendreae remains a problem. The American Epidendreae show a straightforward pattern of relationships, with clear links to the Sobraliinae and between the *Epidendrum* and *Pleurothallis* seed types. The *Elleanthus* seed type, or something similar, is also found in the Old World, especially in the Podochilinae, but the pattern of relationships in the Old World phylad is less clear, in part because we have too little information on seed structure.

The interrelationships between the Podochileae, the Dendrobieae, and the Vandeae seem quite clear and convincing, but it is probable that *Eria* will need to be resolved into two or more groups.

The Arethuseae, as treated here, may be poorly resolved. One is tempted to limit the Arethuseae to the seemingly primitive genera *Arethusa* and *Eleorchis*, but what to do with the other genera that combine terminal inflorescence and soft pollinia is not clear (see discussion under Bletiinae). The Sobraliinae and Arundinae, with reed-stem habit and terminal inflorescence, both have been classified in the Arethuseae. I assign the Sobraliinae to the Epidendreae because *Arpophyllum* and *Meiracyllium*, each with *Elleanthus* seed type and *Epidendrum* velamen type, seem to link these groups. This seems to delimit a natural group, even though no uniquely derived features connect the Sobraliinae with the other Epidendreae. The Arundinae might, similarly, be assigned to the Old World Epidendreae, but this relationship is less clear, so I list them under Misfits and Leftovers in Chapter 11.

Tribe **Arethuseae**

DISCUSSION. I keep this tribe with essentially the same delimitation as in 1981, except for the exclusion of the Sobraliinae and Arundinae. I have interpreted the *Bletia* seed type as derived relative to the *Eulophia* type. The presence of a lateral inflorescence and relatively firm pollinia might be taken to distinguish the Bletiinae and the Chysiinae, for which the name Bletieae is available.

Subtribe **Arethusinae**

DESCRIPTION. Habit: terrestrial with a fleshy corm. Velamen of the *Calanthe* type. Leaf: convolute, plicate, nonarticulate. Inflorescence: terminal, simple, of one or few spiral flowers. Flowers: medium-large, resupinate, lip arched, with yellow hairs; a cavity at the base of the lip seems to lack nectar; column slender, arched, flattened and petaloid apically; anther ventral, incumbent; four pollinia, soft and mealy, weakly sectile; stigma entire, emergent, without a viscidium.

DISTRIBUTION. Temperate Asia and North America.

POLLINATION. *Arethusa* is pollinated by bumblebee queens in early spring; the bees soon learn that the yellow hairs on the lip are not pollen-bearing anthers and then ignore the *Arethusa* flowers.

CHROMOSOME NUMBERS. 40, 44.

SEED STRUCTURE. *Bletia* type.

SPECIES. 3.

GENERA. 2: *Arethusa, Eleorchis.*

DISCUSSION. The Arethusinae have terminal inflorescence, nonarticulate leaves, and rather soft pollinia, all seemingly primitive features, though the soft pollinia could also be a "reversion" reflecting pollination by bristly bumblebees. In habit and seed type they agree with the Bletiinae. The somewhat sectile pollinia of *Arethusa* may reflect a relationship with the Gastrodieae or Nervilieae. An artificial hybrid between *Arethusa* and *Calopogon* was registered recently, but more crosses should be attempted and the results recorded before we can clarify this complex.

Subtribe **Bletiinae**

DESCRIPTION. Habit: terrestrial, epiphytic, or saprophytic, usually with corms or pseudobulbs of several internodes. Velamen of the *Calanthe* type. Leaves: spiral or distichous, convolute, plicate, usually articulate. Inflorescence: lateral or sometimes terminal, simple, with several to many spiral flowers. Flowers: medium-small to large, resupinate or nonresupinate, lip sometimes saccate or with a prominent spur; column short or long, often with a prominent column foot; anther terminal, incumbent, usually eight-celled; eight pollinia or rarely four, hard or relatively soft, laterally flattened or clavate, with ventral or terminal, intralocular caudicles; stigma entire, emergent or not, with a viscidium in some cases. (See Fig. 10-2.)

DISTRIBUTION. Pantropical, extending into temperate areas of Asia and North America.

POLLINATION. Pollination by bees is reported for *Bletia* and *Phaius*, and is to be expected for several genera. The hairs on the lip of *Calopogon* mimic anthers, and pollen-seeking bees occasionally visit the flowers. The bee's weight causes the lip to tip over so that the bee falls against the column. Lepidopteran pollination is to be expected in *Calanthe*, which seems to be a floral analog of the American *Epidendrum* in its adaptations to pollination by moths or butterflies.

CHROMOSOME NUMBERS. 26, 28, 30, 32, 36, 38, 40, 42, 44, 46, 48, 50.

SEED STRUCTURE. *Bletia* and *Eulophia* types, both sometimes occurring in the same genus.

SPECIES. 388.

GENERA. About 21: *Acanthephippium, Ancistrochilus, Anthogonium, Aulostylis, Bletia, Bletilla, Calanthe, Calopogon, Cephalantheropsis, Eriodes, Gastrorchis, Hancockia, Hexalectris, Ipsea, Mischobulbon, Nephelaphyllum, Pachystoma, Phaius, Plocoglottis, Spathoglottis, Tainia.*

DISCUSSION. *Bletilla* has been considered primitive on the basis of a terminal inflorescence and four soft pollinia. The inflorescence may be either terminal or lateral (Tan 1969), however, and some plants show eight distinct pollinia, quite like those of *Bletia. Calopogon*, also, has a terminal inflorescence and soft pollen masses, but the last may be an adaptation to pollination by bristly bees.

Mischobulbon and *Nephelaphyllum* resemble the Collabiinae in habit, but their floral features agree with the Bletiinae.

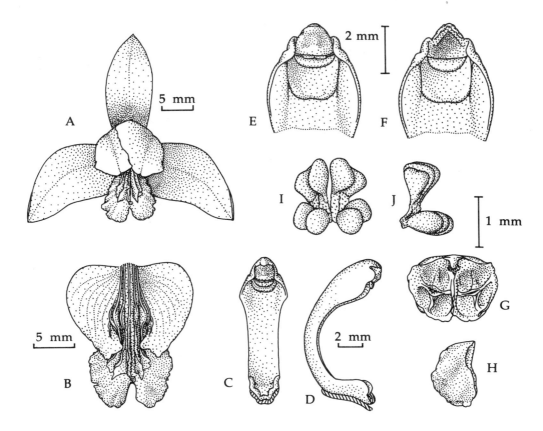

Figure 10-2. *Bletia purpurea* (Epidendroideae: Arethuseae). (A) Flower, front view. (B) Lip, flattened. (C) Column, ventral view. (D) Column, side view. (E) Apex of column, with anther in place. (F) Apex of column, after removal of anther. (G) Anther, ventral view. (H) Anther, side view. (I) Pollinarium, top view. (J) Pollinarium, side view. (After Dressler 1981.)

Plocoglottis, with four pollinia, prominent caudicles, and a large viscidium, and with a motile lip in most species, is very distinctive, and could be the basis of a distinct subtribe, but it does not seem out of place in the Bletiinae.

However the Bletiinae are delimited, the group includes species with the *Bletia* seed type and species with the *Eulophia* seed type, for both seed types occur in *Calanthe* and *Bletia* (if *Crybe* is included). The presence of the *Bletia* seed type in the relatively primitive *Arethusa*, *Bletilla*, and *Calopogon* suggests that this type might be primitive for the subtribe, but outgroup comparison indicates the *Eulophia* seed type as primitive.

REFERENCES. Cribb and Tang 1982; Howcroft 1986; Ishida 1990; Tanaka et al. 1981.

Subtribe **Chysiinae**

DESCRIPTION. Habit: epiphytic, with club-shaped pseudobulbs. Velamen not of a named type. Leaves: distichous, convolute, plicate, scattered on the pseudobulb. Inflorescence: lateral, of few spiral flowers. Flowers: large, resupinate, fleshy, column thick, with a prominent column foot; anther terminal, incumbent, eight-celled; eight pollinia on platelike caudicles, four pollinia parallel, four oblique; stigma entire, without a viscidium.

DISTRIBUTION. Tropical America.

POLLINATION. Not known, but probably by large bees.

CHROMOSOME NUMBER. Not known.

SEED STRUCTURE. *Elleanthus* type.

SPECIES. 7.

GENUS. *Chysis*.

DISCUSSION. *Chysis* was placed in a separate subtribe by Schlechter. I earlier placed *Chysis* in the Bletiinae, but it seems discordant there. The habit is distinctive and rather unlike any other American orchid. The flower structure is reminiscent of either *Laelia* or *Bletia*, but the pollinia are unlike either.

Tribe **Coelogyneae**

DISCUSSION. As delimited here, the Coelogyneae make a very natural group characterized by pollinia with massive caudicles and the *Dendrobium* seed type. I had placed the Coelogyneae with the reed-stem complex, yet, except for *Thunia*, the Coelogyneae have pseudobulbs or corms of a single internode. Though most Coelogyneae have seeds of the *Dendrobium* type, it is clear that this type evolved independently in the Coelogyneae and Dendrobieae. Unpublished work on the seed structure of *Pleione* (mentioned in Cribb and Butterfield 1988) indicates that some groups of *Pleione* have seeds similar to those of *Bletilla* (the *Bletia* type), thus strengthening the evidence that the Coelogyneae may share a common ancestry with the Arethuseae. The flowers of the Coelogyneae resemble those of the Arethuseae in the petaloid column and emergent, clam-shell stigma. These, however, appear to be primitive features, and further evidence might show the Coelogyneae to be more closely allied to the Old World Epidendreae than to the Arethuseae.

Subtribe **Thuniinae**

DESCRIPTION. Habit: terrestrial or lithophytic; with thick, fleshy stems. Velamen of *Coelogyne* type. Leaves: distichous, duplicate, articulate. Inflorescence: terminal, of several spiral flowers, simple, pendant. Flowers: large, resupinate, lip trumpet-shaped, parallel with and enfolding the column; column slender, somewhat petaloid terminally, anther ventral, incumbent, four-celled; eight or four pollinia with massive caudicles; stigma entire, emergent.

DISTRIBUTION. Tropical Asia.

POLLINATION. Not known; flower structure suggests bee pollination.

CHROMOSOME NUMBERS. 38, 40, 42, 44.

SEED STRUCTURE. *Dendrobium* type.

SPECIES. 5.

GENUS. *Thunia*.

DISCUSSION. *Thunia* has been grouped with several different genera with slender stems, though *Thunia*, itself, has quite fleshy stems. Different species have either four or eight pollinia; when there are eight pollinia, these are quite unlike those of the Bletiinae or the Laeliinae. Both flower structure and the velamen and seed types indicate a close alliance between *Thunia* and the Coelogyninae, though the elongate pseudobulbs of *Thunia* are unlike the short pseudobulbs of the Coelogyninae. These elongate pseudobulbs might be a reversal of the condition in the Coelogyninae or may represent early divergence of the subtribes.

Subtribe **Coelogyninae**

DESCRIPTION. Habit: epiphytic or terrestrial; with pseudobulbs or corms of a single internode. Velamen of the *Coelogyne* type (*Calanthe* type in *Pleione*). Leaves: convolute or duplicate, plicate or conduplicate, articulate. Inflorescence: terminal, often produced before growth of pseudobulbs (sometimes borne on rudimentary pseudobulbs and appearing lateral), simple, of few to many flowers, spiral or distichous. Flower: small to large, resupinate; base of lip may be saccate; column short or elongate, apex often petaloid and hooded over anther; anther terminal or ventral, incumbent; four or two pollinia, these superposed or ovoid, with prominent caudicles, sometimes with a viscidium, with a stipe in *Geesinkorchis*; stigma entire, often emergent. (See Fig. 10-3.)

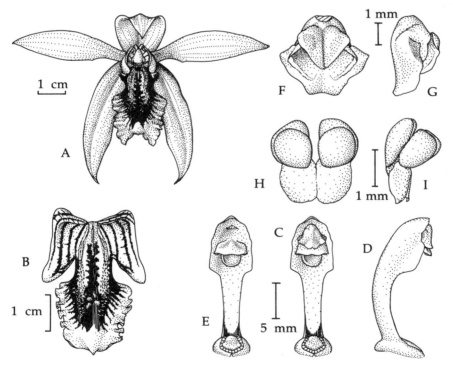

Figure 10-3. *Coelogyne pandurata* (Epidendroideae: Coelogyneae). (A) Flower, front view. (B) Lip, flattened. (C) Column, ventral view. (D) Column, side view. (E) Column, after removal of anther. (F) Anther, ventral view. (G) Anther, side view. (H) Pollinia, top view. (I) Pollinia, side view. (After Dressler 1981.)

DISTRIBUTION. Australasia, Tropical Asia, and into China.

POLLINATION. The flower structure of *Coelogyne* suggests bee pollination; wasp pollination is reported for *Coelogyne fragrans*. According to Cribb and Butterfield (1988), cultivated *Pleione* are often pollinated by bumblebees, and bumblebees are the probable pollinators in nature.

CHROMOSOME NUMBERS. 38, 40, 42, 44.

SEED STRUCTURE. *Dendrobium* type.

SPECIES. 285.

GENERA. 20: *Basigyne, Bracisepalum, Bulleyia, Chelonistele, Coelogyne, Dendrochilum, Dickasonia, Entomophobia, Forbesina, Geesinkorchis, Gynoglottis, Ischnogyne, Nabaluia, Neogyne, Otochilus, Panisea, Pholidota, Pleione, Pseudacoridium, Sigmatogyne.*

DISCUSSION. The Coelogyninae are an unusually clear group. The column resembles that of some Bletiinae, though this may be a suite of primitive features retained in this group.

REFERENCES. Cribb and Butterfield 1988; Cribb et al. 1985–1986; Lund 1987a; Stergianou 1989; de Vogel 1986, 1988.

Tribe **Epidendreae I** (New World Groups)

I treat here the American genera that seem to form a natural group, with distinctive velamen types and distinctive seed types in most.

Subtribe **Sobraliinae**

DESCRIPTION. Habit: terrestrial or epiphytic, stems slender, usually elongate. Velamen not of a defined type. Leaves: distichous or subdistichous, usually convolute and plicate, but duplicate in some cases, articulate. Inflorescence: terminal or lateral, usually of few to many spiral or distichous flowers. Flowers: small to large, membranous, resupinate or nonresupinate; lip more or less enfolding the column, often trumpet-shaped, simple, often with prominent calluses, sometimes basally saccate; column short or elongate, often with armlike wings; anther terminal, two- to eight-celled; eight soft pollinia, superposed or ovoid; stigma usually emergent, with a distinct viscidium in some cases.

DISTRIBUTION. Tropical America.

POLLINATION. Most *Sobralia* species are pollinated by bees, often by Euglossini, while a few appear to be pollinated by hummingbirds. Hummingbird pollination has been observed in *Elleanthus*, and *Sertifera* also appears to be hummingbird-pollinated.

CHROMOSOME NUMBER. 54.

SEED STRUCTURE. *Elleanthus* or (in *Sobralia*) *Bletia* type.

SPECIES. 219.

GENERA. 4: *Elleanthus, Epilyna, Sertifera, Sobralia.*

DISCUSSION. *Elleanthus* has a distinctive seed type and seems clearly allied to the Laeliinae. *Sobralia,* on the other hand, has the *Bletia* seed type, as far as known. Most species of *Sobralia* have very distinctive pollinia, quite unlike either those of *Elleanthus* or *Bletia;* but some species of *Sobralia* have ovoid pollinia, much like those of *Elleanthus.* Most Sobraliinae have clearly plicate leaves, but conduplicate leaves occur in *Epilyna* and some of the smaller *Elleanthus,* which thus approach the Laeliinae (and especially *Isochilus*) in habit. Both *Arpophyllum* and *Meiracyllium* seem to link the Sobraliinae to the Epidendreae.

Subtribe **Arpophyllinae**

DESCRIPTION. Habit: epiphytic, stems slightly thickened, with a single, terminal leaf. Velamen of the *Epidendrum* type. Leaves: conduplicate, leathery or very fleshy, articulate. Inflorescence: terminal, of few to many spiral flowers. Flowers: small or medium-small, membranous, nonresupinate; lip more or less enfolding the column, basally saccate; column short; anther terminal, eight-celled; eight soft, ovoid pollinia; stigma simple, with a distinct viscidium.

DISTRIBUTION. Tropical America.

POLLINATION. Not known, but probably by hummingbirds.

CHROMOSOME NUMBER. Not known.

SEED TYPE. *Elleanthus* type.

SPECIES. 5.

GENUS. *Arpophyllum.*

DISCUSSION. Though the flowers of *Arpophyllum* are very like those of *Elleanthus,* the habit would be discordant in the Sobraliinae, and the velamen structure indicates a closer alliance to the Laeliinae, where the ovoid pollinia would be rather out of place. A distinct subtribe seems best for this genus.

Subtribe **Meiracylliinae**

DESCRIPTION. Habit: epiphytic, stems short, slightly thickened. Velamen of the *Epidendrum* type. Leaves: one per shoot, duplicate, fleshy, articulate. Inflorescence: terminal, simple, of few spiral flowers. Flowers: small, resupinate, lip basally saccate; column elongate, anther erect, dorsal, eight-celled; eight pollinia, clavate, with a distinct viscidium; stigma entire.

DISTRIBUTION. Mexico and Central America.

POLLINATION. Not known; the perfume of methyl cinnamate suggests pollination by male euglossine bees.

CHROMOSOME NUMBER. Not known.

SEED STRUCTURE. *Elleanthus* type.

SPECIES. 2.

GENUS. *Meiracyllium.*

DISCUSSION. *Meiracyllium* has been compared with the Pleurothallidinae and the Old World Podochilinae but may not be close to either one. The seed structure is similar to that of the Podochilinae. If spherical silica bodies were found in *Meiracyllium,* a close alliance to the Podochilinae would be credible, but the velamen type suggests a closer alliance to the Laeliinae.

Subtribe **Coeliinae**

DESCRIPTION. Epiphytic, with smooth, globose pseudobulbs and several terminal leaves. Leaves: distichous, narrow, subconduplicate. Inflorescence: lateral, of several to many spiral flowers. Flowers: small to medium, resupinate or nonresupinate, lip shallowly saccate to deeply spurred at base; column short or long, with a column foot; anther terminal, incumbent, eight-celled; eight ovoid pollinia, with terminal caudicles; stigma simple, with a viscidium.

DISTRIBUTION. Tropical America.

POLLINATION. Not known.

CHROMOSOME NUMBER. 40.

SEED STRUCTURE. *Epidendrum* type.

SPECIES. 5.

GENUS. *Coelia.*

DISCUSSION. *Coelia* superficially resembles the Bletiinae, and I had tried to "shoehorn" it into that group, but it fits poorly. Pridgeon (1978) accepted my idea, but the seed structure is of the *Epidendrum* type. Species with spurs have been separated as *Bothriochilus*, but the difference between *Coelia*, with a "shallowly saccate" lip, and a short-spurred *Bothriochilus* seems no greater than the difference between a short-spurred *Bothriochilus* and a long-spurred one.

REFERENCE. Pridgeon 1978.

Subtribe **Laeliinae**

DESCRIPTION. Habit: epiphytic or terrestrial, stems slender or forming pseudobulbs, these usually of several internodes. Velamen of the *Epidendrum* type. Leaves: distichous or terminal on pseudobulb, duplicate, usually articulate. Inflorescence: terminal or rarely lateral, simple or branched, of one to many flowers, spiral or distichous. Flowers: tiny to large, resupinate or not; flowers may have a cuniculus type of nectary; column short or elongate, often winged, may have a column foot; anther terminal and incumbent or erect; pollinia laterally flattened or ovoid, eight, six, four or two, with prominent caudicles; stigma entire, sometimes with a viscidium. (See Fig. 10-4.)

DISTRIBUTION. Tropical America.

POLLINATION. Most species of *Cattleya*, *Laelia*, *Schomburgkia*, and *Encyclia* subgenus *Encyclia* appear to be bee-pollinated, though we have only a few observations for these genera. The "cockle-shell" group of *Encyclia* subgenus *Osmophyta* seems, however, to be pollinated by wasps; why the wasps visit these flowers is not clear. *Myrmecophila* does not offer any reward for the bees that visit the flowers (Rico-Gray and Thien 1987), and such false advertisement may occur in other bee-pollinated genera. *Caularthron* is pollinated by carpenter bees, and *Brassavola* and *Rhyncholaelia* are surely pollinated by sphingid moths. Most species of *Epidendrum* are pollinated by either diurnal or nocturnal Lepidoptera. *Alamania*, *Neocogniauxia*, *Sophronitis*, and *Hexisea* all seem adapted to hummingbird pollination, as are some species of *Laelia* and *Epidendrum*. Norris Williams (personal communication) has observed the pollination of *Scaphyglottis* by stingless bees (*Trigona*).

CHROMOSOME NUMBERS. 24, 38, 40, 42, 56. Forty is the usual chromosome

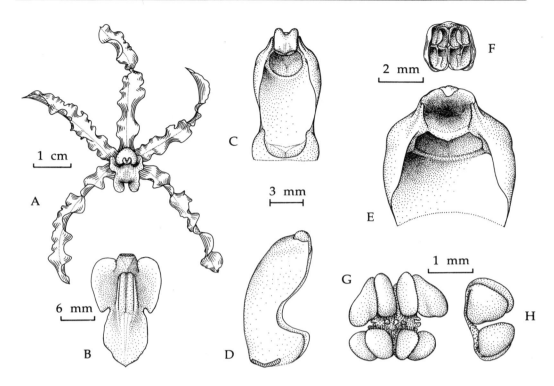

Figure 10-4. *Schomburgkia undulata* (Epidendroideae: Epidendreae). (A) Flower, front view. (B) Lip, flattened. (C) Column, ventral view. (D) Column, side view. (E) Apex of column, with anther and pollinarium removed. (F) Anther. (G) Pollinarium, top view. (H) Pollinarium, side view. (After Dressler 1981.)

number for this subtribe, but the members of the *Epidendrum secundum* complex show both lower and higher numbers.

SEED STRUCTURE. *Epidendrum* type or variants in most genera, with *Pleurothallis* type in a few genera.

SPECIES. 1466.

GENERA. 43: *Acrorchis, Alamania, Artorima, Barkeria, Basiphyllaea, Brassavola, Broughtonia, Cattleya, Caularthron, Constantia, Dilomilis, Dimerandra, Domingoa, Encyclia, Epidendrum, Hagsatera, Helleriella, Hexisea, Homalopetalum, Isabelia, Isochilus, Jacquiniella, Laelia, Leptotes, Loefgrenianthus, Myrmecophila, Nageliella, Neocogniauxia, Nidema, Oerstedella, Orleanesia, Pinelia, Platyglottis, Ponera, Pseudolaelia, Psychilus, Quisqueya, Reichenbachanthus, Rhyncholaelia, Scaphyglottis, Schomburgkia, Sophronitis, Tetramicra.*

DISCUSSION. The old distinction between the Ponerinae (with a column foot, however small or imaginary) and the Laeliinae (without such a column foot) has broken down, leaving the Laeliinae as a natural but diverse group. Many genera have been connected by artificial hybrids, though some of the smaller and less showy groups remain untried, as far as I know. *Dilomilis*, with eight pollinia and a slender reedlike stem, may be the most primitive genus of this group or the sister group of the Pleurothallidinae. Dr. Barthlott has kindly sent photographs of the seed of *Dilomilis scirpoidea*, which seems to reflect relationships with *Elleanthus*, the Pleurothallidinae, and the Old World Epidendreae.

Cattleya is *the* orchid in the eyes of the public and of some orchid growers. The great interest in this group, combined with the traditional emphasis on pollinia number in classification, has resulted in an artificial classification. The distinction between *Cattleya* and *Laelia* is quite imaginary, especially with respect to *Laelia* section *Cattleyodes*, which is surely more closely allied to the Cattleyas than to the other Laelias. It would be more natural to include *Laelia* and *Schomburgkia* (but not necessarily *Myrmecophila*) in *Cattleya*, though horticulturists would surely reject such a change, rather than rename all of the Laeliocattleya hybrids.

Sauleda (1988, 1989) has separated *Psychilus* from *Encyclia*. In spite of the superficial similarity in plant and flower, it is clear that the species of *Psychilus* were misplaced in *Encyclia*. Natural hybrids with members of the *Broughtonia* complex are known in the West Indies, and the flowers of some *Psychilus* are strikingly like those of *Tetramicra*, which includes at least one species with pseudobulbs. In fact, *Basiphyllaea*, *Broughtonia*, *Cattleyopsis*, *Laeliopsis*, *Psychilus*, *Quisqueya*, and *Tetramicra* would appear to form a natural West Indian phylad, though this might be shown quite as clearly with fewer genera.

In terms of species, *Epidendrum* is a major element in this subtribe. Though long used as a receptacle for generic left-overs, *Epidendrum* is quite sharply defined by the union of the lip with the column and its distinct, though usually semiliquid, viscidium. *Dimerandra*, *Jacquiniella*, and *Oerstedella* are clearly distinct from *Epidendrum*, but *Amblostoma*, *Epidanthus*, *Lanium*, *Physinga*, and even the distinctive *Diothonaea* and *Neowilliamsia* are cladistically subgroups of *Epidendrum*.

The *Scaphyglottis* complex is still something of a problem, but the case of *Hexisea* is interesting (see Fig. 10-5). For some time, the genus was made up of all members of the *Scaphyglottis* complex with a prominent nectary at the base of the flower, this more or less closed by the lip. I argued that *Reichenbachanthus reflexus* and *Scaphyglottis amparoana*, with green and white flowers and hinged lips, should be excluded from *Hexisea*, thus restricting *Hexisea* to species with the hummingbird pollination syndrome (Dressler 1974). Still, the details of lip and column are so diverse that even this version of *Hexisea* is almost certainly a polyphyletic grade, derived from different clades within *Scaphyglottis*. Adams (1988) restricts *Hexisea* to *H. bidentata* and *H. imbricata*. This version of *Hexisea* is surely monophyletic, but it is still a small subclade of *Scaphyglottis* and scarcely more deserving of generic status than several others.

The presence of the *Pleurothallis* type seed in some members of the *Ponera* complex suggests that the Laeliinae may be paraphyletic. At present, though, I see no clear resolution of the group. The "Ponerinae," as defined by the presence of a prominent column foot, is unlikely to be a natural group.

REFERENCES. Braem 1986a,b; Dressler 1984; Rico-Gray and Thien 1987; Sauleda 1988, 1989; Sauleda and Adams 1989; Siegerist 1986; Withner 1988, 1990.

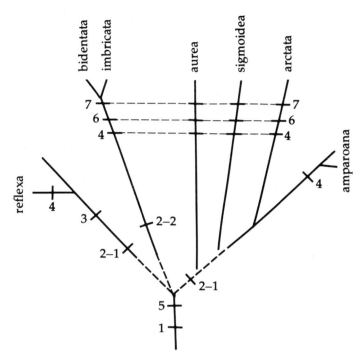

Figure 10-5. A diagram showing the relationships of the species that have been treated as *Hexisea* to *Scaphyglottis*. The relationships within *Scaphyglottis* are not known in detail, as indicated by the *dashed* lines near the base. Unless limited to *H. bidentata* and *H. imbricata, Hexisea* is a polyphyletic grade. Features used in diagram as follows:
1. Stem arising from: 0. base of older shoot; 1. upper part of older shoot (0 → 1).
2. Pseudobulbs: 1. slender; 2. ellipsoid (?).
3. Leaf type: 0. conduplicate; 1. cylindrical (0 → 1).
4. A deep nectary partially closed by lip: 0. lacking; 1. present (0 → 1).
5. Column foot: 0. lacking; 1. prominent (0 → 1).
6. Base of lip: 0. hinged; 1. fixed (0 → 1).
7. Flower color: 0. green and white (or flushed pink); 1. orange or red (0 → 1).

Subtribe **Pleurothallidinae**

DESCRIPTION. Habit: epiphytic or terrestrial; without pseudobulbs, stems unifoliate (exc. *Frondaria*). Velamen of the *Pleurothallis* type. Leaves: duplicate, often fleshy, articulate. Inflorescence: terminal or, rarely, lateral, simple or fascicled, distichous or secund. Flowers: resupinate or not, with a joint between ovary and pedicel; column short or elongate, often with a distinct foot, anther apical and incumbent or dorsal and erect; pollinia clavate, eight, six, four, or two, often with a tiny viscidium (especially when anther is erect); stigma entire or two-lobed; capsule may have two unequal valves.

DISTRIBUTION. Tropical America.

POLLINATION. The Pleurothallidinae are basically a fly-pollinated group, involving fungus gnats and carrion flies in some cases. There are a few cases of hummingbird pollination, especially in *Masdevallia*.

CHROMOSOME NUMBERS. 20, 30, 32, 34, 36, 38, 40, 42, 44.

SEED STRUCTURE. *Pleurothallis* type.

SPECIES. 3021.

GENERA. 28: *Acostaea, Barbosella, Barbrodria, Brachionidium, Chamelophyton, Condylago, Dracula, Dresslerella, Dryadella, Frondaria, Lepanthes, Lepanthopsis, Masdevallia, Myoxanthus, Octomeria, Ophidion, Platystele, Pleurothallis, Porroglossum, Restrepia, Restrepiella, Restrepiopsis, Salpistele, Scaphosepalum, Stelis, Teagueia, Trichosalpinx, Trisetella.*

DISCUSSION. The Pleurothallidinae are a diverse but·very natural group that parallels the Old World Bulbophyllinae in floral evolution (both are pollinated by flies). The presence of the *Pleurothallis* seed type in the *Ponera* complex suggests an origin of the Pleurothallidinae from an ancestor similar to *Ponera*. This group could well be derived from something similar to *Dilomilis*.

REFERENCES. Chase 1985; Luer 1986a–c 1987, 1988, 1990, 1991; Nakata and Hashimoto 1983; Pridgeon 1982a,b; Stern and Pridgeon 1985.

Tribe **Epidendreae II** (Old World Groups)

The Old World Epidendreae share conical silica cells with their New World relatives. Most have four pollinia, except for *Agrostophyllum*, with eight. There are no clear derived features to delimit the group, and some of the species now placed in *Eria* may need to be reclassified in this group, so the classification of this group is quite tentative. Whether the Old World groups form a sister group of the New World Epidendreae or rather a subgroup is not clear, but the dendrobioid sub-clade certainly appears to be a derivative, or subgroup, of the Old World Epidendreae. I had earlier treated these Old World elements as the Glomereae, but they are not clearly distinguished from the New World Epidendreae.

Figure 11-3 suggests that the Glomerinae in the strict sense might be separated from *Agrostophyllum, Earina,* and the other subtribes treated here, but we have too little information on these groups.

Subtribe **Glomerinae**

DESCRIPTION. Habit: epiphytic or terrestrial; with slender, reedlike stems or sometimes with pseudobulbs. Velamen of the *Calanthe* type. Leaves: distichous, duplicate, articulate. Inflorescence: terminal, of few to many spiral flowers, often dense. Flowers: small to medium, resupinate or not; lip often basally saccate or spurred; column short, often with a prominent foot; anther terminal, incumbent or more or less dorsal and erect; eight or four pollinia, laterally flattened, with small caudicles and with or without viscidium; stigma entire.

DISTRIBUTION. Tropical Asia and Australasia, with *Agrostophyllum* also in the Seychelles (Africa).

POLLINATION. Not known, though flower structure suggests moth pollination in many cases.

CHROMOSOME NUMBERS. 38, 40, 46.

SEED STRUCTURE. Similar to the *Elleanthus* type.

SPECIES. 230.

GENERA. 7: *Aglossorhyncha, Agrostophyllum, Earina, Glomera, Glossorhyncha, Ischnocentrum, Sepalosiphon.*

DISCUSSION. I had thought *Agrostophyllum* to be closer to the Podochilinae, because of its eight pollinia, but its silica bodies are conical, so it appears to be out of place in the Podochilinae. It is possible, of course, that *Agrostophyllum* is out of place here, also. *Earina* is rather unlike the other Glomerinae, and may be misplaced here. Its velamen is unlike that of the other genera (Porembski and Barthlott 1988). The presence of elaters suggests a possible alliance among *Agrostophyllum, Earina,* and *Polystachya* (Hallé 1986)

Subtribe **Adrorhizinae**

DESCRIPTION. Habit: small epiphytes without pseudobulbs but with very fleshy roots, velamen of the *Calanthe* type; vegetative stems very short. Leaves: duplicate, articulate. Inflorescence: terminal or lateral, of one or few spiral flowers. Flowers: small, resupinate; column slender, anther terminal, incumbent; four superposed pollinia; stigma entire.

DISTRIBUTION. Tropical Asia (southern India and Ceylon).

POLLINATION. Not known.

CHROMOSOME NUMBER. 48.

SEED STRUCTURE. *Vanda* type.

SPECIES. 3.

GENERA. 2: *Adrorhizon, Sirhookera.*

DISCUSSION. These small orchids have pollinia like those of the Glomerinae. They also resemble *Polystachya*, and their seeds resemble those of that genus. Conical silica cells are reported by Møller and Rasmussen (1984).

REFERENCE. Senghas 1984.

Subtribe **Polystachyinae**

DESCRIPTION. Habit: epiphytic or terrestrial; stems slender or forming pseudobulbs, these usually of several internodes, sometimes of a single internode. Velamen of an undefined type. Leaves: distichous, duplicate, articulate. Inflorescence: terminal or lateral, simple or branched, of several to many spiral flowers. Flowers: small to medium-small, resupinate or not, lip commonly with mealy hairs (pseudopollen) on upper surface; column with a prominent foot, short or somewhat elongate; anther terminal, operculate, with reduced partitions; four or two pollinia, with a small but definite stipe and a viscidium; stigma entire. (See Fig. 10-6.)

DISTRIBUTION. Pantropical, but primarily African, with a number of species in tropical America.

POLLINATION. The flowers of some *Polystachya* species are pollinated by small bees that gather mealy pseudopollen from the lip.

CHROMOSOME NUMBER. 40.

SEED STRUCTURE. *Vanda* type.

SPECIES. 154.

GENERA. 4: *Hederorkis, Imerinaea, Neobenthamia, Polystachya*.

DISCUSSION. These genera have small but definite stipes, so they have been classified with the "vandoid" orchids. In fact, they are "just barely vandoid." In some *Polystachya* species, the pollinia are not superposed in the anther, as they are in most vandoids, but they spread and then appear superposed when removed from the anther. The predominantly terminal inflorescence of this group suggests that it is not closely allied to the Vandeae. Silica bodies are not known in the Polystachyinae, but only one species was examined by Møller and Rasmussen (1984). It is quite possible that silica bodies will be found in other species. Conical silica bodies would convincingly exclude the Polystachyinae from the Dendrobioid subclade, but spherical silica bodies might indicate a closer relationship to *Eria* than to the Vandeae.

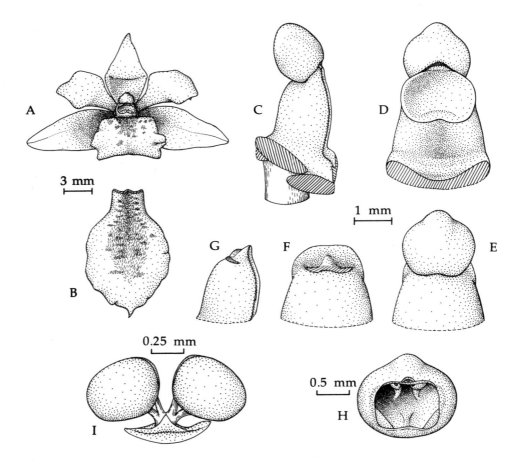

Figure 10-6. *Neobenthamia gracilis* (Epidendroideae: Epidendreae). (A) Flower. (B) Lip, flattened. (C) Column, side view. (D) Column, ventral view. (E) Apex of column with anther in place, dorsal view. (F) Apex of column with anther and pollinarium removed. dorsal view. (G) Apex of column with anther and pollinarium removed, side view. (H) Anther, basal (clinandrial) view. (I) Pollinarium. (After Dressler 1981.)

THE DENDROBIOID SUBCLADE

The dendrobioid subclade appears to be a subgroup of the Old World branch of the Epidendreae. It is potentially one of the clearest and best defined phyletic lines in the Orchidaceae, though its exact boundaries are not yet clear. It is based especially on spherical silica bodies, which are almost certainly a uniquely derived feature, but a feature that is not yet well sampled (see discussion under Eriinae). Aside from the silica bodies, the features of this phylad are essentially those of the Epidendreae, except that upper lateral inflorescences are much more frequent in this complex, and diploid chromosome numbers are largely limited to 38, 40, and 42.

Tribe **Podochileae**

DISCUSSION. The Podochileae includes the least derived subtribes of the dendrobioid subclade, generally with eight pollinia and sometimes without a viscidium. They are, then, very similar to the Epidendreae in most features.

Subtribe **Eriinae**

DESCRIPTION. Habit: epiphytic, rarely terrestrial; stem slender or forming pseudobulbs, usually of several internodes. Velamen of the *Calanthe* type. Leaves: distichous (secondarily spiral in some dwarf species), duplicate or rarely convolute and plicate, articulate. Inflorescence: terminal, or usually upper lateral, simple, of several to many spiral flowers. Flowers: resupinate or not, small to medium-small, sepals free or united, sometimes with a prominent spur; column usually with a prominent column foot; anther terminal, incumbent, eight-celled; eight pollinia, laterally flattened or clavate, with intralocular caudicles; stigma entire or somewhat two-lobed, often with a distinct viscidium. (See Fig. 10-7.)

DISTRIBUTION. Tropical Asia and Australasia, *Stolzia* in Africa.

POLLINATION. Not known; *Cryptochilus* and *Mediocalcar* have features suggestive of bird pollination. Many species of *Eria* have pseudopollen on the lip and are probably pollinated by small bees.

CHROMOSOME NUMBERS. 18, 20, 24, 34, 36, 38, 40, 42, 44, 46.

SEED STRUCTURE. Two distinctive seed types occur in this group (and within *Eria*); neither has been assigned to a defined type.

SPECIES. 701.

GENERA. 10: *Ascidieria, Ceratostylis, Cryptochilus, Epiblastus, Eria, Mediocalcar, Porpax, Sarcostoma, Stolzia, Trichotosia.*

DISCUSSION. The Eriinae are fairly uniform in having eight pollinia. Otherwise, the group is very diverse in nearly every respect, and this is especially true of *Eria*, itself, which includes a few species with plicate leaves. Spherical silica bodies are reported from *Ceratostylis* and four species of *Eria* (Møller and Rasmussen 1984), but conical silica bodies have been found in *Eria javanica* (Dressler and

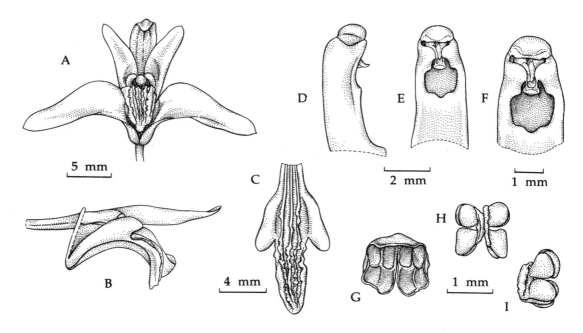

Figure 10-7. *Eria* near *bractescens* (Epidendroideae: Podochileae). (A) Flower, front view. (B) Flower, side view, with near sepals and petal removed. (C) Lip, flattened. (D, E) Column, side and ventral views. (F) Apex of column. (G) Anther. (H, I) Pollinarium, dorsal and side views.

Cook 1988). Until the group has been sampled more thoroughly, one cannot really delimit either the Eriinae or the Dendrobioid subclade with any confidence. Also, the seed structure of this group needs to be more thoroughly sampled.

The rostellum of *Eria* section *Hymeneria* projects downward from the column, and the caudicles of the pollinarium project downward next to the rostellum, so that a pollinator withdrawing from the flower may easily contact first the rostellum and then the caudicles (see Fig. 10-7).

REFERENCES. Andersen et al. 1988; Hashimoto and Tanaka 1983; Seidenfaden 1982.

Subtribe **Podochilinae**

DESCRIPTION. Habit: epiphytic or terrestrial, with slender, reedlike stems or sometimes with pseudobulbs. Velamen of the *Calanthe* type. Leaves: distichous, duplicate, sometimes laterally flattened, articulate or not. Inflorescence: terminal or lateral, of few to many spiral flowers. Flowers: small, resupinate or pendant, lip often basally saccate or spurred; column short, often with a prominent foot; anther terminal and incumbent or dorsal and erect; eight, six, or four pollinia, clavate, often with one or two prominent caudicles or with abortive pollinia at base; stigma entire, with one or two viscidia.

DISTRIBUTION. Tropical Asia and Australasia.

POLLINATION. Not known.

CHROMOSOME NUMBERS. 38, 40.

SEED STRUCTURE. Similar to the *Elleanthus* type.

SPECIES. 140.

GENERA. 6: *Appendicula, Chilopogon, Chitonochilus, Cyphochilus, Poaephyllum, Podochilus*.

DISCUSSION. The Podochilinae seem clearly to be a sister group of the Eriinae, but more advanced in the long-beaked column and anther and in the elongate pollinia. The abortive pollinia of some species have been confused with a stipe. Seidenfaden (1986) illustrates a curious, stipe-like structure in *Appendicula floribunda* that is quite unlike the abortive pollinia of other drawings. Seidenfaden considers the structures in question to be caudicles.

Subtribe **Thelasiinae**

DESCRIPTION. Habit: epiphytic, with or without pseudobulbs; stems, if slender, either short or elongate. Velamen of the *Calanthe* type. Leaves: distichous, duplicate, often laterally flattened and fleshy, articulate or not. Inflorescence: lateral, of few to many spiral flowers. Flowers: tiny, resupinate or not; column short, with or without a foot; anther erect, dorsal; pollinia eight or four, ovoid, with a common caudicle, this sometimes very long; stigma entire, with a viscidium.

DISTRIBUTION. Tropical Asia and Australasia.

POLLINATION. Not known.

CHROMOSOME NUMBERS. 30, 32.

SEED STRUCTURE. Not assigned to a defined type.

SPECIES. 234.

GENERA. 6: *Chitonanthera, Octarrhena, Oxyanthera, Phreatia, Rhynchophreatia, Thelasis*.

DISCUSSION. This group is made up of small plants with tiny flowers, so its members are rarely seen in cultivation. No silica bodies have been observed in this group, but the inflorescence is lateral, and often upper lateral, and seed structure seems to fit well in the Podochileae.

Subtribe **Ridleyellinae**

DESCRIPTION. Habit: epiphytic, with pseudobulbs. Leaves: conduplicate, articulate. Inflorescence: lateral, branched, of many spiral flowers. Flowers: tiny, column short, with a slight foot; anther terminal, operculate; pollinia eight, clavate-ovoid, with short caudicles; stigma entire, without a viscidium (?). Capsule globose.

DISTRIBUTION. Australasia (Papua New Guinea).

POLLINATION. Not known.

CHROMOSOME NUMBER. Not known.

SEED STRUCTURE. Not known.

SPECIES. 1.

GENUS. *Ridleyella*.

DISCUSSION. Schlechter compared the flowers of *Ridleyella* to the Thelasiinae, and it may be a member or ally of that group. The blue color of the flowers is unusual, and the pollinia agree with the Thelasiinae in shape and number, though the caudicles are unusually short for that group.

Tribe **Dendrobieae**

DISCUSSION The Dendrobieae are characterized by naked pollinia, without caudicles or other appendages, by a prominent column foot, and by the *Dendrobium* seed type. *Bulbophyllum* is quite distinct from most Dendrobiinae in its habit (pseudobulbs of a single internode and basal inflorescence), and silica cells have not been found in *Bulbophyllum*, but the resemblances suggest that the Dendrobiinae and Bulbophyllinae are sister groups.

Subtribe **Dendrobiinae**

DESCRIPTION. Habit: epiphytic or occasionally terrestrial, stems slender or forming pseudobulbs, these usually of several internodes, sometimes of a single internode. Velamen of the *Dendrobium* type. Leaves: distichous, duplicate, sometimes laterally flattened or cylindrical, articulate. Inflorescence: lateral or terminal, usually upper axillary, simple or branched, of few to many spiral flowers. Flowers: small to large, resupinate; lip often jointed basally, flowers often with a spur formed by the column foot or by the lip and column foot; column with a prominent foot; anther terminal and incumbent, two-celled; four pollinia, naked, in two pairs, without caudicles or viscidia; rostellum usually with a nonsticky "scraper" behind, the rostellar glue usually covered by a membrane. (See Fig. 10-8.)

DISTRIBUTION. Tropical Asia and Australasia.

POLLINATION. Most of the records for *Dendrobium* indicate bee pollination, though syrphid flies, thynnid wasps, and even a weevil (Lister 1987) are also mentioned. Recent data on several Australian species indicate that the orchids are not markedly specific, but may be pollinated by various bee species in the correct size range. One case of bird pollination is recorded, and a number of the species of higher elevations show the bird-pollination syndrome. Kjellsson and Rasmussen (1987) suggest that *Dendrobium unicum* attracts pollinators by mealy pseudo-pollen on the lip. In northern Thailand, *Dendrobium infundibulum* and *Cymbidium insigne* each appear to mimic the frequent *Rhododendron lyi*; the orchids offer no reward and are only occasionally visited by the bumblebees that pollinate the *Rhododendron* (Kjellsson et al. 1985).

CHROMOSOME NUMBERS. 36, 38, 40.

SEED STRUCTURE. *Dendrobium* type.

SPECIES. 1147.

GENERA. 6: *Cadetia, Dendrobium, Diplocaulobium, Epigeneium, Flickingeria, Pseuderia.*

DISCUSSION. The Dendrobiinae is one of the most distinctive and natural orchid subtribes. The rostellum of *Dendrobium* usually has a thin membrane over the glue, and a very elegant mechanism of pollinia presentation has evolved in

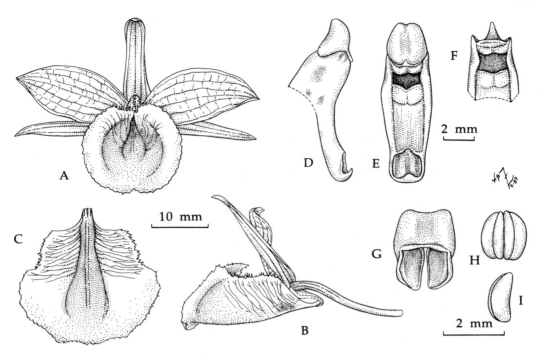

Figure 10-8. *Dendrobium aphyllum* (Epidendroideae: Dendrobieae). (A) Flower, front view. (B) Flower, side view, with near sepal and petal removed. (C) Lip, partially flattened. (D, E) Column, side and ventral views. (F) Apex of column, with anther and pollinia removed. (G) Anther. (H, I) Pollinia, ventral and side views.

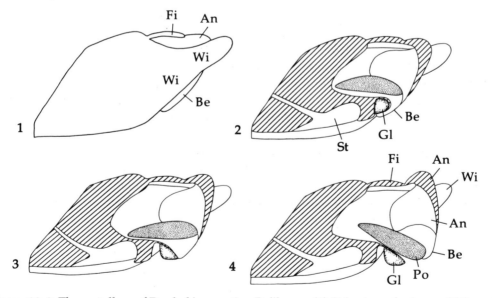

Figure 10-9. The rostellum of *Dendrobium* section *Pedilonum*. (1) Side view of column. (2) Longitudinal section of column, with anther in place. (3) The anther has been moved, breaking the rostellar membrane. (4) Further movement of anther pushes rostellar glue toward pollinator and then ejects pollinia. An, anther; Be, beak of anther; Fi, filament; Gl, rostellar glue; Po, pollinium; St = stigma; Wi, column wing. (After Dressler 1989c.)

section *Pedilonum*. In these species, the anther beak becomes attached to the rostellar membrane (Fig. 10-9). Then, anything that brushes the anther tips it back, rupturing the membrane. As the anther swivels out, it first pushes against the upper side of the rostellum, pushing the freshly exposed glue toward the pollinator, and then ejects the pollinia onto the fresh glue. In these species the rostellar glue solidifies very quickly. The membrane over the rostellar glue may be the reason that *Dendrobium* has not evolved a viscidium, though *Bulbophyllum* has apparently evolved viscidia in several groups.

A number of *Dendrobium* species have ephemeral flowers and flower gregariously, as in the well-known *Dendrobium crumenatum*.

REFERENCES. Adams and Lawson 1987, 1988; Cribb 1983a,b, 1986; Dressler 1989c; Hashimoto 1981, 1987; Jones et al. 1982; Kjellsson and Rasmussen 1987; Kjellsson et al. 1985; Lim 1985b; Lister 1987; H. Rasmussen 1982; Seidenfaden 1985; Slater and Calder 1988.

Subtribe **Bulbophyllinae**

DESCRIPTION. Habit: epiphytic, with pseudobulbs of a single internode, these often widely separated on the rhizome, and sometimes reduced in size. Velamen of the *Bulbophyllum* type. Leaves: duplicate, often fleshy, articulated, sometimes reduced to scales. Inflorescence: lateral, simple, spiral or distichous, of one to many flowers. Flowers: small to large, resupinate; lip often hinged at base; column with a prominent foot; anther terminal, incumbent, two-celled; four or two pollinia, naked, sometimes with a viscidium or viscidia, or even a hamular stipe; stigma entire.

DISTRIBUTION. Pantropical, especially in the Old World tropics.

POLLINATION. Most *Bulbophyllum* species seem to conform to one or another of the fly pollination syndromes, and the available records from Asia all indicate fly pollination, including both carrion flies and nectar seeking flies of other groups. **Three West African species are reportedly pollinated by wasps and stingless bees (Johansson 1974).**

CHROMOSOME NUMBERS. 36, 38, 40, 42.

SEED STRUCTURE. *Dendrobium* type.

SPECIES. 1116.

GENERA. 10–15: *Bulbophyllum, Chaseella, Codonosiphon, Dactylorhynchus, Drymoda, Genyorchis, Hapalochilus, Jejosephia, Monomeria, Monosepalum, Pedilochilus, Saccoglossum, Sunipia, Tapeinoglossum, Trias.*

DISCUSSION. The Bulbophyllinae resemble the Dendrobiinae in naked pollinia, chromosome number, seed type, and general floral form, but are markedly different in habit and lack the rostellar membrane of *Dendrobium*.

Even though the pollinia of the Bulbophyllinae lack caudicles, some Bulbophyllinae have evolved definite viscidia, and a few have even evolved hamular stipes (F. N. Rasmussen 1985, 1986a).

Sunipia has very distinctive double stipes, resembling the hamular stipes of *Bulbophyllum ecornutum* (F. N. Rasmussen 1986a). Both the velamen type and the seed type indicate that *Sunipia* is a member of the Bulbophyllinae. The stipe of

Genyorchis simulates a tegular stipe more closely than does that of *Sunipia*, but it is probably hamular. *Genyorchis* has the *Vanda* seed type but the *Bulbophyllum*-type velamen and seems best classified in the Bulbophyllinae.

Though the Bulbophyllinae are vegetatively very unlike the Pleurothallidinae, there are striking convergences in floral form, reflecting adaptations to fly pollination in both groups.

REFERENCES. Comber 1984; Jones 1985b; Lim 1985a; F. N. Rasmussen 1985; Rivett 1979; Seidenfaden 1979; Vermeulen 1987.

Tribe **Vandeae**

DISCUSSION. Once Møller and Rasmussen's paper (1984) pointed out the spherical silica cells shared by the Eriinae, Dendrobiinae, and Vandeae, it became clear that this feature correlated with chromosome numbers and habit to make the Vandeae a distinctive and "robust" phylad. Most other "vandoid" orchids have pseudobulbs, making them unlikely relatives for the monopodial Vandeae. A reed-stem habit with upper lateral inflorescence, like that of some *Trichotosia* species or *Pseuderia*, makes a very logical starting point for the evolution of the monopodial Vandeae. Only continued apical growth and rooting at the nodes is needed to convert such a plant to monopodial growth.

At first glance, the juxtaposition of the Dendrobieae, with quite naked pollinia, and the Vandeae, with complex pollinaria, seems strange. That they are sister groups, however, seems unlikely. Each group probably evolved independently from more *Eria*-like ancestors, and such a view is supported by seed structure. The *Vanda* and *Dendrobium* seed types are distinct, but each is approached by seed types found in the *Eria* complex.

Nearly all of the African and Madagascan genera show adaptations to pollination by moths, and the term "angraecoid" has been used for them. At one time, chromosome number seemed to divide the angraecoids into two clear subtribes. The two groups are distinguishable (see Fig. 10-11), but the difference in chromosome numbers is by no means as clear as it once seemed. Arends and van der Laan (1986) show a nearly complete series of diploid numbers from 34 to 54. They suggest that these numbers were derived through aneuploidy from an ancestor with $2n = 40$, though the predominance of 38 in the Asiatic Vandeae might suggest that number as the starting point.

The independent evolution of a leafless habit with photosynthetic roots in Asia, Africa, and tropical America is noteworthy. In fact, this remarkable habit probably evolved more than once in Asia, and may have done so in Africa. Benzing and Ott (1981) suggest that this "shootless" condition may have evolved primarily to enhance nutrient economy (see also Benzing et al. 1983).

This group, and especially the Aeridinae, appear to be very finely split, as compared to most other orchid groups.

REFERENCES. Arends and van der Laan 1986; Senghas (1986–1989) in Schlechter 1970–.

Subtribe **Aeridinae**

DESCRIPTION. Habit: monopodial, stem short to elongate. Velamen of the *Vanda* type. Leaves: distichous, rarely secondarily spiral, duplicate, sometimes cylindric, laterally flattened, or lacking. Inflorescence: lateral, simple or branched, of one to many flowers, flowers spiral, distichous or secund. Flowers: tiny to very large, lip may be jointed, saccate, or deeply spurred; column may have a prominent foot; anther terminal, operculate, with reduced partitions; four or two pollinia, with definite stipe or stipes and viscidium; stigma entire. (See Fig. 10-10.)

DISTRIBUTION. Mainly tropical Asia and Australasia, with one *Taeniophyllum* and one *Acampe* in Africa.

POLLINATION. We have few records for this huge group. Carpenter bee pollination is recorded for *Phalaenopsis* and *Vanda*, and bird pollination is certainly to be expected in *Ascocentrum* and *Renanthera*, for example. Jones (1981) reports the pollination of *Sarcochilus* and *Pomatocalpa* species by bees of the genus *Carbonaria*. Beetle pollination is reported in *Vanda cristata* (Pradhan 1983) and *Peristeranthus* (Wallace 1980).

CHROMOSOME NUMBERS. 24, 36, 38, 40, 56. The predominant number is 38; 24 is reported for *Taeniophyllum*.

SEED STRUCTURE. *Vanda* type and *Thrixspermum* variant.

SPECIES. 1253

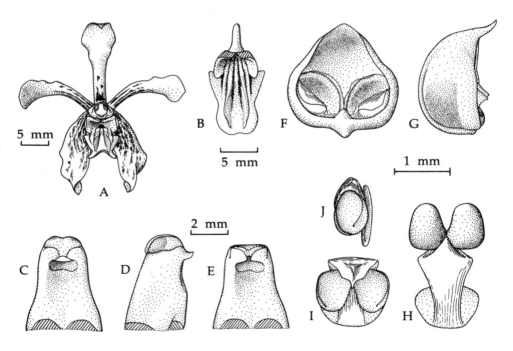

Figure 10-10. *Vanda lamellata* (Epidendroideae: Vandeae). (A) Flower, front view. (B) Lip and spur. (C) Column, ventral view. (D) Column, side view. (E) Column, with anther and pollinarium removed. (F) Anther, ventral view. (G) Anther, side view. (H) Pollinarium, freshly removed. (I) Pollinarium, after stipe has curled, top view. (J) Pollinarium, after stipe has curled, side view. (After Dressler 1981.)

GENERA. About 103: arranged in series, following Senghas (in Schlechter 1970–), with slight modifications.

Series I (pollinia 2, solid): *Ascochilopsis, Ceratocentron, Chamaeanthus, Chroniochilus, Grosourdya, Hymenorchis, Malleola, Megalotis, Microtatorchis, Omoea, Parapteroceras, Pennilabium, Porrorachis, Saccolabium, Tuberolabium.*

Series II (pollinia 2, porate): *Amesiella, Ascocentrum, Ascolabium, Biermannia, Cryptopylos, Dyakia, Eparmatostigma, Gastrochilus, Haraella, Holcoglossum, Luisia, Neofinetia, Seidenfadenia.*

Series III (pollinia 2, cleft): *Aerides, Ascochilus, Brachypeza, Dimorphorchis, Dryadorchis, Macropodanthus, Papilionanthe, Paraphalaenopsis, Phalaenopsis, Phragmorchis, Pteroceras, Rhynchostylis, Robiquetia, Sedirea, Smithsonia, Trudelia, Uncifera, Vanda.*

Series IV-A (pollinia 4, unequal; column with a foot): *Abdominea, Acampe, Ascoglossum, Ceratochilus, Cleisomeria, Cleisostoma, Cottonia, Diplocentrum, Diploprora, Drymoanthus, Loxoma, Micropera, Ornithochilus, Pelatantheria, Plectorhiza, Pomatocalpa, Porphyrodesme, Renanthera, Renantherella, Saccolabiopsis, Sarcanthopsis, Sarcoglyphis, Schistotylus, Schoenorchis, Smitinandia, Staurochilus, Stereochilus, Trichoglottis, Vandopsis, Ventricularia.*

Series IV-B (no column foot, lip fixed): *Arachnis, Armodorum, Esmeralda, Hygrochilus.*

Series IV-C (no column foot, lip movable): *Bogoria, Chiloschista, Cleisocentron, Cordiglottis, Doritis, Gunnarella, Lesliea, Mobilabium, Papillilabium, Peristeranthus, Proteroceras, Rhinerrhiza, Rhynchogyna, Sarcochilus, Thrixspermum.*

Series V (pollinia 4, equal): *Adenoncos, Calymmanthera, Microsaccus, Nothodoritis, Phragmorchis, Sarcophyton, Taeniophyllum, Xenicophyton.*

DISCUSSION. The well-known name Sarcanthinae must be replaced by Aeridinae, because Lindley used the name *Sarcanthus* for two different genera, the first, unfortunately, not corresponding to recent usage.

The orchid world is indebted to Senghas for bringing some kind of order into this difficult group. His series will doubtless be useful in identification, but they may not be phyletic groups. Note that he would place the Asiatic *Amesiella* and *Neofinetia* among the angraecoids. *Neofinetia* has 38 chromosomes that pair readily with those of *Ascocentrum* and *Vanda* (Kamemoto 1964; Shindo and Kamemoto 1962), and one may predict a similar condition in *Amesiella*. Note, too, that *Kingidium* (series IV) has been reduced to *Phalaenopsis* (series III) by Christenson (1986). There are, I am sure, other cases of close cross-series relationships, but we now have a system to work with, even if further study must change it.

REFERENCES. Cheng and Tang 1986; Christenson 1987a,b; Jonsson 1979; Seidenfaden 1988; Senghas 1988; Sweet 1980; Wallace 1980.

Subtribe **Angraecinae**

DESCRIPTION. Habit: monopodial, stem short to elongate. Velamen of the *Vanda* type. Leaves: distichous, duplicate, sometimes laterally flattened, cylindric or lacking. Inflorescence: lateral, simple or branched, of one to many flowers, flowers spiral, distichous or secund. Flowers: tiny to very large, lip usually deeply spurred;

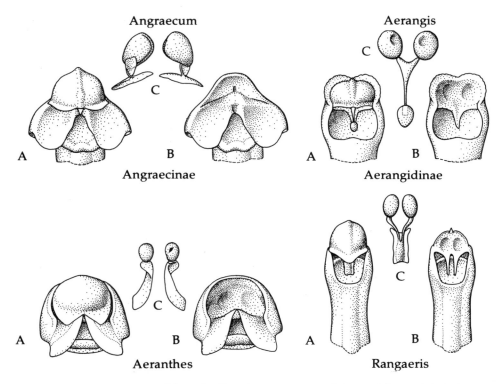

Figure 10-11. A comparison of the subtribes Angraecinae and Aerangidinae. (A) Column with anther in place. (B) Column, with anther and pollinarium removed. (C) Pollinarium, or pollinaria. (After Stewart 1976.)

anther terminal, operculate, with reduced partitions; two pollinia, with definite stipe or stipes and viscidium or viscidia; rostellum deeply divided; stigma entire.

DISTRIBUTION. Africa and especially Madagascar, with a few genera in tropical America and one species of *Angraecum* in Ceylon.

POLLINATION. Most angraecoids are probably moth-pollinated, but we have few records. Sphingid moth pollination is reported in *Angraecum* (Nilsson et al. 1985).

CHROMOSOME NUMBERS. 34, 36, 38, 40, 42, 44, 46, 48, 50.

SEED STRUCTURE. *Vanda* type and *Thrixspermum* variant.

SPECIES. 408.

GENERA. 19: *Aeranthes, Ambrella, Angraecum, Bonniera, Calyptrochilum, Campylocentrum, Cryptopus, Dendrophylax, Harrisella, Jumellea, Lemurella, Lemurorchis, Neobathiea, Oeonia, Oeoniella, Ossiculum, Perrierella, Polyradicion, Sobennikoffia.*

DISCUSSION. In the Angraecinae, the rostellum forms a sort of apron at the apex of the column and is deeply slit, usually with a distinct viscidium at each side of the slit. This system seems admirably suited for placing pollinaria on the tongues of Lepidoptera.

REFERENCES. Hillerman and Holst 1986; Nilsson et al. 1985.

Subtribe **Aerangidinae**

DESCRIPTION. Habit: monopodial, stem short to elongate. Velamen of the *Vanda* type. Leaves: distichous, duplicate or lacking. Inflorescence: lateral, simple or branched, of one to many flowers, flowers spiral, distichous or secund. Flowers: tiny to large, lip usually deeply spurred; anther terminal, operculate, with reduced partitions; two pollinia, with definite, usually long, stipe or stipes and viscidium or viscidia; rostellum elongate, beaklike, stigma entire.

DISTRIBUTION. Tropical Africa.

POLLINATION. There are a few records of visitation by beetles and moths. Moth pollination is to be expected in most cases. In a study of *Aerangis,* Nilsson and Rabakonandrianina (1988) introduce a new and useful technique for the study of nocturnal pollination, the analysis of hawk-moth scales to identify the pollinators.

CHROMOSOME NUMBERS. 42, 44, 46, 48, 50, 52, 54. Fifty is the predominant number.

SEED STRUCTURE. *Vanda* type and *Thrixspermum* variant.

SPECIES. 307.

GENERA. 36: *Aerangis, Ancistrorhynchus, Angraecopsis, Azadehdelia, Beclardia, Bolusiella, Cardiochilus, Chamaeangis, Chauliodon, Cyrtorchis, Diaphananthe, Dinklageella, Distylodon, Eggelingia, Encheiridion, Eurychone, Holmesia, Listrostachys, Margelliantha, Microcoelia, Microterangis, Mystacidium, Nephrangis, Plectrelminthus, Podangis, Rangaeris, Rhaesteria, Rhipidoglossum, Sarcorhynchus, Solenangis, Sphyrarhynchus, Summerhayesia, Taeniorhiza, Triceratorhynchus, Tridactyle, Ypsilopus.*

DISCUSSION. Unlike the Angraecinae, the Aerangidinae have a narrow, beaklike rostellum. Nilsson and Rabakonandrianina (1988) indicate that the pollinaria of *Aerangis ellisii* are probably placed on the frons or the palpi of hawk-moths. The narrower stipes and viscidia of the Aerangidinae would seem less well adapted for placement on the tongue than the pollinaria of the Angraecinae.

REFERENCES. Jonsson 1981; Nilsson and Rabakonandrianina 1988; Senghas 1965; Stewart 1986.

11

Misfits, Parallelisms, and Miscellaneous Problems

I n this chapter I treat the groups that fit poorly in the system and discuss a miscellany of other subjects, including parallelism and more general problems.

MISFITS AND LEFTOVERS

There are misfits in most classifications, though they may be swept under the rug or hidden in some way. In some cases, we need to know more about them; in all cases, they are the groups that should receive special attention in future studies. In my earlier book (Dressler 1981), I appended my misfits to Chapter 8, just after the Orchidoideae, and some readers thought these orphans were being treated as Orchidoideae, which was certainly not my intention. Placing the misfits in a separate chapter may avoid this confusion.

Subtribe **Arundinae**

DESCRIPTION. Habit: terrestrial or epiphytic, with slender stems slightly thickened at the bases. Velamen of the *Calanthe* type. Leaves: distichous, conduplicate, articulate. Inflorescence: terminal, simple or branched, with several to many spiral flowers. Flowers: medium to large, resupinate; column elongate, with a prominent column foot; anther terminal, incumbent, eight-celled; with eight relatively soft pollinia, laterally flattened, with ventral, intralocular caudicles; stigma entire, emergent, without a viscidium.

DISTRIBUTION. Tropical Asia.

POLLINATION. Pollination by bees is suggested by the form of the flowers.

CHROMOSOME NUMBERS. 40, 42.

SEED STRUCTURE. *Bletia* type.

SPECIES. 7.

GENERA. 2: *Arundina, Dilochia*.

DISCUSSION. *Arundina* and *Dilochia* are distinctive in their slender stems, distichous, conduplicate leaves, and terminal inflorescences, yet *Arundina* has *Bletia*-type seeds, and it is not clear that these genera can be separated from the Arethuseae. Tanaka (1971) reports crosses between *Arundina* and *Bletilla*, though I am not sure that these were grown beyond the seedling stage. I have no information on the seed of *Dilochia*, but this genus could well be a close ally of *Arundina*, as generally believed. It is also possible that these genera are basal members of the Old World Epidendreae.

Subtribe **Collabiinae**

DESCRIPTION. Habit: terrestrial, with slender pseudobulbs. Leaves: broad, convolute, plicate or soft herbaceous, articulate. Inflorescence: either terminal on a leafless pseudobulb, or lateral and with the base of the inflorescence thickened (perhaps a semantic difference), simple, with several to many spiral flowers. Flowers: small to medium, resupinate; lip spurred or saccate at the base; column medium to long, sometimes winged; anther terminal, incumbent; pollinia two, angular; stigma entire, with a prominent viscidium.

DISTRIBUTION. Tropical Asia.

POLLINATION. Not known.

CHROMOSOME NUMBER. 36.

SEED TYPE. Not known.

SPECIES. 28.

GENERA. 3: *Chrysoglossum, Collabium, Diglyphosa*.

DISCUSSION. When I first read of the Collabiinae, I felt that a subtribe with either eight or two pollinia (but not four or six) was improbable, and I eventually divided the group between the Bletiinae and the Cyrtopodiinae (then more inclusive than now). Seidenfaden (1983) gives more information on the genera with two pollinia, shows that they lack a stipe, and considers them misplaced in the Cymbidieae. His drawings show angular pollinia quite unlike those of the Eulophiinae.

I restrict the Collabiinae to the distinctive genera with two pollinia, leaving the genera with eight pollinia and quite distinct floral features in the Bletiinae. The Collabiinae may be allied to the cymbidioid phylad, but we have no clear evidence.

Claderia

Claderia was treated as a separate subtribe by Mansfeld, though the name Claderiinae was not validly published. Even if the subtribe were validly named, I would not be sure where to put it. The plant is a creeper almost without corm or pseudobulb, and the leaves are plicate. The rather fleshy flower is reminiscent of Coelogyne. It has two deeply cleft pollinia with a large viscidium, but no stipe. The viscidium develops behind a pair of rostellar flaps similar to those of Bromheadia (see Fig. 11-1). Claderia has seed of the Eulophia type, according to W. Barthlott (personal communication), and may be allied to Bromheadia and/or the Collabiinae.

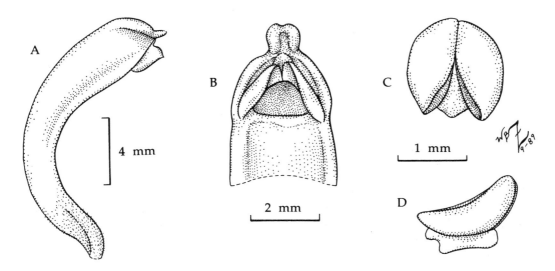

Figure 11-1. *Claderia viridiflora* (Epidendroideae: Cymbidieae?). (A) Column, side view. (B) Apex of column, anther and pollinarium removed. (C, D) Pollinarium, dorsal and side views. A and B drawn from material preserved in liquid.

Eriopsis

The tropical American *Eriopsis* has pollinia very like those of the Cyrtopodiinae, but the warty pseudobulbs and leathery, terminal leaves are anomalous, and the flower is quite distinctive. Mark Chase (personal communication) finds its seed to be of the *Maxillaria* type and considers it a member of the Maxillarieae, rather than the Cymbidieae. *Eriopsis* may well merit a separate subtribe. Its diploid chromosome number is 40. (See Fig. 11-2.)

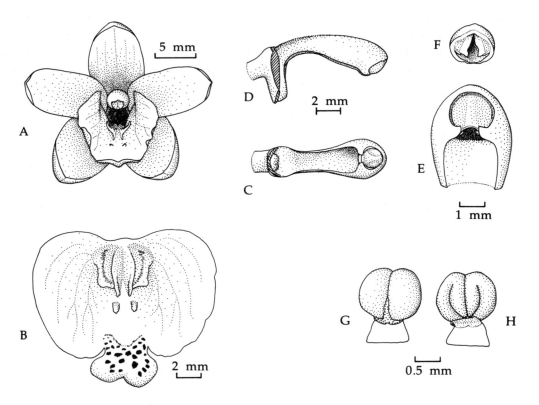

Figure 11-2. *Eriopsis biloba* (Epidendroideae: Maxillarieae?) (A) Flower, front view. (B) Lip, flattened. (C) Column, ventral view. (D) Column, side view. (E) Apex of column, with anther and pollinarium removed. (F) Anther. (G) Pollinarium, dorsal view. (H) Pollinarium, ventral view. (After Dressler 1981.)

SPECIAL PROBLEM AREAS

Primitive Groups

We almost always have problems resolving primitive groups, and phylogenetic theory explains much of the problem. Such groups are those with few derived features, and they are likely to be paraphyletic or unresolvable. Primitive orchids add additional problems. Many are difficult to cultivate and some have very delicate flowers. Thus, it is difficult for the student to find adequate material of these groups. These are the orchid groups of which we most need studies of anatomy, cytology, and other aspects, and they are the most poorly represented in such studies. Further, as pointed out by Seidenfaden (1978), these terrestrial groups are especially sensitive to environmental change, and many may become extinct in the near future.

The groups treated here as primitive Epidendroideae are especially trouble-some. We do not know as much as we should of their anatomy, cytology, or biochemistry, but they might still be difficult to classify even if we had much more detailed knowledge of their features. Molecular systematics may offer the best hope for better understanding of this grade.

Diurideae

In the case of the Diurideae, our problem is basically one of polarity. The curious column of *Diuris* has prominent staminodia free from the style nearly to the base. Is this condition primitive or secondarily derived? It may be more parsimonious to consider it a secondary derivation, but I am not sure which polarity is the more difficult to accept. If the *Diuris* type column is a primitive condition, I find it more difficult to visualize the monandrous orchids as a monophyletic group.

Podochileae

Though the Dendrobioid subclade seems one of the clearest in the family, its "roots" are not well resolved. The genus *Eria* is diverse in most features and is held together primarily by ancestral features, especially the presence of eight pollinia. The few species sampled show two rather different seed types; the first species sampled for stegmata show spherical silica bodies, but we find conical silica bodies in *Eria javanica* (Dressler and Cook 1988). Considering these factors and the great vegetative variation, it is quite probable that *Eria*, itself, will be resolved into two or more phylads, at least one with spherical silica bodies and one or more with conical silica bodies. The position of the Eriinae and Glomerinae, with resemblances to the Epidendreae and the Dendrobieae, make this complex espe-cially critical.

Since *Eria stellata* (= *javanica*) is the conserved type of *Eria*, it cannot be "removed from the genus *Eria*" without a major nomenclatural overhaul. Seiden-faden (1982) suggests that a species of the section *Hymeneria* would be much more appropriate as nomenclatural type, as this would permit retention of the name *Eria* for most of the species now known under that name. We need to resolve the nom-enclatural problems and to know more about the features of *Eria* in the broad sense before the lower level of the Dendrobioid phylad can be resolved.

Old World Epidendreae

The problems of the Old World Epidendreae are closely tied to the problems of the Podochileae, since any rejects from the "*Eria*" complex might be reclassified in the Epidendreae. *Glomera* and its close allies appear rather derived in their perianth, having a prominent column foot and often a spur, but *Earina*, the Adrorhizinae, and the Polystachyinae have simpler perianth structure. It is quite possible, of course, that some of these groups are closer to elements now classified in *Eria* than to *Glomera*. Silica bodies have not been found in *Polystachya*, but spherical silica bodies might be found there. A possible resolution of the Eriinae and Old World Epidendreae is shown in Figure 11-3. The complex around the Glomerinae will probably require revision when we have sufficient information.

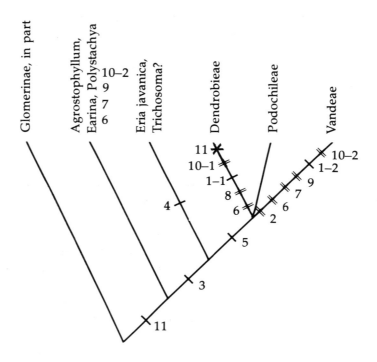

Figure 11-3. A tentative diagram of the dendrobioid subclade and its closest allies. This suggests a possible resolution of the poorly delimited Old World Epidendreae and stresses the awkward position of *Eria javanica* (the conserved type of *Eria*). Features as follows:
1. Velamen: 0. *Calanthe* type; 1. *Dendrobium/Bulbophyllum* types; 2. *Vanda* type.
2. Growth habit: 0. sympodial; 1. monopodial (0 → 1).
3. Inflorescence: 0. terminal; 1. upper lateral (0 → 1).
4. Leaf sheaths: 0. free; 1. thick and adnate to stem (0 → 1).
5. Silica cells: 0. conical; 1. spherical (0 → 1).
6. Pollinia: 0. eight; 1. reduced to four or two (0 → 1).
7. Pollinia: 0. parallel; 1. superposed (0 → 1).
8. Caudicles: 0. present; 1. lacking (0 → 1).
9. Tegular stipe: 0. lacking; 1. present (0 → 1).
10. Seed type: 1. *Dendrobium*; 2. *Vanda*.
11. Elaters: 0. lacking; 1. present (0 → 1).

Arethuseae

Almost any of the major groups may prove to be paraphyletic when we know more about its features. The Arethuseae seem especially likely to be paraphyletic. The Arethusinae appear distinctly less advanced than most other Arethuseae. *Bletilla* and *Calopogon* share some of the primitive features of *Arethusa*, yet they may not be close allies of *Arethusa* and they seem closely allied to *Bletia*. A recent artificial hybrid between *Calopogon* and *Arethusa* suggests, though, that these genera may be more closely allied than their appearance suggests. Even without considering *Bletilla* and *Calopogon*, the Bletiinae are rather diverse, but I see no way to divide them into two or more clear groups. Here, as in many other groups, we need more information.

INFORMATION WE NEED

In many of the problems discussed above, we need more good derived features, of whatever sort, preferably features that are uniquely derived or show only limited parallelism. Ideally, we need students to work from the species upward in many different genera and subtribes. I have tried to work from the family downward, but in too many cases the current taxonomic units are neither monophyletic nor well enough known.

I outline briefly something of what we need to know in several areas.

Anatomy

Anatomical work on the Orchidaceae has been rather spotty, and may well be one of our best hopes for additional derived features. Both the silica bodies and the velamen types have proven useful in classification. The orchid volume for Metcalfe's *Anatomy of the Monocotyledons* is in preparation and should be very useful, both as a source of information and an indication of areas where more work is needed.

Cytology

Orchid chromosome numbers are relatively well known, and better known than they were ten years ago. There are still some groups about which we know nothing, but the main need is probably for careful studies of tribes and subtribes. Cytology includes a good deal more than just chromosome numbers. The types of resting nuclei studied by Tanaka (1971) appear to have some usefulness in systematics, and there may well be other kinds of information that are yet to be utilized.

Embryology

Quite a bit of work has been done on orchid embryology (Abe 1972), but some of the features of embryo and embryo sac development vary so much within groups that they seem to hold little promise for systematics. Still, a comprehensive review of this area might be valuable.

Floral Ontogeny

The works of F. N. Rasmussen (1982, 1986b) and Kurzweil (1987a,b, 1988, 1989a,b) are quite useful. Our greatest need is probably for more information on some of the Australian Diurideae, of which neither author had enough material.

Pollen Structure

Schill and Pfeiffer (1977) have given us a general survey of pollinaria structure. There are a few detailed studies of selected groups (Ackerman and Williams 1980, 1981; Burns-Balogh and Hesse 1988). Detailed studies of other groups would undoubtedly be rewarding, and the pollen structure of some groups remains quite unknown. The sculpturing of the exine is of interest mainly in those groups with soft pollinia. In the advanced Epidendroideae, the surface of the pollinia is generally quite smooth. Though the ontogeny of the embryo-sac may have little of systematic value to offer, the ontogeny of the pollen and pollinia may prove to be rewarding (Wolter and Schill 1985).

Seed Structure

A good deal of work has been done on seed structure, but only a portion of this work has been published to date (see Chapter 2). The extensive work of Barthlott and his collaborators will, I hope, be published eventually. We clearly need to know more about seed structure and variation in the Eriinae and Glomerinae, for example, and careful studies of tribes or subtribes will usually be rewarding, as in Chase and Pippen (1988, 1990).

THE PREVALENCE OF PARALLELISM

The great deal of parallel evolution within the orchids has been emphasized before (Dressler 1981; F. N. Rasmussen 1982) and is of special interest if the pattern of evolution in the family is one of our concerns. Detailed discussion of each case of parallelism might be profitable, but could easily fill a fair-sized book. Some of the parallelisms are summarized in Table 11-1. I discuss a few of the features and some more general ideas. Features that appear to be uniquely derived are listed in Table 3-4.

Some of the parallelisms in seed structure are of special interest. In the Bletiinae one finds two rather distinct seed types, the *Bletia* type and the *Eulophia* type. One's first impression is that this might give the clue to dividing the subtribe into two more homogeneous groups, but both seed types may occur within a seemingly natural genus such as *Calanthe*. Similarly, the *Diuris* and *Goodyera* types both occur within some genera of the Caladeniinae. In this case, the seed types may reflect ecological differences. In both cases, the pattern suggests that some plants have the capacity to produce either of two rather different seed types, with the seed type controlled by some sort of genetic "switch." This hypothesis could be tested with interspecific crosses of *Calanthe*, for example.

Table 11-1. Features showing parallelism within the Orchidaceae. The estimate of parallelism is minimal, and the number of times that a given feature may have evolved is listed as 1 to 5, more than 5, and more than 10.

Monopodial growth	>5
Pseudobulbs	>5
Soft herbaceous leaves	5
Conduplicate leaves	>5
Cylindrical leaves	>5
Distichous leaves	4
Leaflessness (without saprophytism)	4–?
Lateral inflorescence	>10
Sepals imbricate	2 (?)
Loss of median anther	2
Separate stigmatic areas	>5
Viscidium	>10
Tegular stipe	>5
Hamular stipe	4
Sectile pollinia	4
Firm pollinia	4
Double pollinaria (separate viscidia)	2 (–3?)
Eight pollinia	2
Reduction 8–4–2	4
Clavate pollinia	5
Superposed pollinia	5
Spheroid pollinia	5
Unilocular ovary	3
Capsule	4
Loss of crustose seed coat	5
Vanda type seed	3
Dendrobium type seed	2
Twig-epiphyte seeds	3 (?)
Ecological features	
Epiphytism	>5
Saprophytism ("neotenic")	>10
Deceit pollination	>10
Pseudocopulation	4
Flower mimicry	5
Sapromyophily	2
Myrmecophily	4

There is no sharp distinction between parallelism and convergence; as commonly used, convergence is essentially parallelism in several different features, or "parallel parallelism," guided by ecological factors in the classical cases. Convergence between distantly related organisms should be resolved by careful analysis, but convergence between closer allies may be harder to detect. As Simpson (1953) has stressed, however, convergence among close relatives may be common. Similar organisms under similar conditions offer the most favorable material for either parallelism or convergence.

Ecological Convergence

We are familiar with ecologically guided convergence, and several good cases of this occur among the orchids. There are striking parallelisms between *Bulbophyllum* and the Pleurothallidinae, and these surely reflect pollination by Diptera in both groups. The very striking parallels between the African angraecoids and *Amesiella philippinense* or *Neofinetia falcata* (both once classified as *Angraecum*) show the moth pollination syndrome in relatively close allies. The American genus *Brassavola* shows many of the same features, but with a different growth habit and a different kind of nectary.

Integrational Parallelism

The concepts of integrational and contingent parallelism (discussed below) are neither clearly distinct nor independent of ecological factors, but they may help to understand some of the parallelism we find in the orchids. By integrational parallelism, I refer to cases in which a new feature or structure functions best only after associated features are modified. One can imagine a blunt, snub-nosed column with a terminal viscidium (*Pachyplectron* comes close), but few such orchids exist. The terminal viscidium functions best if the column is beaklike, with the viscidium placed high relative to the anther. In many cases, elongate caudicles or a slender hamulus permit a narrow, elongate rostellum. The interrelationships of anther, stigma, and rostellum have been interpreted to support a much broader Spiranthoideae (Burns-Balogh and Funk 1986). That these relationships reflect a functional and structural convergence is indicated by the similarities of epidendroid genera, such as *Meiracyllium*, *Podochilus*, or *Thelasis*, all of which show very "spiranthoid" columns.

Similarly, there are a number of features that characterize the more advanced "vandoid" orchids (see Chapter 8), and it would not be too difficult to make credible cladograms showing the Vandoideae to be a natural subfamily. It appears, however, that these strongly correlated features are all features that improve the efficiency of the viscidium in pollinia transfer.

Contingent Parallelism

By "contingent parallelism," I refer to features that may evolve only after another feature is present (the other side of the coin of integrational parallelism). Thus, feature A permits the evolution of feature B, and B may, in turn, permit the evolution of a succession of other features. I strongly suspect that many of the "tendencies," "underlying synapomorphies," and "orthogenetic trends" discussed in the literature are, in fact, contingent parallelism. The concept of contingent parallelism is close to that of "preadaptation," but without, I hope, the purposive implications that term seems unable to shed.

The Orchidaceae demonstrate a beautiful sequence of contingent parallelism. The loss of the ventral stamens left a dorsoventral flower with one anther directly over the stigma. This facilitated the loss of the lateral anthers. Then the median stigma lobe took over a function in pollen transfer, that is, evolved a rostellum. Once this had happened, a distinct viscidium evolved in many different groups. Different types of stipe have evolved a number of times. On the vandoid level, we find reduced anther partitions, the anther incumbent very early in ontogeny, and superposed pollinia, all ultimately contingent on the symmetry of the primitive orchid flower.

In the Vandeae there is apparently some feature, whether structural or physiological, that permits the evolution of leaflessness, and leaflessness has arisen independently in the American tropics, Africa, and Asia, probably more than once in both Africa and Asia.

Contingent parallelism within phylads may be quite widespread. There has been some debate over the polyphyletic origin of vessels in the early angiosperms. Carlquist (1987) argues that vessel specialization has been quite polyphyletic, and that a polyphyletic origin of vessels in the dicotyledons is highly probable. From another perspective, once vascular tissues had evolved to a certain level in the ancestral angiosperms, the polyphyletic origin of vessels was almost inevitable.

Dunstervillea mirabilis is a fine example of convergence in orchid evolution. Garay described this orchid as a member of the Sarcanthinae (= Aeridinae), though I believe it should be placed in the Ornithocephalinae. I can find no feature, however, that will prove either of us right or wrong. Tiny monopodial plants with laterally flattened leaves and globose pollinia occur in both groups. Such tiny plants may not have silica cells and commonly show reduction in chromosome number. We find the same type of seed in twig epiphytes of both groups (Chase 1987c). If we could show that these tiny flowers produce oil, rather than nectar, this might support my viewpoint on the position of *Dunstervillea*, but otherwise the convergence is so thorough that it is difficult to prove or disprove either position.

The prevalence of deceit pollination in the Orchidaceae is noteworthy. Pseudocopulation is a remarkable form of deceit that has evolved independently in three or four different groups, though this form of mimicry is not known in other plant families. The bilateral symmetry of the orchid flower is doubtless a factor. I earlier mentioned that a lack of nectaries in the primitive orchids might have been a factor in the evolution of mimicry, or deceit pollination. Another possibility is that the orchids can make a success of such a system because of their pollinia. That is, a low percentage of pollination produces many seeds and can maintain a relatively large population. Mimicry may be favorable in causing a high percentage of cross-pollination or achieving pollination over relatively great distances, but only a small proportion of the flowers are pollinated in a mimetic system. Another form of mimicry, carrion fly pollination or sapromyophily, is especially well represented among the asclepiads, the other major group with pollinia.

FALSE ADVERTISEMENT
AND THE EVOLUTION OF INEFFICIENCY

Mimicry is fascinating and bizarre, but where mimicry is clear and specific, the continued attraction of pollinators is easily understood. There are, however, many cases of "generalized food flower mimicry" that do not involve a clear and recognizable model. Flowers with clusters of yellow hairs or yellow calli may be said to mimic the pollen-bearing anthers of a generalized model, just as flowers with an empty spur or spurlike extension may mimic nectaries, but generalized food flower "mimics" do not have specific models. Indeed, this system is termed "nonmodel mimicry" by Dafni (1986). Surprisingly, it is not easy to draw a clear line between specific mimicry and "generalized mimicry." *Orchis caspia*, for example, does not closely mimic any other flower, yet it commonly grows with *Asphodelus microcarpus*, *Bellevalia flexuosa*, and/or *Salvia fruticosa* (Dafni 1983). When *O. caspia* grows with all three species, about 86 percent of its flowers are pollinated, but populations that grow with none of these species have only 11 to 16 percent of their flowers pollinated. It would seem, then, that the presence of these three species facilitates the pollination of *O. caspia*, even if they are not truly models.

In generalized food flower mimics, the pollinators soon learn that the flowers offer no reward, and these systems are characterized by a low percentage of pollination. Floral ecologists write of "pollinator limitation." In these cases, pollinators may be abundant, but pollination is limited by the availability of gullible pollinators, and gullibility is short-lived. *Orchis spitzelii*, for example, seems to have no specific model, and fruit set per spike varies from 7 to 44 percent in smaller populations and 10 to 16 percent in a large population (Fritz 1990). In such a system, we may expect lower percentages of pollination in larger or denser populations because the pollinators are proportionately fewer and quickly learn that the flowers offer no reward. These generalized food flower mimics may show from 2 to 50 percent pollination, though related species that offer nectar may have 80 to 95 percent pollination.

In terms of fruit set, mimicry is relatively inefficient, and this is especially true of generalized food flower mimicry, yet generalized food flower mimics may make up as much as a quarter of the orchid family. Gill (1990) suggests a third of the family, citing van der Pijl and Dodson (1966), but their calculation refers to orchids without nectar, including other systems, such as male euglosine pollination and specific mimicry. After reviewing the data on generalized food flower mimicry, Gill (1990) suggests that these pollination systems cannot possibly be "evolutionarily stable strategies." He suggests that mutants that cause self-pollination or the production of a reward should rapidly take over the entire populations.

Though "pollinator limited" species may set numerous fruits when hand pollinated, the plants may be smaller or produce fewer flowers in succeeding years (Ackerman 1986a; Ackerman and Montalvo 1990; Zimmerman and Aide 1989). In the long term, then, the plants may be resource limited, and heavy fruit set may weaken the plants.

The energetic cost of producing nectar may be a factor that favors the evolution of pollination through deceit. I know of no data on the energetic costs of nectar production in orchids. Southwick (1984) calculates that nectar production accounts for 4 to 37 percent of the daily net photosynthetic production in a milkweed, *Asclepias syriaca*. Similarly, he calculates that nectar production in cultivated alfalfa may represent twice the energy investment of the seed crop. It is quite possible that the energetic cost of producing nectar is significant, especially in plants that grow in the shade or in epiphytes that are generally adapted to survival with minimal resources.

Orchis coriophora, one of the two *Orchis* species that produces nectar, may achieve a high percentage of pollination, but according to Dafni (1987), "The pollinator usually visits several flowers on the same plant, which leads to a high frequency of geitonogamy." In the food flower mimics, on the other hand, the pollinator typically leaves the inflorescence as soon as it finds that there is no nectar. Though there is much less pollination, most of the fruits result from cross-pollination.

Because orchid seeds are numerous and only a small proportion can germinate or survive, it is often assumed that survival is a matter of chance. I suggest, however, that, except for autogamous plants, the large number of seeds permits high recombination and the maximum testing of recombinants in nature. Gill (1990) suggests that *Cypripedium acaule* must have very low genetic diversity. I suspect, rather, that it is usually cross-pollinated and that the mature plants with a life expectancy of nearly 24 years may be genetically superior to most of the younger plants whose median life expectancy is 5.6 years. This hypothesis could easily be tested in the Orchidinae. With enzyme chromatography, one could sample the genetic diversity of *Orchis* species that produce nectar and others that practice false advertisement. If the latter have greater genetic diversity, it would indicate that a high percentage of cross-pollination may be one of the advantages of this seemingly inefficient pollination system.

BREEDING SYSTEMS

Breeding systems cover a spectrum from apomixis, with no sexual recombination, to obligate cross-pollination (see Table 11-2), and strongly influence variation and the pattern of evolution. I have generalized, saying that orchids are usually self-compatible but have mechanisms that favor cross-pollination (Dressler 1981). This may be a valid generalization, but like all generalizations, it is deceptive if not quite false. Breeding systems are often discussed as though each were an absolute, but many plants are not 100 percent apomictic, autogamous, self-compatible or self-incompatible. Thus we find, for example, the dandelion pattern of variation: innumerable apomictic "microspecies" without variation, but with infrequent sexual reproduction producing new micro-species. There is a broad spectrum of breeding systems, and the plants that exactly fit one of our labels may be in the minority.

Table 11-2. Major breeding systems and their genetic consequences.

Breeding system	Recombination and variation
Apomixis	none
Autogamy	little
Self-compatibility	moderate or high
Self-incompatibility	high
Dioecy (separate male and female plants)	very high

Apomixis

In apomixis, the plant produces flowers and seeds but actually bypasses sexual reproduction. That is, the embryos are not the result of sexual recombination but are maternal tissue. Thus, apomixis is essentially vegetative reproduction in the guise of sexual reproduction. This system has been reported in a few orchids, mainly among the Cranichideae (Fryxell 1957).

Autogamy

Autogamy, or automatic self-pollination, occurs in most orchid groups and involves many mechanisms (Catling 1990). As noted above, it is not necessarily absolute. In southern Mexico I observed *Epidendrum nocturnum*, whose flowers normally opened and presumably required pollination to set seed (in that area), and *E. carpophorum*, whose flowers did not open but set fruit automatically. Still, I found two plants that appeared to be hybrids between *E. nocturnum* and *E. carpophorum*; this suggests that the flowers of *E. carpophorum* must occasionally open.

Autogamy is clearly more frequent than the literature indicates (Catling 1990), and may occur in 10 to 15 percent of orchid species. Systematically, autogamy is most frequent in relatively primitive groups and least frequent in the advanced, "vandoid" groups (Table 11-3). Catling notes that autogamy is more frequent in terrestrials than in epiphytes, and it is especially frequent in saprophytes. It is relatively frequent on isolated islands or peninsulas, where immigrants may not find their customary pollinators. Autogamous populations have obvious advantages as immigrants or invaders. Autogamy is also frequent at higher elevations, where pollination is unreliable. In the Australian Diurideae, it is interesting that the Diuridinae depend on false advertising to achieve pollination and about 23 percent are autogamous, but the related Prasophyllinae offer nectar and only about 6 percent of the species are autogamous.

Table 11-3. Percentages of self-pollinating species in different orchid groups and grades. Larger percentages are rounded to whole numbers.[a]

Cypripedioideae	4.0%
Diurideae	8.0%
Primitive epidendroids	7.0%
Calypsoeae	17.0%
Cranichideae	3.0%
Orchideae	1.3%
Diseae	1.2%
Malaxideae	1.0%
Cymbidieae (except Catasetinae)	3.0%
Advanced epidendroid grade	2.0%
Polystachyeae	3.0%
Vandeae	0.3%
Maxillarieae	0.4%

[a] Based on Catling 1990.

Self-compatibility

Autogamous plants are obviously self-compatible, but this term usually refers to plants that are not automatically self-pollinated but may be either self-pollinated or cross-pollinated by flower visitors. In many ways, this is the most flexible system, in that an isolated plant has at least the potential of being pollinated and establishing a population. This would appear to be the common condition in the Orchidaceae, but in fact we have no firm statistics. Even in plants that are technically self-compatible, fruits resulting from self-pollination may abort or have fewer or smaller seeds than those resulting from cross-pollination.

Self-incompatibility

In self-incompatible plants, some biochemical mechanism prevents the germination or growth of pollen from the same plant or clone, so that self-pollination is prevented. There are scattered reports of self-incompatibility in various orchid groups. Self-incompatibility is the rule in *Cryptostylis* (Stoutamire 1975), and is general in most species of *Oncidium* and probably in the subtribe. Charanasri and Kamemoto (1977) found self-incompatibility in 73 percent of the Oncidiinae studied, with both self-compatible and self-incompatible plants in two of the species studied. In the Vandeae, Agnew (1986) found about 30 percent of the species sampled to be self-incompatible, with about the same percentage needing further study. Johansen (1990) found self-incompatibility in 72 percent of the *Dendrobium* species that he studied, and found this to be a new type of incompatibility, apparently mediated by specialized, detached cells in the stigma. These small samples of the Dendrobiinae, Oncidiinae, and Vandeae suggest that at least 10 percent of the orchids are self-incompatible, and systematic sampling might show a much greater percentage.

Species-incompatibility

Though species incompatibility is not necessarily an element of breeding system, it is of special interest here, because of the high degree of species compatibility known in some groups of orchids. Johansen (1990) found that the incompatibility system of *Dendrobium* caused a high degree of species incompatibility. Species incompatibility is also found in Australian *Cryptostylis*, which are all pollinated by the same species of wasp.

Dioecy

True dioecy is not reported from the orchids. The staminate and pistillate flowers of the Catasetinae may occur on the same plant, so that there is at least the possibility of infrequent self-pollination (geitonogamy). Nevertheless, cross-pollination is clearly the rule in the Catasetinae, and probably also in *Satyrium ciliatum* (Chen 1979).

LIFE HISTORY STRATEGIES

There is a generally recognized spectrum in plant "behavior" from the "weedy" plants of disturbed habitats to the plants of stable habitats. This corresponds in general to the r- and K-selection theory that has been much discussed in the ecological literature (Stearns 1977). An organism that experiences density-independent mortality or an organism of fluctuating enviroments should experience "r-selection," and one would expect many offspring with little energetic investment in the individual offspring, early maturity, short lifespan, and often little genetic recombination. An organism of stable environments ("K-selection"), on the other hand, should produce relatively few offspring but invest heavily in each one, have later maturity and longer lifespan, and its breeding system should favor high recombination.

At first glance, the production of many tiny seeds would seem to fit the characteristics of r-selection, but in other respects, most orchids fit this pattern poorly. Only in the case of autogamous orchids do we find a close approach to the r-selection syndrome, with enormous seed production, reduced recombination, and sometimes early maturity or short lifespan. Many orchids, on the other hand, favor or require cross-pollination (high recombination), and tend to have a long lifespan, even though the number of seeds is relatively high.

Grime (1979) has suggested a three-cornered spectrum of life history patterns, in which plants may be tolerant of disturbance (the "weedy" pattern) or of stress, or may be highly competitive (the stable habitat pattern). In general, orchids approach the stress-tolerant pole of the triangle. Many are tolerant of low nutrient supply, some are tolerant of low light, and most epiphytes must be somewhat tolerant of drought.

12

Classification and Phylogenetic Analysis

D iscussion of classification and, especially, of phylogenetic analysis tends to bog down in abstruse terminology. I have tried to weed out unnecessarily complicated terms, but some readers may wish to review the Glossary, where words marked with asterisks (*) are especially relevant to this chapter.

From earliest times, people have classified at least the plants and animals that are useful or dangerous. Systematics as a science has developed gradually and rather unevenly. Early systematists considered species to be immutable and sought God's plan in the diversity of nature. Not surprisingly, many early classifications were essentially identification schemes, formulated with little thought of phylogenetic relationships. The doctrine of immutability in species lost support when local and geographic variation had been studied in detail. At the same time, the idea developed that some groups are "natural," while others, especially those in identification schemes, are "artificial," or arbitrary. Darwin's work later supplied a theoretical basis to explain the naturalness of groups and classifications. A natural group is one that evolved from a single ancestral species and inherited a number of different features from that common ancestor; it is what we would now call a monophyletic group. An artificial group, on the other hand, is not descended from a single common ancestor, but is polyphyletic. Such a group may be characterized by one or few obvious features, but the other features of its members are not well correlated with each other or with the supposed key features on which it was based.

What Is to Be Classified?

One might suppose that classification deals with classes, but etymology is deceptive. Classes are defined groups of objects or entities. A class may be defined to include all three-legged dogs, all white dogs with brown ears, or all short-haired dogs. Clearly, a single individual *Canis familiaris* might claim membership in all three classes, but it cannot be a member of *C. lupus* (wolf) or *C. latrans* (coyote). Philosophically, the biological groups we classify are not classes, but rather "historical entities" or "historical individuals." For special purposes, of course, one may group plants into classes in the strict sense of the term. Field manuals often class wild flowers by color to facilitate their identification, in spite of the occasional chromatic nonconformists that occur in almost every species. One may form all sorts of ecological or structural classes for special studies, but in each case, their usefulness is limited to special purposes.

Biological groups differ fundamentally from classes in important ways. Classes are defined and may be quite arbitrary, but biological groups are notoriously difficult to define, even when quite natural and clearly delimited. Horses are characterized by having one toe on each foot, but a colt with three toes on each foot would still be *Equus caballus*. One may prepare a diagnosis to distinguish most members of a taxon, but not a definition, which all members of a class must fit. Note, for example, that a species is not rejected merely because the original author chose the wrong delimiting features. Rather, the original name is retained, but the diagnosis is revised.

Though we must study individual plants and their features, we should not classify either features or individual plants. The individual plants we study are samples of populations, and it is the populations that are classified and named. Mistakes are made when samples are inadequate, but when a better sample is available, the classification must change to reflect our improved knowledge.

The Goals of Classification

Classification is useful for students of anatomy, cytology, and ecology; for agronomists, conservationists, horticulturists, librarians, and natural products chemists; and for those who prepare field manuals or data bases. Understandably, any of these people may think of classification as a service activity and feel that it should be tailored to their specific needs. Nevertheless, systematics has its own goals, and I will argue that the goals have not changed from pre-Darwinian times. The main goal is to develop a natural classification, which is, in fact, the classification with the greatest general utility. A natural classification is a system that is based on, or at least consistent with, phylogeny, but an artificial, or arbitrary, classification is discordant with phylogeny. In a natural classification, the groups are monophyletic, each derived from a common ancestor, but the groups in artificial classifications are often polyphyletic. A phylogenetic classification may predict unsampled features, but an arbitrary classification can predict only the key features on which it is based.

A classification that is intended to be natural is, in many ways, a hypothesis about relationships. Ideally, of course, it will represent the best hypothesis that can be reached with the available information.

Clades and Grades

As we learn more about plants and animals, we find that some seeming "groups" are not phylogenetic units at all. A clade is, by definition, monophyletic and includes all descendants of a common ancestor (and, in theory, the ancestor). A "grade" is usually defined by similar adaptive features or similar evolutionary level. The genus *Hexisea* has been a polyphyletic grade during much of its history (see Fig. 10-5). In many cases, such a grade is made up of two or more units that are similar in some few features, but each unit is phylogeneticly most closely related to other units not included in the grade. The vandoid orchids once appeared to be a very natural subfamily, but we now realize that they are almost certainly three different clades, each derived from a different ancestral "epidendroid" group. We may use the terms "epidendroid grade" and "vandoid grade." Indeed, the concept of grade can be quite useful, but grades are not necessarily phylogenetic units. As commonly used, the term "reptile" refers to a grade, though the reptiles might become a natural taxonomic group by excluding the ancestral mammals and including the surviving dinosaurs, that is, the birds.

The vandoid orchids demonstrate the practical problems in treating a polyphyletic grade as a natural taxon. There are a number of features that are loosely correlated in the vandoid genera, but no two of these features are always correlated. If we make an arbitrary diagnosis of the grade, we will separate close allies, and the features will permit only an arbitrary diagnosis.

Features in Classification

Any classification is based on features, especially on the features shared by different members of the group. Many authors have attempted to distinguish anatomical details, or other classes of features, as being more "fundamental," but the features that are consistent and useful are not necessarily the same in different groups. Some features are much easier to see or to sample than others, but the important differences between "classes" of features reflect their history, rather than their physical nature.

Primitive versus Derived

Primitive, or ancestral, features are those inherited from remote ancestors. A primitive feature may give some hints as to ancestry, but shared primitive features are not reliable evidence of close relationship. A primitive feature shared by species A and B may reflect descent from a remote common ancestor, but it does not necessarily indicate a close relationship between A and B.

Derived features, that is those that are inherited from a relatively close common ancestor, are much more informative in determining relationship. One must emphasize that primitive and derived are relative terms. The genus *Epidendrum* is characterized by having the whole length of the column united with the lip. In the subtribe Laeliinae, or in orchids in general, this is a derived feature. For a subgroup of *Epidendrum*, however, it would be viewed as an ancestral feature. Evolution is not necessarily unidirectional. Many features may evolve in different ways in different groups, or the pattern of evolution may even be reversed. Thus, there is no *a priori* way to determine polarity, or the direction of evolution, in a given feature. There has been much discussion of this problem, and "outgroup comparison" seems to be the best approach. A study group is compared with its sister group, its closest relative, presumably descended from a common ancestor. If we find states A and B in our study group and only state A in its sister group, then B is considered the derived state. We may be able to identify the sister group only after careful study, but we may survey the pattern in several potential sister groups for an estimate of polarity.

One should beware of the old and discredited "common equals primitive." Features that are general in closely related groups may be primitive features for a study group, but this should not be used uncritically. If a given feature shows extensive parallelism in several related groups, such a comparison can give a false polarity, especially if the primitive state has become rare. One colleague argues that dust seeds must be primitive in the orchids, because they predominate in all subfamilies.

Transformation Series

Outgroup comparison considers two states of a given feature to determine their polarity. Two or more such states may be termed a "series." Pollen texture in the orchids, for example, may be powdery, sticky, soft but coherent, divided into packets (sectile), brittle, or hard. In general, this list goes from the ancestral to the most advanced, but brittle or hard pollinia are not necessarily derived from sectile pollinia. Transformation series and their states are usually numbered, as, for example:

15. Pollen texture: 0. powdery; 1. sticky; 2. soft but coherent; 3. sectile; 4. brittle; 5. hard ($0\rightarrow1\rightarrow2\rightarrow3$, $2\rightarrow4$, $2\rightarrow5$, $3\rightarrow5$).

The numbers in parentheses indicate probable patterns of evolution for these states. These may also be shown by arrow diagrams connecting the different states. The states may be shown in phylogenetic diagrams as 15–1, 15–2, etc.

Reversals

It has been suggested that evolution cannot reverse itself, or that a feature once lost is lost forever, but this is not strictly true. A plant may still carry the genetic program for a feature that is not normally developed. In such a case, the loss or misfunction of a suppressor gene may be enough to bring back the "lost" feature. A number of "monandrous" orchid species include self-pollinating forms with three fertile anthers. These represent a sort of reversion to the state of a remote triandrous ancestor. The radially symmetrical flowers of *Thelymitra* show a seemingly ancestral feature. In this case, though, it is a secondary adaptation. The flowers of *Thelymitra* mimic other, radially symmetrical flowers and attract pollen-seeking bees. Both *Bletilla* and *Calopogon* resemble *Bletia* in many features but have very soft pollinia. Whether their pollinia are primitively soft or represent something of an adaptive "reversion" to softer pollen masses is difficult to determine. Small reversals undoubtedly do occur, but it is still true that a major and complicated feature is rarely lost without trace. We find self-pollinating forms in the Maxillarieae and the Vandeae, yet these would never be mistaken for primitive orchids of the neottioid grade. Indeed, when the complicated pollinary apparatus of the vandoid grade breaks down, the flowers function only in self-pollination.

Parallelism and Convergence

Parallelism refers to those cases in which similar, or even indistinguishable, features evolve independently in two or more different groups, including closely allied groups. Parallelism in single, isolated features is usually easy to recognize. There are, however, many cases of convergence, that is, parallelism in a suite of features, usually guided by ecological factors. (See alternative definitions of convergence in Glossary.) Plants that grow in deserts often resemble each other in a suite of vegetative features, including even anatomical details, yet these plants are often members of different families and not at all closely related. When I visited Papua New Guinea I "recognized" several genera of American Pleurothallidinae, but all were species of *Bulbophyllum*. Both groups are pollinated by flies, and similar pollination systems have led to striking convergence in flower structure between the two groups. Orchid flowers that are pollinated by birds commonly show a suite of correlated features: fleshy brightly colored parts, tubular lip or flower, dark colored pollinia, and some structure that forces the bird's beak toward the rostellum. Compare, for example, *Sophronitis*, *Neocogniauxia*, *Hexisea*, and *Alamania*. These show some of the same features that one finds in distantly related bird-pollinated orchids of New Guinea. At one time, some genera of American Gesneriaceae were based on flower shape. Now we realize that such "genera" each included members of several different natural groups and were united only by their pollination systems. In this case, chromosome number, crossability, and obscure anatomical features are much more meaningful than flower shape or color (Wiehler 1983).

It should be clear that parallelism is most likely to occur where similar (or related) organisms occur under similar ecological conditions. Thus, parallelism between close relatives may be quite common, and this parallelism will be the most difficult to distinguish from inheritance due to common ancestry. There seems to be no formula or methodology that will dependably detect parallelism. The best procedure is to analyze as many features as possible for a given group and to look for conflict or discordance between different features or kinds of features.

Homology

Homology is a basic concept in biology, but there is some confusion in the use of the term. Some recent authors define homology simply as referring to features derived from a common ancestor. From the morphologist's or anatomist's viewpoint, though, features that are similar in structure, topology, and development are often considered homologous. These are the features that *could* be derived from a common ancestor. We may distinguish the two viewpoints as "morphological homology" and "phyletic homology." Morphological homology is a prerequisite for features to be used in phylogenetic analysis; at the same time, the analysis determines the probability of phyletic homology for each feature used.

By either definition, homology is a relative term. A classic example of non-homologous features would be the wings of butterflies and birds. The two types of wing are so different in both structure and development that they could not have been derived from a common ancestor. If we compare the wings of bats and birds, however, they are homologous *as forelimbs*, but they are not homologous as wings; each uses the ancestral forelimb in a different way.

Correlations

Ideally, classification is based upon derived features that are correlated because they are inherited from common ancestors. Features change during evolution, and different stages in a transformation series may be more informative than unchanging features. Even "lost" features may be significant. We would prefer to find several strongly correlated features at each level of the classification. When we can do this, we feel confident that we have a natural, phylogenetic classification.

Correlated features may arise from causes other than common ancestry. Features that are ecologically related are suspect, as they may have evolved independently; their correlation may reflect ecology, rather than inheritance. Suites of features that are structurally or developmentally related also call for caution. A suite of interdependent features should not be treated as independent features in phylogenetic analysis and might better be treated as a single feature. Aside from these factors, parallelism, by itself, should not cause correlation so close that it would be confused with common ancestry.

Identification and Key Characters

Classification and identification are closely related, and often practiced by the same people, but the activities should not be confused. Plants or animals may often be identified by a few key features, but a classification should be based on as much information as possible. Once a group has been classified, we may be lucky enough to have a few, consistent features that permit easy identification, but taxa that are based on a few key features are often arbitrary and unnatural.

The idea of "monothetic" groups that can be defined by single features is intuitively very attractive, but this idea corresponds to classes in the philosophical sense rather than to biological groups. Biological groups are usually polythetic at any level. One usually finds some exceptions to virtually every delimiting feature, so species and especially genera and other inclusive groups must be delimited by combinations of features. Groups or individuals that lack any one of the delimiting features may still be members in good standing of their respective groups or species.

One viewpoint holds that if X/Y separates two genera in a given group, then every pair of species that differ by X/Y should be placed in different genera. This has a clear, algebraic logic, but plant evolution does not necessarily follow algebraic logic. When all other features indicate that two species are close allies, a supposed key feature does not necessarily mean that they should be placed in different genera. In other words, the significance of a given feature in determining phylogenetic relationships must be determined independently for each group.

Modern workers rarely present their classifications as identification schemes, but some still think of classification in that way. There may be several different and equally valid identification schemes for any group of plants. We can objectively choose one classification as more natural than another, but quite different systems may function equally well as identification schemes.

Monophyly, Polyphyly, and Paraphyly

A monophyletic group is one that is descended from a single ancestral species, but a polyphyletic group reflects descent from at least two, relatively unrelated ancestral species. Figure 12-1 illustrates a polyphyletic group, B, in which two species are derived from a common ancestor shared with group A, while the other two share a common ancestor with group C. In this case, the two ancestral species are relatively close allies. In many cases, the two ancestral species are not at all close allies, but they or their offspring have, through parallelism, evolved one or a few close resemblances that led to their descendants being classified together. The monophyletic group, then, is "natural," but the polyphyletic group is artificial, or arbitrary.

In traditional systematics, workers have not insisted that groups include all descendants of their common ancestor, but the cladistic school of classification demands that all descendants of a common ancestor be included in a single taxon. There may, of course, be subgroups within that group. The term "holophyly" has

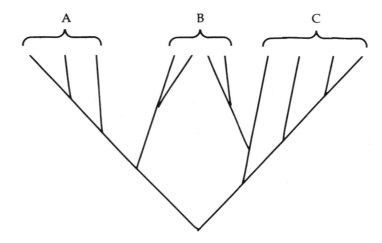

Figure 12-1. A diagram showing a polyphyletic group (B). Some of the taxa assigned to "B" share a common ancestor with group A, while others are a subgroup of group C.

been used to distinguish this concept from the traditional idea of monophyly.

Cladists have introduced another concept, that of paraphyly. A group that is descended from a common ancestor but does not include all descendants of that common ancestor is termed "paraphyletic." There is typically a subclade that is clearly distinguished by one or several derived features, as C in Figure 12-2, and often classified at the same rank as the corresponding paraphyletic group (AB). If a feature can be found that is shared by B and C (at X in the diagram) to resolve AB into two groups, then B and C are considered sister groups, and together form the sister group of A. Paraphyletic groups may also be called "remnant groups," as they are species or groups left over when distinctive derived groups are delimited. In many cases, the paraphyletic group would be monophyletic and quite acceptable if there were no strongly divergent subclade. Indeed, it might be useful to use a special term, such as "exophyletic," for the strongly divergent subclades that render their relatives "paraphyletic."

How to treat paraphyletic or unresolved groups is a major point of conflict. Many traditional taxonomists see nothing wrong with unresolved groups, but the cladists prefer to eliminate all such groups from the classification. Mayr (1982) argues that classification should reflect degrees of differentiation, as well as branching pattern, and that paraphyletic taxa are thus acceptable. In practice, even the cladists must treat some groups as paraphyletic or unresolved, simply because no shared derived features have been found to resolve them. Further study, of course, may find anatomical or other features that permit the complete resolution of such a group. The cladistic ideal of a fully hierarchical classification without paraphyletic groups is intellectually very attractive, but not necessarily practical or achievable.

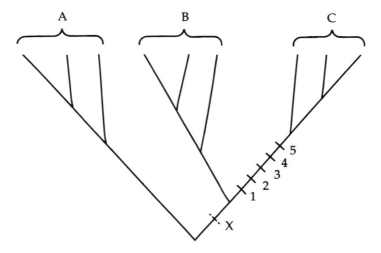

Figure 12-2. Diagram of a paraphyletic group AB. C represents a clearly defined subclade (exophyletic group). If one or more features can be found at X (shared by B and C), then AB can be resolved into A and B.

THE SPECIES IN BIOLOGY AND PHYLOGENY

In many respects, the basic element in biological classification is the species. The species is either a population or a system of populations with relatively continuous variation within the species, but there are discontinuities in variation between species. A genus is a group of species, ideally a group of species that could have descended from a single ancestral species. Similarly, other taxonomic categories above the species are groups of species, groups of genera, or groups of groups, ideally monophyletic at each level.

The Concept of Species

In spite of the reams of paper that have been written on the definition (or indefinability) of species, species is the most objective and definable category in biological classification. In recent years, three major species concepts have been much discussed: the biological species concept, the evolutionary species concept, and the ecological species concept. As Andersson (1990) points out, these are really models, rather than definitions, and it is difficult to state a universal species concept more exact than "the smallest groups that are consistently and persistently distinct, at least in a given area." Andersson favors the ecological species concept, and in practice the biological and ecological concepts come to about the same thing. Consistently distinct kinds that occur in the same areas are good species. In case of doubt or geographic isolation, the fact that each kind occurs in a distinctive habitat suggests that they should be considered species. The evolutionary species concept seems unassailable as a concept but contributes little in the way of concrete criteria for delimiting species.

Some recent authors stress genetic reorganization in peripheral or founder populations as a fundamental aspect of speciation. Even so, the biological species concept remains useful. Species do not normally lose their identity by hybridization with other species. Hybrids may occur, and they may be locally abundant, but they do not obscure the species. When species are sympatric, or occur together, the Mayrian criterion—that sympatric species maintain their identities in a given area—is both objective and practical. We have no really objective criteria for species in plants or animals that reproduce asexually, and there is a "catch 22" for species that do not occur together. If we put them together, they are not under natural conditions. With geographically separate (allopatric) populations and organisms that reproduce asexually, classification may have to be somewhat arbitrary.

The biological species concept stresses interbreeding, but this does not mean that taxonomists must spend their time studying breeding patterns in nature. One may infer interbreeding or the lack of interbreeding by an analysis of variation. In practice, if A and B occur together and are always easily distinguished, they may be considered distinct species (ignoring for the moment the problems of gender and polymorphism). Occasional hybrids that are easily identified as such do not prevent the treatment of their parent populations as species. If, however, A and B occur in adjacent areas, and one finds a gradual transition from A to B, they are surely not distinct species; some would use the subspecies concept in such a case. If we find that X and Y, though superficially very similar, never interbreed, many taxonomists would say, "They are biologically distinct species, but it's not practical to treat them as species if one can't tell them apart." Such populations may be termed "cryptic" species, and pairs of cryptic species often prove to be abundantly distinct when they are carefully studied.

We should not be disturbed by the many difficulties in defining species. We would only expect them to be neat and clearly defined if they had been created at the beginning of time. If they developed over millions of years, we should find a spectrum of not species, almost species, just barely species, and clearly defined species, and that is exactly what occurs in nature.

Polyphyly and Paraphyly in Species

Most evolution occurs at (or below) the species level. Genera or families may appear to change over time, but the changes really occur in the constituent species. Wheeler and Nixon (1990) argue that the terms monophyly and polyphyly are simply not relevant at the species level. Species do not descend from individuals, but from populations. Whether or not a species is derived from a single, "monophyletic" population, or from the merging of two or more distinctive populations cannot be determined *a posteriori* because of interbreeding.

Wheeler and Nixon (1990) also argue against the use of paraphyly at the species level, but it may be useful to stress that species can be paraphyletic by any reasonable criteria (Donoghue and Cantino 1988; de Queiroz and Donoghue 1988). Hennig, himself, recognized that one daughter species would sometimes be indistinguishable from the parent species. Mayr (1954, 1982) suggests that many widespread species are paraphyletic and that speciation occurs mainly in small, peripheral populations. Most such peripheral isolates are doubtless short-lived in geological terms, but some achieve greater success and coexist with, or replace, their parent species.

The cladistic assumption of strict dichotomy may be the best procedure to analyze groups of species. At the species level, though, insistence on strict dichotomy may distort the pattern of phyletic relationships.

Categories within Species

In botany, the rules of nomenclature permit the use of several categories below the species: subspecies, variety, and forma (with subcategories sometimes added). "Subspecies" is usually interpreted as a geographic subpopulation within a species, and *varietas*, or "variety," may be used in the same way. These, then, can refer to populations and may have some objective reality. At the same time, "variety" is often used without regard to geography, in the same way that *forma* is usually used. For example, any individual plant that is very hairy might be considered a member of "var. *villosa*" or "forma *villosa*," even though some of its siblings might be hairless. Such taxa are nearly meaningless and quite out of place in a biological classification. A plant with hairy leaves and white flowers might claim simultaneous membership in "var. *alba*" and "var. *villosa*." In orchids, an individual *Cattleya labiata* "var. *alba*" may be only distantly related to another "var. *alba*." Even the genes that result in white flowers may be different in the two plants. The names of clonal (horticultural) varieties are useful, and it would be much better if we could stop mixing different things under the term "variety." Many botanists (myself included) would prefer to abolish both variety and form as formal taxonomic units. There are better ways to describe the variation within a population.

THEORY AND PHILOSOPHY OF CLASSIFICATION

Several authors have attempted lengthy theories of classification, for the most part with only limited success. Here I discuss briefly the viewpoints that have dominated the literature in the last few decades.

Traditional Systematics

It is difficult to discuss other philosophies of classification, without contrasting them with more traditional procedures; however, "traditional" systematics is not easily characterized. By analogy with biological groups, the traditional systematists are a paraphyletic or remnant group, without clear delimiting features. Perhaps the major feature of traditional systematics is a relative lack of theory. Most systematists recognize organic evolution as the basis for their work, but beyond this, many do not advocate any particular theory of classification. In general, traditional workers survey the distribution and correlation of features in their study groups and select the features to delimit their groups and to identify the groups and species without following any explicit formula. Some traditional systematists do this very well. In retrospect, we can say that Lindley or J. J. Smith were careful, competent systematists, or that F. Kränzlin simply did not have "an eye for relationships." The best of traditional systematists explain their choice of delimiting features in detail, while others simply offer their classification with the implication that "this is correct because I say it is."

Lacking a clear theory of systematics, however, even the best of traditional systematists have had little success communicating their good judgment to others. Mayr, Simpson, and Stebbins all have made important contributions, especially to evolutionary theory, but a clear and coherent theory of classification has eluded us. In recent decades the acrimonious debates between different schools of thought have often cast more heat than light, yet these debates may have brought us closer to a clear theory of systematics. The Uniter program being developed by Hall (1991) would seem to simulate good traditional systematic procedure much better than any other methodology now available. Hall's theory and program appear to merit careful attention and further evaluation.

Phenetics

Phenetics was much touted in the 1950s and early 1960s; Sokal and Sneath, its major proponents, proposed to measure "overall resemblance" by enumerating and coding as many different features as possible, giving equal weight to each feature, and calculating a percentage of resemblance between each pair of species, or "operational taxonomic units." Phenetic procedure had the virtue that it forced students to observe the features of their study groups in detail. Too often, one "knows," from the literature or personal bias, what features to expect in one's group and fails to check carefully.

At first, phenetics seemed a scientific and unbiased approach to systematics. What was intended to avoid bias proved to be its fatal weakness. Pheneticists did not distinguish between primitive and derived features, or between uniquely derived features and parallelisms. In other words, "overall similarity" could not be distinguished from superficial similarity.

Cladistics

The cladistic school of thought evolved from the work of Willi Hennig, who published especially in the 1950s and 1960s. Hennig's writings are somewhat turgid and laborious, and parts of the English translation of his major work are almost unintelligible. The Spanish edition (1968) is clearer but by no means light reading. Hennig created a complex and rather dysphonious terminology; some of this has come into general use, especially "apomorphic" for derived and "plesiomorphic" for ancestral. At first, Hennig's ideas were little known outside of entomology, but they have become much better known within the last 15 or 20 years.

Some have taken up Hennig's ideas with religious zeal; indeed, cladistics has given rise to a number of sects (or heresies, depending on one's viewpoint). For a time, the journals were so filled with convoluted and contentious discussion of the finer points of cladistic theory that it was hard to see how anyone could keep up with the literature and still find time to study plants. In common with the pheneticists, many cladists seek not so much a theory of classification as a technique, or methodology, to derive phylogenies and classifications. Dahlgren and Rasmussen (1983) give an unusually clear explanation of cladistic methodology, and Donoghue and Cantino (1988) give a clear and straightforward discussion of cladistic concepts, while distinguishing clearly between method and concept. One hopes that the futile controversies of the last couple of decades need not be repeated.

A basic tenet of Hennig and his followers is that only derived features may be used in determining phyletic relationships, and this seems to be generally accepted. Similarly, the cladistic doctrine is that no modern group can be derived from any other modern group, though two modern groups may share a common ancestor. Above the species level, at least, this is simply common sense if one is dealing with holophyletic groups. Several other cladistic tenets may be less universally accepted:

1. Many cladists feel that classification should reflect phylogeny in detail, and derive their classifications directly from cladograms, or diagrams derived through phylogenetic analysis. Hennig and especially his early followers tended to make classification reflect every branching point in the corresponding phylogeny, but such a procedure leads to an inordinately complicated classification. There is also a tendency to view the branching patterns that emerge from cladistic analysis as measures of time, but a cladogram is, at best, a relative measure

of time. Evolutionary rates vary within and between groups, as does the amount of extinction.

2. Cladists stress branching pattern (cladogenesis) rather than degree of difference in classification. Degree of resemblance or difference is considered only as it is reflected in the derived features available for analysis.

3. Cladists insist that groups must be not only monophyletic, but holophyletic. That is, each group should include all descendants of its common ancestor. In other words, if the common ancestor of group B (if it were known) would be assigned to group A, then the derivative group B should be treated as a subgroup of A, no matter how drastically it may deviate from the features of the parent group.

4. In cladistic procedure, each group should not only be descended from a common ancestor, but it should include the common ancestor. In other words, each group should be a clade. This is not so much a theory of evolution as a procedural convention in cladistic analysis. The ancestors derived through cladistic analysis are quite hypothetical and unnamed. In the relatively few cases that we can identify ancestors, some may prove to be unlike their hypothetical counterparts.

5. Cladists assume that speciation is dichotomous and that when there is speciation, the parent species disappears, as such, and is replaced by two daughter species. This simple assumption may have generated more opposition than all the rest of cladistic methodology. Now it seems clear that even the cladists cannot expect the real biological world to fit this assumption, though it makes the drawing of cladograms much simpler. As biology, the assumption is simply senseless (Andersson 1990).

6. Though scarcely mentioned by Hennig, the principle of parsimony has become important in cladistics. That is, one should accept the simplest explanation (or phylogeny) that is consistent with the available data. This usually means the phylogeny that involves the fewest parallelisms or reversals. Some cladists have even suggested accepting taxa without unique derived features in the name of parsimony, a conclusion Hennig would have rejected (see Duncan 1986).

An advantage of the cladistic system is that the author must indicate clearly what features he is using and how he interprets them, so that the work is more easily evaluated or criticized. Cladists recognize that features may be lost, and a feature that is lost in part of a taxon may still have value in determining phylogeny. In effect, polarity is more important than frequence or constancy. In any method of analysis, the results depend on the features chosen, how they are polarized, and how carefully the analysis is done. Obviously, the inclusion of additional features may lead to a different result.

Cladistic Classification

Phylogenetic analysis is the best, if not the only, method of determining phylogeny with any confidence. To what degree the diagrams derived from phylogenetic analysis should be converted directly into classifications is still subject to debate.

The conflicts between cladistic and other approaches are especially evident when we consider classification through time. Many cladists have dismissed evidence from fossils as irrelevent, but Donoghue et al. (1989) show that fossils may be critical in the interpretation of polarity and phylogeny. Lidén and Oxelman (1989) hold that applying temporal aspects to taxa can only lead to unsolvable paradoxes. If this is the case, we should perhaps revise our philosophy of classification (assuming that there is one). We do classify fossil as well as living organisms, and in some cases we know a good deal about extinct taxa. To be sure, only the gaps and imperfections in the fossil record let us classify taxa through time without being quite arbitrary, but that is a practical problem rather than a theoretical one.

Cladistic procedure demands that common ancestors must be included in the taxa that develop from them. In theory, then, we would place quite similar Cretaceous sister species in different families if each gave rise to a clade that was eventually treated as a distinct family. Cladists also insist that highly divergent, or exophyletic, groups should be treated as members of the paraphyletic (or unresolved) groups from which they appear to be derived. In many cases, these exophyletic taxa must be early members of clades that systematists of the remote future might treat as distinct families or orders. An alternative and more consistent view might be to consider each taxon to be made up of all members of its respective clade or subclade during a period of time. Thus, a group might be treated as a family at the present, while its common ancestor could be treated as a member of a genus in a different, Cretaceous family. These objections to cladistic classification are philosophical, rather than practical, but we should have a philosophy of classification that avoids unsolvable paradoxes.

PHYLOGENETIC ANALYSIS

There are several different methods of phylogenetic analysis. The ground-plan divergence method of Wagner (1980) is one of the oldest, and continues to be relatively free of controversy. In recent years, most cladists have used methods that stress parsimony. In some respects, the cladist's parsimony methods parallel phenetics in putting all available (polarizable) features into the data base and accepting the most parsimonious pattern that emerges from analysis as the best hypothesis of phylogeny. Compatibility analysis, or clique analysis (Duncan 1980; Meacham and Estabrook 1985), however, gives greater weight to "compatible" features, that is, those that agree with each other completely, and gives less weight to features that may show parallelism. The advocates of parsimony accuse the compatibility school of discarding useful information, and the debates between the two schools are acrimonious, with discussion at cross purposes and some fatalities among straw men. Nordal (1987) defends the compatibility viewpoint, and considers the difference one of weighting, rather than parsimony. As Fink (1986) stresses, each method involves some assumptions, and one should try to understand the arguments and be willing to accept the assumptions before using any method.

Any method of analysis may be done "by hand" for small groups and small data sets, but computer programs must be used for larger analyses. Several programs are available (Fink 1986; Platnick 1988). The computer is totally unbiased and may calculate a most parsimonious cladogram that involves quite unacceptable "reversals." In all cases, the student must evaluate the results and decide which cladograms are credible, as well as reasonably parsimonious.

Both Dupuis (1984) and Saether (1986) suggest that parsimony methods may be appropriate at or near the species level, while other approaches may be better at higher systematic levels. Felsenstein (1983) suggests that parsimony methods are not appropriate where evolution is rapid or there is much parallelism. Alfred North Whitehead (1957) suggests: "The guiding motto in the life of every natural philosopher should be 'seek simplicity and distrust it.' " This motto seems especially apt for those doing phylogenetic analysis. For a given data set, the most parsimonious cladogram is a sort of systematic null hypothesis; all other things being equal, it is acceptable. Nevertheless, considerations of polarity, reversibility, morphological homology, and probability may be more important in determining the validity of a hypothesis than strict numerical parsimony.

Bias and Objectivity

The cladistic approach tries to minimize bias in analysis, certainly a worthwhile goal. Bias, however, is not easily exorcized. Bias is insidious, may be quite unconscious, and can influence one's choice and coding of features, even when the method is carefully mechanical. One possible expression of bias is to code several different aspects of a single feature as independent characters. One author distinguishes a pair of genera by features of column, column foot, sepals, and lip, when, in fact, all of these features are different aspects of a spur at the base of the lip, and could be summed up as "lip with a deep or shallow spur" versus "lip base shallowly saccate, but not quite a spur."

Hypotheses about phylogeny, polarity, or relationships may also introduce some bias. Outgroup comparison tries to avoid bias, but even the choice of an outgroup may influence the result of an analysis. Saether (1986) holds that objectivity is simply a myth, and it is, at least, a very relative concept. Certainly, the cladistic format is no guarantee against bias. One systematist has said, "Anything can be proven by cladistics." The cladists, similarly, ridicule traditional procedure as picking only the data that support a pet idea. Perhaps the critical question is not how to prove one's pet idea, but how to disprove it, a question every scientist should be willing to ask.

Pitfalls and Perils

Christenson (1987b) notes that phylogenetic analysis is especially subject to the computer axiom, "garbage in, garbage out." One must be especially careful about the data that go into an analysis. One should use only truly independent features, and determine morphological homology and polarity as carefully as possible.

Some cladists seem confident that parallelisms will not outnumber uniquely derived features and create a false hypothesis of phylogeny. This attitude seems close to "ignore it, and maybe it will go away." My own reservations about cladistics are surely conditioned by work with the Orchidaceae, where parallelisms could easily outnumber uniquely derived features and give a false phylogeny in an uncritical analysis.

One may also be misled by functional or developmental correlations. That is, one can easily treat different aspects of a single adaptive complex as independent features. The features of bird-pollinated flowers might be treated as independent, to prove that *Alamania* and *Hexisea* or *Sophronitis* were closely related, even though they are not. When the relative positions of anther and stigma are considered, one might code the same feature independently for anther base, anther apex, stigma, and rostellum, though these are structurally connected.

Norris Williams and I ran a computer program with data sets in which I had tried already to cut out duplication. The results were lovely, with every major group delimited by several derived features. On careful inspection, though, it was clear that many features were structurally interrelated. When these dependent features were factored out, the resulting cladograms were little better than what had been done "by hand."

CONCLUSIONS

There is much common ground in systematics. Phylogenetic analysis has greatly improved the objectivity and explicitness of systematic biology, and few would argue with its virtues. If a group is shown to be clearly polyphyletic, most systematists would agree to its dismemberment. I suspect, too, that most would agree to the division of groups that are "positively paraphyletic" (Donoghue 1985). Groups that are merely unresolved, however, may be with us forever. We should strive for a classification that is consistent with phylogeny, as far as that is feasible, but I doubt that we can show all details of phylogeny in a usable system of classification. Hypotheses about phylogeny, however strong they may be, are communicated more effectively by diagrams and discussion than by incorporating them into a classification.

Many biologists are using phylogenetic analysis with common sense and a minimum of dogma. There are currently several methods and philosophies of phylogenetic analysis, and this diversity is desirable. I doubt that there is any magic formula, but a phylogenetic point of view is essential. Clearly polyphyletic groups should be divided, but the fact that one may divide a group into two clear subgroups does not necessarily mean that the group is polyphyletic. Recognition and some degree of communication are necessary elements in classification.

Ultimately, the argument over phylogeny versus similarity seems a bit silly, though it may help to clarify our ideas. In most cases we can only infer phylogeny through the study of similarity. Similarity in primitive features does not mean the same as similarity in derived features, and we should not be deceived by superficial similarity. Finally, we should consider Stebbins' clever parable (1969). The perfect classification, like absolute truth and beauty, may be an ideal to be sought, but not always an achievable goal.

NOMENCLATURE

Classification without names would be difficult, but nomenclature is distinct from classification. The International Code of Botanical Nomenclature is reviewed at each International Botanical Congress and has become relatively stable after years of use with periodic revision. The Code lists the categories of taxa that may be named but does not define them; it merely specifies their order (Table 12-1). For example, a genus may not include a tribe, and a variety may not include a subspecies.

A name must be published to be valid. The publication of a new taxon must cite a type specimen and where it is deposited, and must include a diagnosis in Latin. If a known taxon is to be renamed in any way, the publication must include a clear and direct reference to the original publication and description. When two or more names of the same rank are available for a single taxon, the first validly published name is correct. If, however, two names are spelled identically, the second usage is forever invalid (a later homonym). Some names are single words, but the names of species have two parts. *Cattleya labiata* is the name of a species; the first part, *Cattleya*, is also the name of the genus that includes that species. The specific epithet, *labiata*, by itself, is not the name of a plant but merely a Latin adjective.

For taxa described before the "type" method was developed, type specimens may have to be selected after the fact, preferably from material studied by their authors before the taxa were published. Now, the type specimen should be selected by the author, himself, and deposited in an herbarium where others may refer to it. The first plants to be recognized as new may not be representative of their populations. In other words, type specimens are not necessarily typical. They are simply name-bearing specimens, bench marks, or reference points that determine the application of their respective names. All available material should be studied before delimiting a species (or other taxon). Once it is delimited, the correct name is the earliest valid species name based on one of the included type specimens. If no type specimen falls within the limits of a species, the species has no valid name and must be considered as "new."

Table 12-1. Hierarchy of formal nomenclatural categories.

Category	Examples
Order	Liliales
Family	Orchidaceae, Liliaceae
Subfamily	Orchidoideae, Epidendroideae
Tribe	Orchideae, Vandeae
Subtribe	Orchidinae, Laeliinae
Genus (pl. genera)	*Orchis, Vanda, Epidendrum*
Species	*Orchis purpurea, Cattleya labiata*

Botanists try to keep nomenclature as stable as possible, but better information often leads to name changes. When we know more about the variation of populations, we may find that two names really refer to the same species; if the well-known name is the younger one, it becomes a synonym. If we find that a single name has been applied to several distinct species, some of the names we had thought to be synonyms may be resurrected and used. Validly published names are stable, in that they remain in the literature, but the delimitation or circumscription of taxa often changes with further study. Thus, neither classification nor nomenclature can be truly stable as long as we continue to learn more about plants.

Because priority gives preference to the older names but delimitation changes through time, names of higher taxa, such as "tribe Epidendreae," do not mean much without qualification. One really needs to specify something like "Epidendreae as used by Dressler 1981" (not quite the same as "Epidendreae as used by Dressler 1993"). Note, however, that the name is cited as "Epidendreae Lindley," never "Epidendreae Dressler."

Glossary

Terms that may be needed to follow the discussion in Chapter 12 are marked with asterisks (*). Some readers may wish to review these terms before reading that chapter. Terms followed by (c) are much used by the cladistic school of systematics.

Abaxial—With reference to leaves and other organs, the side away from the stem, normally the lower surface.

Abscission—The process by which leaves (or other organs) fall from a plant; a special layer of easily broken cells is formed at the base and the organ falls off.

Acetolysis—Treatment of pollen grains with acetic acid (or acetic anhydride) and concentrated sulfuric acid. The treatment is necessary for studying fossil pollen grains, and is commonly used for modern materials.

Acrocentric—Of chromosomes, having the centromere (the point that appears to be attached to the spindle in cell division) near or at the end of the chromosome.

Acrotonic—Having the rostellum or viscidium associated with the apex of the anther.

Adaxial—With reference to leaves and other organs, the side toward the stem, normally the upper surface.

Adventitious—Refers to organs that develop in unexpected places: stems or buds from roots, leaves, or between axils, or to roots from stems or leaves.

Allele—Refers to genes that occupy the same positions in homologous chromosomes and affect the same features but differ in structure and effect.

Allopatric—Refers to species or populations that occur in different areas (not overlapping).

Anagenesis—Change through time without phyletic branching (cladogenesis).

Aneuploid—The pattern of evolution in which chromosome numbers change by small steps (see polyploidy).

Anther—The part of the flower that produces pollen.

Anticlinal—In seed coat cells, the lateral cell walls that are in contact with each other, as contrasted with the inner and outer (periclinal) cell walls.

Apomorphic (c) —Refers to derived features, as opposed to primitive, or ancestral features. A derived feature may be termed an **apomorphy.**

Articulate—Jointed, said of leaves or other organs that have a clear abscission layer or joint, usually at the base.

Asparagoid—Resembling asparagus, referring to the complex of Liliiflorean families that are similar to the Asparagaceae, as contrasted with the colchicoid families.

Atectate—Of pollen wall structure, the exine without columellae and thus without a tectum.

Auricle—A small ear-like structure; in the orchids especially a lateral organ on the column; apparently part of the filament or anther in the Orchideae, though similar structures may occur on the staminodia of other groups.

Autapomorphic (c) —Refers to derived features that are each limited to a single monophyletic group.

Baculate—Covered with rods that are higher than wide and not constricted at their bases; used especially with reference to pollen grains.

Basifixed—With reference to anthers, attached at the base.

Basitonic—Having the rostellum or viscidium associated with the base of the anther.

Bract—A scale-, sheath- or leaflike structure; leaflike, but not a normal foliage leaf.

Bursicle—A sack-like membrane or covering over the viscidium of some orchids.

Caducous—Deciduous, or soon falling from the plant.

Calyculus—A small cup or circle of bractlike structures outside of the sepals.

Caudicle—A slender, mealy, or elastic extension of the pollinium, or a mealy portion at one end of the pollinium; produced within the anther.

Chloroplast—The intracellular body that contains chlorophyll; the site of photosynthesis.

Chromosome—One of the rodlike bodies that contain genetic material. Sexually reproducing organisms normally receive one set of chromosomes from each parent.

*****Clade**—A natural phyletic line or group of any size or rank; essentially a twig or a branch of the family tree (see **Grade**).

Cladogenesis—Phyletic branching, that is, the division of a single phyletic line into two or more, commonly associated with speciation.

*****Cladogram**—A diagram of a hypothesis of phylogeny showing the features on which it is based, especially a diagram derived from cladistic analysis (see Chapter 12).

Clavate—Club-shaped, thick above and tapering to the base.

Clinandrium—The anther bed; that portion of the column under, or surrounding, the anther.

Colchicoid—Referring to the Liliales in the narrow sense, as contrasted with the Asparagales.

Colpate—Referring to pollen grains, having one or more furrows in the exine, these reaching from pole to pole.

Columellate—Of pollen wall structure, the exine with rodlike structures that may branch or fuse distally to form a roof, or tectum.

Column—The central structure of an orchid flower, made up of the style united with the filaments of one to three anthers.

Column foot—A ventral extension at the base of the column; the lip is attached at its tip.

Conduplicate—Of leaflike organs, with a single median fold, each half being flat or curved.

Congeneric—Referring to species, members of the same genus.

Connective—The tissue that connects the two lobes of an anther.

Convergent evolution, or **Convergence**—Several meanings have been given to convergence: (1) A pattern in which members of different groups independently become similar to each other, usually reflecting adaptation to similar habitats or conditions. In effect, this is parallelism in a number of features, though the resemblances may be superficial. (2) Parallelism in unrelated or distantly related groups. (3) A pattern in which a specific feature has become more similar in two taxa, whether related or not.

Convolute—Refers to leaves that are rolled up during early development.

Corm—A thick stem, usually of several internodes, as in *Gladiolus* or *Bletia*, usually underground or near the surface.

Cortex—In monocotyledonous roots, the tissue beneath the epidermis (or exodermis) and outside of the vascular tissue.

Crustose—With reference to seeds, having a hard seed coat.

Cuniculus—A tubular nectary that is concealed within the "stem" (actually a floral tube above the ovary) of the flower.

Cytology—The science of cell structure, often referring especially to chromosomes or chromosome numbers.

*Delimit—To determine the limits of something.

Dendrogram—A diagram of tree-like form, representing the phylogeny of a group of organisms.

*Derived—Referring to any state or feature that has evolved from another; a relative term, like primitive. **Advanced** is used in much the same way.

Dicotyledon—Refers to the large class of flowering plants that normally has two seed leaves and net-veined leaves; contrasts with **monocotyledon.**

Dioecy—Having male and female flowers always on different plants; not found in the orchids.

Diploid—Having two sets of chromosomes in each nucleus, the normal condition in most animals and many plants.

Distichous—Having leaves or other parts in two opposite rows.

DNA—Desoxyribonucleic acid, the material in which the genetic code is written.

Duplicate—Refers to leaves that are once folded during early development.

Elaiophore—A gland producing oil.

Electrophoresis—The separation of particles or molecules in a gel or liquid by applying an electric field.

Endothecium—The inner layer of cells of the anther well, lining the anther cell.

Enzyme—A protein that promotes a chemical reaction, and remains unchanged after the reaction.

Epi-—A prefix meaning upon or above, as in epicuticular, upon the cuticle.

Epichile—The terminal part of a lip that is divided into two or three distinct parts.

*Evolution—Basically, long-term change in biological populations: includes (1) speciation, or cladogenesis, one species evolving into two or more distinct species; (2) divergence, two or more species or lines becoming more different from each other (whether giving rise to additional species or not); (3) progressive adaptation or change, a single species or line changing through time without reference to any other.

Exine—The outer wall of a pollen grain or spore.

Exodermis—The outermost cortical layer of the root; a specialized layer of cells just beneath the velamen.

Exophyletic—Refers to clearly delimited groups with several derived features; a term introduced here to contrast with paraphyletic or unresolved.

Filament—The slender stalk of an anther, part of the column in most orchids.

Floral tube—A narrow tube formed of united flower parts, often a nectary.

Foveolate—Marked with small, shallow pits (larger than in perforate), especially in reference to pollen grains.

Freniculum—See **Caudicle.**

Globose—Globe-shaped or spherical.

Globular—Approaching a spherical shape.

*Grade—A group of plants (or animals) that are similar in some features, but not necessarily a phyletic group; especially an unnatural, or polyphyletic, group; also applied to paraphyletic groups.

Hamulate—Having small, hooked processes, especially in reference to pollen grains.

Hamulus—A pollinium stalk developed from the apex of the rostellum.

Haploid—Having a single set of chromosomes in each nucleus, that is, half the usual number.

Hastate—Shaped like an arrowhead.

Herbaceous—Refers to plants with little woody tissue, especially plants whose aboveground shoots persist less than a year. **Soft herbaceous** is used here to refer to thin leaves that are convolute in development but not plicate.

Heterobrochate—With reference to pollen grains, having netlike (reticulate) sculpturing, but with the meshes of the network differing in size.

Hexaploid—Having six sets of chromosomes in each nucleus.

Hierarchic—In a graded or ranked series, with each rank subordinate to, or included in, the one above it.

Homology—This term has long been used in two senses. Recent usage stresses phyletic homology, referring to similar features of different taxa, the similarities reflecting inheritance from a common ancestor. Morphological homology refers to structural and developmental similarity such that the features *could* have been derived from a common ancestor.

Homoplasic (c)—Refers to similar features that have evolved independently in two or more phyletic groups; such features may be called parallelisms or homoplasies.

Hygroscopic—Readily absorbing and retaining moisture.

Hypochile—The basal part of a lip that is divided into two or three distinct parts.

Incumbent—Refers to anthers that bend forward during development.

Intercellular—Between adjacent cells.

Internode—The section of stem between two nodes or joints.

Introrse—Refers to anthers that open toward the center of the flower.

Isozyme—Any of two or more chemically distinguishable enzymes that have the same function (and are produced by allelic genes).

Kernel seed—See *Körnchensame.*

Körnchensame—German term referring to orchid seeds without air space between the embryo and the seed coat. The seed coat may be hard, or crustose.

Laevigate—Smooth, especially referring to pollen grains.

Lenticular—Lens-shaped, equally convex on each side.

*Level—The relative position of a point or a group in a phylogeny.

Lithophytic—Growing on rocks.

Locule—A chamber or cavity, as in an anther or a fruit.

Lumen—The chamber within a cell wall.

*Lumping—A tendency to combine taxa into one (see splitting).

Lunate—Moonlike or crescent-shaped.

Massula (pl. **massulae**)—A packet of pollen in those orchids that have the pollinium subdivided into small packets; see **Sectile.**

Medial—In or near the middle portion of an object.

Mentum—A chinlike extension at the base of the flower, made up of the column foot and the lateral sepals.

Mesochile—The mid portion of a lip that is divided into three distinct portions.

Mesoperigenous—Having one or more subsidiary cells derived from the same parent cell as the guard cells and the other subsidiary cell or cells from a different parent cell or cells.

Mesotonic—See **Pleurotonic.**

Metacentric—In chromosomes, having the centromere (the point that appears to be attached to the spindle in cell division) near the middle of the chromosome.

Micro- —A prefix meaning small.

Microsporogenesis—The development of the microspores, or pollen grains.

Mimicry—A resemblance between different kinds of organisms that is favored by natural selection, as the resemblance is advantageous for one or both.

Mitochondria—Intracellular organelles that produce energy through respiration.

Mitra—In the Diurideae, a hood formed by the staminodia and the filament of the fertile stamen.

Monad—A single pollen grain, unattached to others.

Monandrous—Having a single anther in each flower.

Monocotyledon—Refers to the major class of flowering plants with a single seed leaf (none in most orchids) and usually parallel leaf venation.

*Monophyletic—Derived from a single ancestral species; a "natural" group, united by derived features and without discordant features, is thought to be monophyletic.

Monopodial—A growth habit in which the stem may continue to grow indefinitely at its apex; see Sympodial.

Monothetic—A group that may be characterized or recognized by a single feature.

Mycorrhiza—A symbiotic relationship between fungi and the roots of vascular plants.

Mycotrophic—Refers to vascular plants that obtain some or all of their nutrition from the substrate through mycorrhizal fungi; saprophytic.

Myrmecophyte—Any plant associated with ants.

Nectary—A nectar-producing structure or gland.

Node—Joint, the point on a stem where a leaf is attached.

Nodular—In reference to tuberoids, those that make up only a short section of the root.

Octoploid—Having eight sets of chromosomes in each nucleus.

Ontogeny—The development of an individual or organ.

Operculate—Lidlike or caplike.

Ornate—With reference to pollen grains, a netlike pattern with curved and discontinuous ridges.

Osmophore—A scent-producing gland.

Ovoid—Egg-shaped in three dimensions.

Ovule—An embryonic seed; develops into a seed after fertilization.

Papilla—A small bump or nipple-like structure.

Parallel evolution—The evolution of a similar feature in two or more groups, not necessarily guided by similar life-styles or habitats.

Parsimony—Thriftiness, in science commonly applied to the principle of accepting the simplest explanation or hypothesis that is consistent with the data.

Pedicel—The stem immediately beneath an individual flower.

Peloric—An abnormality with the lip like the other petals, or vice versa; a radially symmetrical mutant of a species that normally has bilaterally symmetrical flowers.

Perianth—A collective term for sepals and petals, together.

Perforate—With reference to pollen grains, the surface having small pits; with reference to pollinia, each one having a single deep pit.

Periclinal—Referring to seed coat cells, the cell walls that are tangential with respect to the seed; the inner and outer cell walls.

Petiole—The narrow, stemlike basal portion of a leaf.

Phenology—The study of seasonality, that is, the relation between growth, flowering, fruiting, etc. and the seasons.

Phylad—A phylogenetic line, essentially equal to clade.

Phyletic—Pertaining to phylogeny.

*Phylogenetic analysis—Any system of analysis designed to infer phylogeny; especially cladistic analysis (see Chapter 12).

*Phylogeny—History and development of a group through geological time; a scheme or diagram representing the phylogeny of a group.

Pilate—Covered with rods that are enlarged at their apices, especially referring to pollen grains.

Plesiomorphic (c)—Refers to features that are primitive, or ancestral; such features may be called plesiomorphies.

Pleurotonic—Having the middle of the anther or pollinia associated with the rostellum or viscidium.

Plicate—Refers to leaves with several or many major longitudinal veins, usually folded at each major vein; pleated.

Polarity—The ordination of different character states with respect to an ancestral condition.

Pollenkitt—A sticky liquid that occurs on all pollen grains, usually in minute quantities.

Pollinarium—The complete set of pollinia from an anther, with associated parts, viscidium, or viscidium and stipe; when there are two viscidia, each half of the set might be termed a pollinarium.

Pollinium (pl. **pollinia**)—A more or less compact and coherent mass of pollen; often the contents of an anther cell.

Polymorphic—A state in which a population may include two or more discontinuous variants, independent of gender. In some plants, for example, each population may include short-styled and long-styled forms.

*Polyphyletic—Refers to an "unnatural" group, that is, one that is derived from two or more ancestral species. This is most obvious when the ancestral species were members of different groups, but even when they were members of the same group but not sister species, the descendant group is, strictly speaking, polyphyletic. In practice, though, we may not be able to distinguish such a case from a monophyletic group.

Polyploidy—Duplication of whole sets of chromosomes.

Polyporate—Refers to pollen grains with several or many pores.

Polythetic—A group that must be characterized by a combination of features.

Porate—Having a pore, or rounded hole.

*Primitive—A relative term used in the sense of ancestral. A feature that is primitive for one group may be an advanced, or derived, feature in another group. A primitive feature is not necessarily a simple one.

Primordium—The first visible trace of a developing organ or structure.

Protandrous—Having the anther(s) fertile or functional before the stigma of the same flower.

Protologue—Everything associated with a name at its first publication. When a type specimen is lost or destroyed, one may be able to determine its identity from the diagnosis, description, or other evidence.

Pseudobulb—A thickened stem, usually aerial.

Pseudocopulation—A type of mimicry, in which flowers resemble female insects and are pollinated by the males when these attempt to mate with the flowers.

Pseudopollen—Refers to granular pollenlike tissue that may be gathered as food by insects.

Rachis—The axis of an inflorescence (above the lowermost flower).

*Rank—Relative position of categories; more inclusive categories are said to have higher rank. In the higher categories, especially, rank is very subjective, and a given rank may not be comparable between different groups.

Raphide—A needlelike crystal, usually of calcium oxalate, often forming bundles in orchid cells.

Recumbent—Refers to anthers that bend backward during development.

Relict (or **Relic**)—May refer to an organism that once had a wider distribution, or to remnants of once larger groups.

Resupinate—In orchids, having the lip on the lower side; the ovary or pedicel twists or bends during development to achieve this position.

Reticulate—Netlike, refers to veins that form a branched network, or to netlike sculpturing of a pollen grain.

Retinaculum—See **Viscidium.**

Retrorse—Projecting back or downward toward the base.

Rhizome—A horizontal stem usually on or in the substrate; in sympodial orchids, made up of the bases of successive shoots.

Ribosome—An intracellular granule that is the site of protein synthesis.

Rostellum—A portion of the median stigma lobe that aids in gluing the pollinia to the pollinator; the tissue that separates the anther from the fertile stigma.

Rugose—Wrinkled.

Rugulate—Surface covered with minute wrinkles, especially referring to pollen grains.

Saccate—Deeply concave, sacklike.

Saprophytic—Refers to any plant with little or no green coloring that depends on organic matter in the substrate; in vascular plants, always in conjunction with mycorrhizal fungi.

Sclerenchyma—Refers to hard or stony cells or tissue.

Scutellum—Refers to the cellular plate that forms part of most viscidia.

Sectile—The condition in which soft, granular pollinia are subdivided into small packets, these usually connected by elastic material.

*Sister groups—Refers to groups of any rank that share an immediate common ancestor.

Soft herbaceous—Refers to soft or thin leaves that are convolute in development, but not plicate.

*Speciation—The achievement of species status. If a subpopulation achieves the capacity to coexist with its parent population without losing its identity, speciation has occurred. Both populations now have species status.

*Splitting—Dividing taxa into smaller groups; a "splitter" is one who consistently recognizes very narrow groups (see also **Lumping**).

Staminode, staminodium (pl. **staminodia**)—A sterile stamen that produces no functional pollen.

Stegmata—Longitudinal series of silica-containing cells adjacent to vascular or nonvascular fibers.

Stigma—The sticky, receptive part of the pistil; pollen grains that reach the stigma germinate and fertilize the embryonic seeds.

Stipe—Any pollinium stalk not derived from the anther.

Stomate (pl. **stomata**)—The opening by which gas exchange occurs between the leaf and the atmosphere, the opening is flanked by a pair of guard cells that can open or close the stomate by changing shape.

Style—The slender part of the pistil, connecting the stigma to the ovary; forms part of the column in orchids.

Subsidiary cell—A cell that is associated with a stomate and is structurally distinct from other surrounding cells.

Subtend—To occur next to and beneath, as a bract always subtends a flower.

Sulcate—Longitudinally furrowed or grooved.

Superposed—One on top of the other; with respect to pollinia, flattened parallel to the clinandrium.

Symbiont—Said of any organism that lives cooperatively with an organism of a different species.

Sympatric—Refers to populations or species that occur in the same area.

Symplesiomorphic (c)—Refers to shared primitive features.

Sympodial—A habit of growth in which each shoot has limited growth, new shoots usually arising from the bases of older ones.

Synapomorphic (c)—Refers to shared derived features.

Synsepal—A compound organ formed by the union of two lateral sepals.

Systematics—The science of classification; the study of organismic diversity and phylogenetic relationships in populations and taxa.

Tabular—Tablelike, having a flat surface.

*Taxon (pl. **taxa**)—Originally intended to refer to any nameable category. Thus, each category from forma to kingdom would be a taxon. The term has also proven useful in discussing plants and plant populations. One may refer to a population or type of plant as a taxon, without specifying that it should be classified as a species, a subspecies, or some other category.

Taxonomy—The science of classifying and naming.

Tectate—Of pollen wall structure, the outer surface of the exine forming a more or less continuous "roof" supported by columellae.

Tegula—A pollen stalk derived from the surface tissues of the clinandrium or column.

Tenuate—With reference to pollen grains, having one or more thin, aperturelike areas, these not as clearly delimited as true apertures.

Tepals—Used for the perianth of lilylike plants in which the outer and inner parts ("sepals and petals") are similar.

Testa—Seed coat, a layer of cells or tissue from the maternal plant, surrounding an embryo; this becomes the outer covering of the seed.

Tetrad—A set of four pollen grains that do not separate before dispersal.

Tetraploid—Having four sets of chromosomes in each nucleus.

Theca (pl. **thecae**)—One of the four segments of the anther, a pollen sac.

Transformation series (c)—In phylogenetic analysis, a series of different states of a single character; ideally arranged in a series from ancestral to most derived.

Trichome—Hair, often microscopic.

Tuberoid—A tuberlike organ; often used for thickened roots (tubers being defined as thickened stems).

Type—In systematics, the reference specimen by which the identity of a taxon is determined. Now selected by the author of new taxa; with reference to early descriptions, the type should be selected from material that was studied by the original author.

Vandoid—Resembling *Vanda*; refers to the orchids with viscidium, stipe and two or four hard pollinia.

Vascular—Pertaining to vessels or water-conducting tissue; those plants with water-conducting tissues.

Velamen—One or more layers of spongy cells on the outside of a root; in origin related to the epidermis.

Verrucose—Warty.

Vibration pollination—A system in which the pollen does not flow freely from the (often tubular) anther, but bees of the right size free the pollen by vibrating on the flower (buzz pollination).

APPENDIX A

Keys to Major Orchid Groups

Because there is so much parallelism in the Orchidaceae, writing a simple key on a world-wide basis is especially difficult. "Natural" keys, such as Schlechter (1926), may be useful as a summary or outline of the classification, but they have serious shortcomings as tools for identification. There is the danger, of course, that the classification is made to fit the key rather than the reverse. For any given region, it is much more practical to write a quite artificial key in which some variable groups may key out under several different alternatives. The present key makes no pretense of being "natural," and it certainly will not serve to identify all plants of all groups.

Key I. Subfamilies

1.   Lateral anthers fertile; filaments only partially united with column . . . 2
1.   Lateral anthers normally sterile or lacking; filaments usually united with column up to apex . . . 3
2(1).   Lip deeply saccate; usually with a shieldlike median staminodium. . . . . Cypripedioideae
2.   Lip similar to petals; median stamen fertile, fingerlike or lacking . . . . Apostasioideae
3(1).   Anther incumbent, caplike at apex of anther; pollinia often hard; plants may have pseudobulbs . . . . Epidendroideae, key IV
3.   Anther erect or nearly so; pollinia soft (rarely brittle); without pseudobulbs . . . 4
4(3).   Leaves conduplicate. . . . . Epidendroideae, key IV
4.   Leaves plicate or soft herbaceous . . . 5
5(4).   Leaves soft-herbaceous, not or weakly plicate . . . 6
5.   Leaves distinctly plicate . . . 10

6(5).   Pollinia sectile *and* viscidia at base of anther; base of anther fused with column
          apex . . . . Orchidoideae, key III
6.   Pollinia mealy *or* viscidia near apex of anther; base of anther attached to a narrow
          filament . . . 7
7(6).   Anther with blunt, fleshy beak . . . . Epidendroideae, key IV
7.   Anther without a prominent beak, usually acute . . . 8
8(7).   Plants with globose root–stem tuberoids (roots dimorphic) . . . .
          . . . . Orchidoideae, key III
8.   Plants without root–stem tuberoids; roots thin or fleshy, but not dimorphic . . . 9
9(8).   Viscidia at or near apex of anther; anther and rostellum more or less tapering . . .
          . . . . Spiranthoideae, key II
9.   Viscidium ventral or basal with respect to anther . . . . . Orchidoideae, key III
10(5).   Pollinia sectile, with apical viscidium . . . . Spiranthoideae, key II
10.   Pollinia mealy or soft; viscidium basal or ventral . . . 11
11(10).   Column with narrow, erect wings that surpass the anther (*Diceratostele*) . . . . .
          . . . . Spiranthoideae, key II
11.   Column wings small, not surpassing the anther (Neottieae, Triphoreae) . . . . .
          . . . . Epidendroideae, key IV

Key II. **Spiranthoideae**

1.   Leaves soft-herbaceous, leaves often in rosette . . . . Cranichideae . . . 2
1.   Leaves clearly plicate, often scattered on elongate stem . . . 8
2(1).   Pollinia sectile; plants usually without a distinct rhizome, the stems basally hori-
          zontal, but similar to the erect stems . . . . Goodyerinae
2.   Pollinia soft or brittle, not sectile; rhizomes short, underground . . . 3
3(2).   Column bent like a door hook (West Africa) . . . . Manniellinae
3.   Column straight or curved, but not sharply bent . . . 4
4(3).   Lip with a prominent external spur (New Caledonia) . . . . . . Pachyplectrinae
4.   Lip without a spur, or the spur partially united with the ovary . . . 5
5(4).   Pollinia clavate, brittle . . . . Cranichidinae
5.   Pollinia soft, mealy . . . 6
6(5).   Flowers resupinate, lip lowermost . . . . Spiranthinae
6.   Flowers nonresupinate, lip uppermost . . . 7
7(6).   Column very short (Australasia) . . . . (Diurideae) Cryptostylidinae
7.   Column moderately long (New World) . . . . Prescottiinae
8(1).   Pollinia sectile; viscidium terminal . . . . Tropidiinae
8.   Pollinia soft, without a viscidium or the viscidium subbasal (West Africa) . . .
          . . . . Diceratosteleae

Key III. **Orchidoideae**

1.   Pollinia sectile *and* viscidia basal; without stipes; base of anther fused with
          column apex . . . 2
1.   Pollinia mealy or sectile; with apical viscidia or hamular stipes if sectile; base of
          column usually narrowly attached to column . . . . Diurideae . . . 7
2(1).   Anther erect or nearly so . . . . Orchideae . . . 3
2.   Anther bent markedly backwards, sometimes 180 degrees . . . . . Diseae . . . 5
3(2).   Anther cells connivent apically, but markedly divergent basally, petals stalked
          and fringed (Africa) . . . . Huttonaeinae
3.   Anther cells usually separated by a prominent connective, not strongly diver-
          gent basally; petals not stalked . . . 4

4(3).	Stigma borne in a depression, surface flat or concave . . . . Orchidinae
4.	Stigma convex, commonly of two stalked lobes. . . . . Habenariinae
5(2).	Column long and slender; anther upside-down; lip with two spurs. . . Satyriinae
5.	Column short; anther not fully reversed; lip without spurs . . . 6
6(5).	Lip free from column or nearly so, spreading; dorsal sepal often spurred . . . . Disinae
6.	Lip erect, united with the face of the column, usually with a prominent appendage; lateral sepals often saccate . . . . Coryciinae
7(1).	Plants saprophytic, underground, the inflorescence scarcely reaching the surface; flowers clustered in a dense head surrounded by prominent bracts . . . . Rhizanthellinae
7.	Plants with green leaves or not wholly underground; inflorescence never capitate . . . 8
8(7).	Column very short; staminodia small and inconspicuous . . Cryptostylidinae
8.	Column much longer or staminodia large and conspicuous . . . 9
9(8).	Staminodia prominent, partially free from column; column relatively short . . . 10
9.	Staminodia prominent or not, united with the full length of column; column relatively long . . . 12
10(9).	Pollinia with hamular stipes, usually sectile; staminodia basally united with column and then spreading or recurved; leaves cylindrical . . . . Prasophyllinae
10.	Pollinia mealy or soft, without stipes; staminodia not spreading or recurved . . . 11
11(10).	Staminodia united with median filament to form a prominent hood about stigma, with a definite staminodial rim surrounding ventral base of style. . . . . Thelymitrinae
11.	Staminodia free nearly to bases, leaflike; without an obvious ventral staminodial rim . . . . Diurideae
12(9).	Leaves cordate to palmately lobed, with reticulate venation . . . . Acianthinae
12.	Leaves elongate, with parallel veins . . . 12
13(12).	Sepals connivent, forming a chamber around column; dorsal sepal hooded; lip motile, may have an appendage (trigger) near base; column wings retrorse; stigma near middle of column . . . . Pterostylidinae
13.	Sepals spreading, dorsal sepal not markedly hooded; stigma near anther . . . 14
14(13).	Lip stalked and hinged or motile, with a prominent insectlike callus. . . . . Drakaeinae
14.	Lip hinged or not, not markedly stalked; calli rarely insectlike . . . 15
15(14).	Plants usually with root–stem tuberoids (Australia) . . . . Caladeniinae
15.	Plants usually without root–stem tuberoids (New Caledonia and southern South America) . . . . Chloraeinae

Key IV. **Epidendroideae**

1.	Pollinia mealy or soft . . . 2
1.	Pollinia firm or hard . . . 17
2(1).	Anther suberect or slightly incumbent, may have a thick, fleshy beak . . . 3
2.	Anther distinctly incumbent, caplike on apex of column; anther beak thin . . . 5
3(2).	Leaves soft herbaceous, paired and subopposite (lacking in saprophytes); rostellum sensitive, ejecting a drop of glue when touched . . . . Neottieae—Listerinae
3.	Leaves plicate, subplicate or herbaceous, never paired and subopposite; rostellum not sensitive . . . 4

4(3).   Lip three-lobed, forming a slender tube with the lateral lobes enfolding the
           column; lip never stalked, spurred, or with a saccate base (New World) .
           . . . . Triphoreae
4.   Lip diverse, but not as in Triphoreae, usually stalked, spurred or with a saccate
           base . . . . Neottieae—Limodorinae
5(2).   Pollinia eight, clearly defined . . . 6
5.   Pollinia four or mealy and ill-defined . . . 8
6(5).   Plants with corms and usually lateral inflorescences . . . . Arethuseae, key VI
6.   Plants with slender stems and terminal inflorescences . . . 7
7(6).   Leaves conduplicate (Asia) . . . . Arundinae
7.   Leaves usually plicate (New World). . . . . Epidendreae—Sobraliinae
8(5).   Plants cormous . . . 9
8.   Plants without distinct corms . . . 10
9(8).   Leaves narrow, erect; pollinia mealy or faintly sectile . . . .
           . . . . Arethuseae—Arethusinae
9.   Leaves cordate to fan-shaped, blades horizontal; pollinia distinctly sectile . . .
           . . . . Nervilieae
10(8).   Leaves thin, plicate (New World) . . . . Palmorchideae
10.   Leaves fleshy, leathery or lacking . . . 11
11(10).   Pollinia usually sectile; always small saprophytes. . . . . Gastrodieae . . . 12
11.   Pollen mealy or pastelike, scarcely forming pollinia . . . . Vanilleae . . . 14
12(11).   Roots fleshy, spindle-shaped; plants with Y-shaped hairs (New World) . . . . .
           . . . . Wullschlaegeliinae
12.   Roots slender or fleshy, but not tapering at each end; without Y-shaped hairs . . .
           13
13(12).   Pollinia with caudicles . . . . Epipogiinae
13.   Pollinia without caudicles . . . . Gastrodiinae
14(11).   Plants shrubs or vines; lip united to column by margins; lip with a cluster of
           scales . . . . Vanillinae
14.   Plants herbs or saprophytes (may be saprophytic vines); lip free from column,
           without cluster of scales . . . 15
15(14).   Massive saprophytes; seeds winged (Asia) . . . . Galeolinae
15.   Small saprophytes or with green leaves; with dust seeds . . . 16
16(15).   Delicate saprophytes with a cupule (calyculus) around base of perianth (Asia) . .
           . . . . Lecanorchidinae
16.   Saprophytic or autophytic; without a calyculus . . . . Pogoniinae
17(1).   Pollinia naked, without caudicles (may have viscidia or rarely stipes) . . . 18
17.   Pollinia with definite caudicles, often with viscidia or stipes . . . 20
18(17).   Leaves condupicate; inflorescence lateral or upper lateral, rarely terminal;
           pollinia oblong. . . . . Dendrobieae . . . 19
18.   Leaves diverse; inflorescence usually terminal; pollinia usually clavate or
           oblique . . . . Malaxideae
19(18).   Stems/pseudobulbs of several to many internodes; inflorescence upper lateral
           or rarely terminal . . . . Dendrobiinae
19.   Pseudobulbs of one internode; inflorescence basal . . . . Bulbophyllinae
20(17).   Pollinia with viscidia and usually stipe, 2–4, superposed if 4 . . . 21
20.   Pollinia without stipe, 2–8, usually laterally flattened (except Coelogyneae) . . .
           25
21(20).   Plants monopodial, each stem with indefinite growth; never forming pseudo-
           bulbs . . . 22
21.   Plants sympodial, each stem with limited growth, with or without pseudo-
           bulbs . . . 24
22(21).   Plants of Old World . . . . Vandeae, key X
22.   Plants of New World . . . 23

23(22). Flowers with definite sepaline spurs...................... Vandeae, key X
23.   Flowers without sepaline spurs....................... Maxillarieae, key V
24(21). Leaves conduplicate; inflorescence terminal .. Epidendreae—Polystachyinae
24.   Leaves plicate *or* inflorescence lateral........... Cymbidioid phylad, key V
25(20). Inflorescence lateral ... 26
25.   Inflorescence terminal ... 27
26(25). Inflorescence usually basal; leaves plicate; plants usually with corms ....
         .... Arethuseae, key VI
26.   Inflorescence usually terminal or upper lateral; rarely cormous; leaves usually
         conduplicate ... 28
27(25). Pseudobulbs of one internode; column rather petaloid; Asia..............
         .... Coelogyneae—Coelogyninae
27.   Stems elongate or usually of several internodes ... 29
28(26). Plants strictly New World.............. New World Epidendreae, key VII
28.   Plants of Old World (except some *Polystachya*) ... 29
29(28). Pollinia four.......................... Old World Epidendreae, key VIII
29.   Pollinia eight ... 30
30(29). Pseudobulbs elongate, thickened, with whitish deciduous leaves; flowers large,
         never densely capitate; column markedly winged above..............
         .... Coelogyneae—Thuniinae
30.   Stems usually slender; leaves not whitish nor deciduous; flowers small, com-
         monly densely clustered; column not strongly winged above ....
         .... Epidendreae (*Agrostophyllum*)

Key V. Cymbidioid Phylad
(Including *Claderia* and Collabiinae—for purposes of identification.)

1.   Pollinia without caudicles or stipes, may have tiny viscidia ..... Malaxideae
1.   Pollinia with caudicles and prominent viscidium, usually with stipes ... 2
2(1).   Plants of the Old World ... 3
2.   Plants of the New World ... 11
3(2).   Column foot hollow; pollinia with viscidium but no stipe .... Thecostelinae
3.   Column foot solid; pollinia usually with stipes ... 4
4(3).   Pollinia 4 ... 5
4.   Pollinia 2 ... 6
5(4).   Leaves thin, clearly plicate (lacking in saprophytes), may be wide and clearly
         stalked... Calypsoeae
5.   Leaves conduplicate or weakly plicate, never wide and clearly stalked ......
         .... Cyrtopodiinae
6(4).   Pollinia with a prominent viscidium but without stipe or prominent cellular cap
         (scutellum) on viscidium ... 7
6.   Pollinia usually with stipe or at least with clear scutellum ... 8
7(6).   Pollinia angular, not grooved............................ Collabiinae
7.   Pollinia oblong, grooved or partially split .... *Claderia*
8(6).   Stems slender; inflorescence terminal; stipe formed behind rostellar flaps, con-
         tinuous with viscidium .... Bromheadiinae
8.   Stem usually somewhat thickened *or* inflorescence lateral; stipe formed from
         surface of clinandrium, usually sharply distinguished from viscidium ... 9
9(8).   Pollinia narrow, thin, and bladelike; lip strongly united with column, forming a
         tube... Acriopsidinae
9.   Pollinia broad, not thin and bladelike; lip not forming a prominent tube with
         column ... 10

10(9).   Pollinia rounded, perforate............................... Eulophiinae
10.   Pollinia angular, long-grooved.......................... Cyrtopodiinae
11(2).   Pollinia two... 12
11.   Pollinia four... 18
12(11).   Pseudobulbs warty; leaves thick and leathery, not distinctly conduplicate ...
            .... Maxillarieae ?—*Eriopsis*
12.   Pseudobulbs smooth; leaves plicate or conduplicate... 13
13(12).   Plants with elongate pseudobulbs and several plicate leaves scattered along
            pseudobulb... 14
13.   Pseudobulbs short; leaves terminal or nearly so, plicate or conduplicate... 15
14(13).   Rostellum sensitive, moving at least slightly when touched; flowers often
            unisexual.. Catasetinae
14.   Rostellum never sensitive; flowers bisexual .... Cyrtopodiinae
15(13).   Inflorescence terminal .... (*Galeandra*) .... Cyrtopodiinae
15.   Inflorescence lateral... 16
16(15).   Plants cormous (pseudobulbs of one internode in *Oeceoclades*).. Eulophiinae
16.   Plants with definite pseudobulbs, usually of one internode... 17
17(16).   Leaves plicate, often stalked............................. Stanhopeinae
17.   Leaves conduplicate, not stalked .... Oncidiinae
18(11).   cormous plants (corms may be elongate or branched) with thin plicate leaves,
            the leaves wide and with a prominent tubular petiole; anther commonly
            with a narrow clawlike beak............................. Goveniinae
18.   Leaves various, but never with tubular petiole... 19
19(18).   Terrestrial, cormous or saprophytic; leaves (when present) 1–2, broad and
            distinctly petiolate................................... Calypsoeae
19.   Epiphytic, pseudobulbous, many-leaved, *or* leaves without distinct petioles..
            .... Maxillarieae... 20
20(19).   Plants small, with conduplicate leaves; pollinia without stipes but with promi-
            nent, translucent caudicles; lip four-lobed or anchor-shaped ....
            .... Cryptarrheninae
20.   Plants usually larger; with distinct stipes *or* without large, translucent
            caudicles... 21
21(20).   Leaves plicate... 22
21.   Leaves conduplicate... 24
22(21).   Stipes very long, more than twice width, with prominent viscidia.... Lycastinae
22.   Stipe usually short; viscidium may be similar to stipe... 23
23(22).   Callus usually prominent and longitudinally ridged or keeled; with or without
            pseudobulbs, may have pseudobulbs of several internodes.. Zygopetalinae
23.   Callus usually low, smooth or without keels; always with pseudobulbs of single
            internodes .... Lycastinae
24(21).   Each flower on a separate peduncle (one-flowered inflorescences) ....
            .... Zygopetalinae
24.   Each peduncle commonly with two to many flowers... 25
25(24).   Callus usually prominent and longitudinally ridged or keeled... Zygopetalinae
25.   Callus usually low, smooth, or without keels... 26
26(25).   Each flower on a separate peduncle (one-flowered inflorescence) ....
            .... Maxillariinae
26.   Each peduncle commonly with two to many flowers... 27
27(26).   Column usually bristly; viscidium hooked................. Telipogoninae
27.   Column not bristly; viscidium rarely hooked; stigma usually at base of column;
            anther often long-beaked .... Ornithocephalinae

Key VI. **Arethuseae**

1. Pendant epiphyte; pseudobulbs club-shaped; flowers fleshy; pollinia flattened, but not all in same plane . . . . Chysiinae
1. Terrestrial or epiphytic, usually with short corms; flowers not very fleshy; if pollinia flattened, then all in same plane . . . 2
2(1). Pollinia four, soft and mealy; column somewhat petaloid . . . . . . Arethusinae
2. Pollinia usually eight, clearly defined; column not markedly petaloid. . Bletiinae

Key VII. New World **Epidendreae**

1. Pollinia soft; leaves usually plicate . . . . Sobraliinae
1. Pollinia firm; leaves conduplicate . . . 2
2(1). Ovary jointed to pedicel, the pedicels persisting on rachis after flowers have fallen; without pseudobulbs, usually one-leaved . . . . . . Pleurothallidinae
2. Ovary not jointed to pedicel, pedicel falling with flower . . . 3
3(2). Pollinia eight, ovoid, not flattened . . . 4
3. Pollinia clavate or laterally flattened . . . 5
4(3). Inflorescence terminal; stem narrow, with one fleshy leaf . . . . Arpophyllinae
4. Inflorescence lateral; stem globose, with several terminal leaves. . . Coeliinae
5(3). Pollinia eight, clavate; each stem with one fleshy leaf . . . . Meiracylliinae
5. Pollinia two to six *or* laterally flattened . . . . Laeliinae

Key VIII. Old World **Epidendreae**

1. Pollinia with small but definite stipes . . . . Polystachyinae
1. Pollinia without stipes . . . 2
2(1). Pseudobulbs short, inconspicuous (Ceylon and southern India). . Adrorhizinae
2. Stems slender, or pseudobulbs elongate . . . . Glomerinae

Key IX. **Podochileae**

1. Pollinia eight, not strongly clavate or with long caudicles. . . . . Eriinae
1. Pollinia four to eight, clavate or with long caudicles . . . 2
2(1). Pollinia clavate; stem slender, with distichous leaves . . . . Podochilinae
2. Pollinia ovoid, with long caudicles; stems usually thickened or short . . . .
. . . . Thelasiinae

Key X. **Vandeae**

1. Plants Asiatic (except one species each of *Acampe, Taeniophyllum*), usually without a spur . . . . Aeridinae
1. Plants of Africa, Madagascar, or tropical America (one in Ceylon); flowers usually whitish or green with a prominent spur . . . 2
2(1). Rostellum beaklike. . . . . Aerangidinae
2. Rostellum deeply notched . . . . Angraecinae

APPENDIX B

Outline of Classification with Lists of Genera

In this appendix, I list most currently recognized orchid genera, arranged according to the classification used here, with the original authors of the names and approximate numbers of species. I also include some synonyms and invalid names that have been in the recent literature, with indication of the correct names where possible. I hope this list will be useful, but I must emphasize that it is only a working list with estimates of the numbers of species. All data listed, and especially the numbers of species, will change as we learn more.

To estimate the numbers of species I have used various sources; especially important are a data base of western hemisphere orchid names by Dr. C. H. Dodson; Clements (1989) for Australian groups; Seidenfaden's many contributions on Asiatic orchids; and Senghas' works on the Vandeae. I have better data on western hemisphere groups than on those of Asia, but, even here, groups in urgent need of revision, such as the Oncidiinae or the Spiranthinae, are difficult to evaluate.

I have not followed the customary abbreviations of authors' names, which are incomprehensible to the uninitiated. When a name originated with one botanist but was published by another, I list only the publishing author, to avoid lengthy combinations such as "Reichenbach in Cransifarb & Glockenspiel."

The synonyms and names that are invalid or doubtfully distinct are given in *Italics*. Such names may be followed by an equal sign (=) and the correct name for the genus in question. They may also be followed by ∿, meaning "similar to or the same as" or ≃, meaning "very similar to." The last two symbols are also used to indicate currently used generic names that are doubtfully distinct from other genera.

The names of orchid subfamilies, tribes, and subtribes are reviewed by Butzin (1971). Such names that were published since Butzin's paper are referenced to the literature cited.

Apostasioideae Reichenbach

Adactylus Rolfe = Apostasia
Apostasia Blume, 8

Neuwiedia Blume, 8

Cypripedioideae Lindley

Cypripedium Linnaeus, 40
Paphiopedilum Pfitzer, 60
Phragmipedium Rolfe, 16

Selenipedium Rolfe, 6
Criosanthes Rafinesque = Cypripedium

Spiranthoideae Dressler (1979)

Diceratosteleae Dressler (1990b)
  Diceratostele Summerhayes, 1

Tropidieae Dressler (1983b)
  Corymborkis Thouars, 8
  Muluorchis Wood = Tropidia
  Tropidia Lindley, 35

Cranichideae Endlicher
     GOODYERINAE KLOTZSCH
  Anoectochilus Blume, 35
  Aspidogyne Garay, 30
  Chamaegastrodia Makino & Maekawa, 1
  Cheirostylis Blume, 25
  Cystorchis Blume, 8
  Dicerostylis Blume, 3
  Dossinia E. Morren, 1
  Erythrodes Blume, 60
  Eucosia Blume, 2
  Eurycentrum Schlechter, 7
  Evrardia Gagnepain, 1
  Gonatostylis Schlechter, 1
  Goodyera R. Brown, 55
  Gymnochilus Blume, 3
  Haemaria Linnaeus = Ludisia
  Herpysma Lindley, 1
  Hetaeria Blume, 27
  Hylophila Lindley, 6
  Kreodanthus Garay, 6
  Kuhlhasseltia J. J. Smith, 6
  Lepidogyne Blume, 3
  Ligeophila Garay, 8
  Ludisia A. Richard, 1
  Macodes Lindley, 14
  Moerenhoutia Blume, 10
  Myrmechis Blume, 6
  Orchipedum Breda, 1
  Papuaea Schlechter, 1
  Platylepis Lindley, 10
  Platythelys Garay, 8
  Pristiglottis Cretzoiu & Smith, 13
  Rhamphorhynchus Garay, 1

Stephanothelys Garay, 4
Tubilabium J. J. Smith, 2
Vrydagzynea Blume, 40
Zeuxine Lindley, 76
    PRESCOTTIINAE DRESSLER (1990b)
Aa Reichenbach, 25
Altensteinia Kunth, 9
Gomphichis Lindley, 23
Myrosmodes Reichenbach, 9
Porphyrostachys Reichenbach, 2
Prescottia Hooker, 21
Stenoptera Presl, 10
    SPIRANTHINAE LINDLEY
Aracamunia Carnevali, 1
Aulosepalum Garay, 4
Beadlea Small = Cyclopogon
Beloglottis Schlechter, 7
Brachystele Schlechter, 18
Buchtienia Schlechter, 3
Centrogenium Schlechter = Eltroplectris
Coccineorchis Schlechter, 4
  (~ Stenorrhynchos)
Cogniauxiocharis Hoehne = Pteroglossa
Cotylolabium Garay, 1
Cybebus Garay, 1
Cyclopogon Presl, 55
Degranvillea Determann, 1
Deiregyne Schlechter, 7
Dichromanthus Garay, 1
Discyphus Schlechter, 1
Dithyridanthus Garay, 1
Eltroplectris Rafinesque, 12
Eurystyles Wawra, 10
Funkiella Schlechter, 8
Galeottiella Schlechter, 5
Gamosepalum Schlechter = Aulosepalum
Greenwoodia Burns-Balogh, 1
Gularia Garay = Schiedeella
Hapalorchis Schlechter, 9
Helonema Garay, 2
Kionophyton Garay, 3

Lankesterella Ames, 10
Lyroglossa Schlechter, 3
Mesadenella Pabst & Garay, 7
Mesadenus Schlechter, 7
Microthelys Garay = Galeottiella
Odontorrhynchos Garay, 5
Oestlundorchis Szlachetko, 11
Pelexia L. C. Richard, 72
Physogyne Garay = Schiedeella
Pseudoeurystyles Hoehne = Eurystyles ?
Pseudogoodyera Schlechter, 2
Pteroglossa Schlechter, 8
Sacoila Garay, 10
Sarcoglottis Presl, 40
Sauroglossum Lindley, 9
Schiedeella Schlechter, 10
Skeptrostachys Garay, 13
Spiranthes L. C. Richard, 30
Stalkya Garay, 1
Stenorrhynchos L. C. Richard, 13

Stigmatosema Garay, 2
Svenkoeltzia Burns-Balogh = Funkiella
Synanthes Burns-Balogh, Robinson & Foster = Eurystyles
Synassa Lindley = Sauroglossum
Thelyschista Garay, 1
MANNIELLINAE SCHLECHTER
Manniella Reichenbach, 1
PACHYPLECTRONINAE SCHLECHTER
Pachyplectron Schlechter, 2
CRANICHIDINAE LINDLEY
Baskervilla Kunth, 7
Cranichis Swartz, 60
Fuertesiella Schlechter, 1
Nothostele Garay, 1
Ponthieva R. Brown, 53
Pseudocentrum Lindley, 6
Pseudocranichis Garay, 1
Pterichis Lindley, 20
Solenocentrum Schlechter 3

Orchidoideae

Diurideae Endlicher
CHLORAEINAE REICHENBACH
Asarca Lindley = Gavilea
Bipinnula Jussieu, 7
Chloraea Lindley, 50
Codonorchis Lindley, 3
Gavilea Poeppig, 11
Geoblasta Barbosa Rodrigues, 1
Megastylis Schlechter, 8
CALADENIINAE PFITZER
Adenochilus Hooker, 2
Aporostylis Rupp & Hatch, 1
Burnettia Lindley, 1
Caladenia R. Brown, 100
Elythranthera A. S. George, 2
Eriochilus R. Brown, 6
Glossodia R. Brown, 2
Leporella A. S. George, 1
Leptoceros Fitzgerald = Leporella
Lyperanthus R. Brown, 5
Petalochilus R. Rogers = Caladenia
Rimacola Rupp, 1
DRAKAEINAE SCHLECHTER
Arthrochilus F. Mueller, 4
Caleana R. Brown, 3
Chiloglottis R. Brown, 9
Drakaea Lindley, 4
Paracaleana Blaxell = Caleana
Spiculaea Lindley, 1

PTEROSTYLIDINAE PFITZER
Pterostylis R. Brown, 100
ACIANTHINAE SCHLECHTER
Acianthus R. Brown, 27
Corybas Salisbury, 110
Cyrtostylis R. Brown, 4
Stigmatodactylus Makino, 4
Townsonia Cheeseman, 2
CRYPTOSTYLIDINAE SCHLECHTER
Coilochilus Schlechter, 1
Cryptostylis R. Brown, 15
Diuridinae Lindley
Epiblema R. Brown, 1
Diuris Smith, 38
Orthoceras R. Brown, 2
THELYMITRINAE LINDLEY
Calochilus R. Brown, 10
Thelymitra Förster, 65
RHIZANTHELLINAE ROGERS
Cryptanthemis Rupp = Rhizanthella
Rhizanthella Rogers, 2
PRASOPHYLLINAE SCHLECHTER
Genoplesium R. Brown, 20
Goadbyella Rogers = Microtis
Microtis R. Brown, 14
Prasophyllum R. Brown, 65

[Orchidoideae continued]

Orchideae
ORCHIDINAE

Aceras R. Brown, 1
Aceratorchis Schlechter, 2 = ?
Amerorchis Hultén, 1
Amitostigma Schlechter, 10
Anacamptis L. C. Richard, 15
Aorchis Vermeulen, 1
Barlia Parlatore, 2
Bartholina R. Brown, 3 (∿ Holothrix)
Blephariglottis Rafinesque = Platanthera
Brachycorythis Lindley, 25
Chamorchis Richard, 1
Chondradenia Maekawa, 2
Chusua Nevski, 17
Coeloglossum Hartman, 1
Comperia C. Koch, 2
Dactylorchis Vermeulen = Dactylorhiza
Dactylorhiza Necker, 30
Fimbriella author? = Platanthera
Galearis Rafinesque, 12
Galeorchis Rydberg = Galearis
Gymnadenia R. Brown, 10
Gymnadeniopsis Rydberg = Platanthera
Hemipilia Lindley, 16
Himantoglossum Koch, 2
Holothrix Lindley, 55
Leucorchis E. Meyer = Pseudorchis
Lysiella Rydberg = Platanthera
Neobolusia Schlechter, 4 (= Brachycorythis)
Neotinea Reichenbach, 2
Neottianthe Schlechter, 7
Nigritella L. C. Richard, 2
Ophrys Linnaeus, 25
Orchis, 33
Piperia Rydberg, 4
Platanthera L. C. Richard, 40
Ponerorchis Reichenbach = Gymnadenia
Pseudodiphryllum Nevski, 1
Pseudorchis Séguier, 3
Schwartzkopffia Kränzlin = Brachycorythis
Schizochilus Sonder, 26
Serapias Linnaeus, 13
Steveniella Schlechter, 1
Symphyosepalum Handel-Mazzetti, 1
Traunsteinera Reichenbach, 1
Tulotis Rafinesque = Platanthera
Vermeulenia Löve & Löve = Orchis
HABENARIINAE BENTHAM
Androcorys Schlechter, 4

Arnottia A. Richard, 2
Benthamia A. Richard, 26
Bonatea Willdenow, 20 ∿ Habenaria
Centrostigma Schlechter, 5
Cynorkis Thouars, 125
Diphylax Hooker, 1
Diplomeris D. Don, 2
Gennaria Parlatore, 1
Habenaria Willdenow, 600
Herminium R. Brown, 30
Kryptostoma Geerinck = Habenaria
Megalorchis H. Perrier, 1
Oligophyton Linder & Williamson, 1
Parhabenaria Gagnepain = ?
Pecteilis Rafinesque, 4
Peristylus Blume, 70
Physoceras Schlechter, 7
Platycoryne Reichenbach, 17 ∿ Habenaria
Porolabium Tang & Wang = ?
Roeperocharis Reichenbach, 5
Smithorchis Tang & Wang, 1
Stenoglottis Lindley, 4
Thulinia Cribb, 1
Tsaiorchis Tang & Wang, 2
Tylostigma Schlechter, 3

Diseae Dressler (1979)
HUTTONAEINAE SCHLECHTER
Huttonaea Harvey, 5
SATYRIINAE SCHLECHTER
Pachites Lindley, 2
Satyridium Lindley, 1
Satyrium Swartz, 100
CORYCIINAE BENTHAM
Anochilus Rolfe = Pterygodium
Ceratandra Lindley, 2
Corycium Swartz, 14
Disperis Swartz, 75
Evota Rolfe = Ceratandra
Evotella Kurzweil, Linder & Chesselet, 1
Pterygodium Swartz, 15
DISINAE BENTHAM
Amphigena Rolfe = Disa
Brownleea Lindley, 6
Disa Bergius, 99
Forficaria Lindley = Herschelia
Herschelia Lindley, 16
Monadenia Lindley, 16
Orthopenthea Rolfe = Disa
Penthea Lindley = Disa
Schizodium Lindley, 6

Epidendroideae Lindley

Neottieae Lindley
    LIMODORINAE BENTHAM
  Aphyllorchis Blume, 15
  Cephalanthera L. C. Richard, 14
  Eburophyton A. A. Heller = Cephalanthera
  Epipactis Swartz, 21
  Limodorum Linnaeus, 1
  Sinorchis S. C. Chen = Cephalanthera
  Tangtsinia S. C. Chen = Cephalanthera
    LISTERINAE LINDLEY
  Archineottia S. C. Chen = Neottia
  Diplandrorchis S. C. Chen = Neottia
  Holopogon Komarov & Nevski = Neottia
  Listera R. Brown, 20
  Neottia Linnaeus, 9

Palmorchideae Dressler (1990c)
  Palmorchis Barbosa Rodrigues, 12

Triphoreae Dressler (1979)
  Monophyllorchis Schlechter, 2
  Psilochilus Barbosa Rodrigues, 7
  Triphora Nuttall, 19
  Vanilleae Blume
    GALEOLINAE GARAY
  Cyrtosia Blume, 5
  Erythrorchis Blume, 4
  Galeola Loureiro, 10

Pseudovanilla Garay, 8
    VANILLINAE LINDLEY
  Clematepistephium Hallé, 1
  Dictyophyllaria Garay,
  Epistephium Humbert, 14
  Eriaxis Reichenbach, 3
  Vanilla Swartz, 100
    LECANORCHIDINAE DRESSLER (1979)
  Lecanorchis Blume, 20

Gastrodieae Lindley
    GASTRODIINAE LINDLEY
  Auxopus Schlechter, 2
  Didymoplexiella Garay, 6
  Didymoplexis Griffith, 10
  Gastrodia R. Brown, 16
  Neoclemensia Carr, 1
  Uleiorchis Hoehne, 1
    EPIPOGIINAE SCHLECHTER
  Epipogium R. Brown, 3
  Silvorchis J. J. Smith, 1
  Stereosandra Blume, 1
    WULLSCHLAEGELIINAE DRESSLER (1983)
  Wullschlaegelia Reichenbach, 2

Nervilieae Dressler (1990c)
  Nervilia Gaudichaud, 65

Cymbidioid Phylad

Malaxideae Lindley
  Hammarbya Kuntze = Malaxis
  Hippeophyllum Schlechter, 6
  Liparis L. C. Richard, 350
  Malaxis Swartz, 300
  Microstylis Eaton = Malaxis
  Oberonia Lindley, 300
  Orestias Ridley, 3
  Risleya King & Pantling, 1

Calypsoeae Dressler (1979)
  Aplectrum Nuttall, 1
  Calypso Salisbury, 1
  Changnienia Chien = ?
  Corallorhiza R. Brown, 15
  Cremastra Lindley, 2
  Dactylostalix Reichenbach, 1
  Diplolabellum Maekawa = ?
  Didiceia King & Pantling = ?
  Ephippianthus Reichenbach, 1
  Hakoneasta Maekawa = ?
  Kitigorchis Mori = Oreorchis

Oreorchis Lindley, 9
Tipularia Nuttall, 3
Yoania Maximowicz, 2

Cymbidieae Pfitzer
    GOVENIINAE DRESSLER (1990c)
  Govenia Loddiges, 20
    BROMHEADIINAE DRESSLER (1990c)
  Bromheadia Lindley, 12
    EULOPHIINAE BENTHAM
  Cyanaeorchis Thouars, 2
  Dipodium R. Brown, 20
  Eulophia R. Brown, 200
  Eulophidium Pfitzer = Oeceoclades
  Geodorum Jackson, 8
  Lissochilus R. Brown = Eulophia
  Oeceoclades Lindley, 31
  Pteroglossaspis Reichenbach, 3
    THECOSTELINAE SCHLECHTER
  Thecopus Seidenfaden, 2
  Thecostele Reichenbach, 2

[Cymbidioid continued]

CYRTOPODIINAE BENTHAM

Acrolophia Pfitzer, 9
Ansellia Lindley, 2
Cymbidiella Rolfe, 3
Cymbidium Swartz, 45
Cyrtopodium R. Brown, 30
Eulophiella Rolfe, 2
Galeandra Lindley, 25
Grammangis Reichenbach, 2
Grammatophyllum Blume, 12
Graphorkis Thouars, 5
Grobya Lindley, 3
Porphyroglottis Ridley, 1

ACRIOPSIDINAE DRESSLER (1979)

Acriopsis Blume, 6

CATASETINAE SCHLECHTER

Catasetum Kunth, 100
Clowesia Lindley, 6
Cycnoches Lindley, 23
Dressleria Dodson, 5
Mormodes Lindley, 60

Maxillarieae Pfitzer

CRYPTARRHENINAE DRESSLER (1971)

Cryptarrhena Lindley, 7

ZYGOPETALINAE SCHLECHTER

Acacallis Lindley = Aganisia
Aganisia Sprengel, 3
Batemania Lindley, 45
Benzingia Dodson, 2
Bollea Reichenbach, 10
Chaubardia Reichenbach, 2
Chaubardiella Garay, 6
Cheiradenia Lindley, 1
Chondrorhyncha Lindley, 24
Cochleanthes Rafinesque, 15
Colax Lindley = Pabstia
Dichaea Schlechter, 55
Dodsonia Ackerman, 2
Galeottia A. Richard, 11
Hoehneella Ruschi, 2
Huntleya Lindley, 10
Kefersteinia Reichenbach, 36
Koellensteinia Reichenbach, 16
Mendoncella Hawkes = Galeottia
Neogardneria Schlechter, 1
Otostylis Schlechter, 3
Pabstia Garay, 5
Paradisianthus Reichenbach, 4
Pescatorea Reichenbach, 16
Promenaea Lindley, 14
Scuticaria Lindley, 7
Stenia Lindley, 8
Vargasiella C. Schweinfurth, 2

Warscewiczella Bentham & Hooker =
  Cochleanthes
Warrea Lindley, 4
Warreella Schlechter, 2
Warreopsis Garay, 3
Zygopetalum Hooker, 15
Zygosepalum Reichenbach, 7

LYCASTINAE SCHLECHTER

Adipe Rafinesque = Bifrenaria
Anguloa Ruiz & Pavon, 10
Bifrenaria Lindley, 24
Horvatia Garay, 1
Lycaste Lindley, 49
Neomoorea Rolfe, 1
Rudolfiella Hoehne, 7 (≃ Bifrenaria)
Stenocoryne Lindley = Bifrenaria
Teuscheria Garay, 6
Xylobium Lindley, 29

MAXILLARIINAE BENTHAM

Anthosiphon Schlechter, 1
Camaridium Lindley = Maxillaria
Chrysocycnis Linden & Reichenbach, 3
Cryptocentrum Bentham, 19
Cyrtidiorchis Rauschert, 4
Cyrtidium Schlechter = Cyrtidiorchis
Maxillaria Ruiz & Pavon, 420
Mormolyca Fenzl, 7
Neo-Urbania Fawcett & Rendle =
  Maxillaria
Ornithidium R. Brown = Maxillaria
Pityphyllum Schlechter, 4
Sepalosaccus Schlechter = Maxillaria
Trigonidium Lindley, 14

STANHOPEINAE BENTHAM

Acineta Lindley, 20
Braemia Jenny, 1
Cirrhaea Lindley, 6
Coeliopsis Reichenbach, 1
Coryanthes Hooker, 20
Embreea Dodson, 1
Endresiella Schllechter = Trevoria
Gongora Ruiz & Pavon, 50
Horichia Jenny, 1
Houlletia Brongniart, 10
Kegeliella Mansfeld, 3
Lacaena Lindley, 2
Lueddemannia Linden & Reichenbach, 1
Lycomormium Reichenbach, 5
Paphinia Lindley, 12
Peristeria Hooker, 15
Polycycnis Reichenbach, 15
Schlimia Lindley & Paxton, 7
Sievekingia Reichenbach, 15
Soterosanthus Jenny, 1

Stanhopea Hooker, 55
Trevoria Lehmann, 6
Vasqueziella Dodson, 1
      TELIPOGONINAE SCHLECHTER
Dipterostele Schlechter = Stellilabium
Hofmeisterella Reichenbach, 1
Stellilabium Schlechter, 20
Telipogon Kunth, 100
Trichoceros Kunth, 5
      ORNITHOCEPHALINAE SCHLECHTER
Caluera Dodson, 2
Centroglossa Barbosa Rodrigues, 6
Chytroglossa Reichenbach, 3
Dipteranthus Barbosa Rodrigues, 2
Dunstervillea Garay, 1
Eloyella Ortiz, 4
Hintonella Ames, 1
Oakes-Amesia Schweinfurth & Allen =
    Sphyrastylis
Ornithocephalus Hooker, 28
Phymatidium Lindley, 10
Platyrhiza Barbosa Rodrigues, 1
Rauhiella Pabst & Braga, 1
Sphyrastylis Schlechter, 8
Thysanoglossa Porto & Brade, 2
Zygostates Lindley, 7
      ONCIDIINAE BENTHAM
Ada Lindley, 15
Amparoa Schlechter, 2
Antillanorchis Garay, 1
Aspasia Lindley, 8
Baptistonia Barbosa Rodrigues =
    Oncidium
Binotia Rolfe, 1 (∿ Gomesa)
Braasiella Braem = Tolumnia
Brachtia Reichenbach, 5
Brassia R. Brown, 35
Buesiella Schweinfurth, 1
Capanemia Barbosa Rodrigues, 14
Caucaea Schlechter, 1
Centropetalum Lindley = Fernandezia
Chaenanthe Lindley = Diadenium
Cischweinfia Dressler & Williams, 7
Cochlioda Lindley, 5
Comparettia Poeppig & Endlicher, 10
Cuitlauzinia Llave & Lexarza, 1
Cymbiglossum Halbinger =
    Lemboglossum
Cypholoron Dodson & Dressler, 2
Cyrtochilum Kunth = Oncidium
Diadenium Poeppig & Endlicher, 2
Dignathe Lindley ?
Erycina Lindley, 1
Fernandezia Ruiz & Pavon, 9

Gomesa R. Brown, 13
Helcia Lindley, 4
Hispaniella Braem = Tolumnia
Hybochilus Schlechter, 1
Ionopsis Kunth, 3
Jamaicella Braem = Tolumnia
Konantzia Dodson, 1
Lemboglossum Halbinger, 14
Leochilus Knowles & Westcott, 10
Leucohyle Klotzsch, 3
Lockhartia Hooker, 24
Lophiaris Rafinesque, 25 ∿ Trichocentrum
Macradenia R. Brown, 12
Macroclinium Dodson, 27
Mesoglossum Halbinger, 1
Mesospinidium Reichenbach, 7
Mexicoa Garay, 1 ∿ Oncidium
Miltonia Lindley, 9 (≃ Oncidium)
Miltonioides Brieger & Lückel = Oncidium
Miltoniopsis Godefroy-Lebeuf, 5
Neodryas Reichenbach, 6
Neoescobaria Garay = Helcia
Neokoehleria Schlechter, 7
Notylia Lindley, 50
Odontoglossum Kunth, 140 (∿
    Oncidium)
Olgasis Rafinesque = Tolumnia
Oliveriana Reichenbach, 4
Oncidium Swartz, 420
Ornithophora Barbosa Rodrigues =
    Sigmatostalix
Osmoglossum Schlechter, 7 (≃
    Palumbina)
Otoglossum Garay & Dunsterville, 7
Pachyphyllum Kunth, 35
Palumbina Reichenbach, 1
Papperitzia Reichenbach, 1
Plectrophora Focke, 8
Polyotidium Garay, 1
Psychopsiella Lückel & Braem, 1
Psychopsis Rafinesque, 4
Psygmorchis Dodson & Dressler, 5
Pterostemma Kränzlin, 1
Quekettia Lindley, 5
Raycadenco Dodson, 1
Rhynchostele Reichenbach =
    Lemboglossum
Rodriguezia Ruiz & Pavon, 40
Rodrigueziella Kuntze, 5 (∿ Gomesa)
Rodrigueziopsis Schlechter, 2
Roezliella Schlechter = Sigmatostalix
Rossioglossum Garay & Kennedy, 4
Rusbyella Rolfe, 1
Sanderella O. Kuntze, 2

[Cymbidioid continued]
Saundersia Reichenbach, 2
Scelochiloides Dodson & Chase, 1
Scelochilus Klotzsch, 35
Sigmatostalix Reichenbach, 35
Solenidiopsis Senghas, 2
Solenidium Lindley, 3
Stictophyllum Dodson & Chase, 1
Suarezia Dodson, 1
Sutrina Lindley, 1

Symphyglossum Schlechter, 5
Systeloglossum Schlechter, 5
Theodorea Barbosa Rodrigues =
  Rodrigueziella
Ticoglossum Rodríguez, 2
Tolumnia Rafinesque, 35
Trichocentrum Poeppig & Endlicher, 30
Trichopilia Lindley, 30
Trizeuxis Lindley, 1
Warmingia Reichenbach, 4

Epidendroid Phylad

Arethuseae Lindley
        ARETHUSINAE LINDLEY
Arethusa Linnaeus 1
Eleorchis Maekawa, 2
        BLETIINAE BENTHAM
Acanthephippium Blume, 12
Ancistrochilus Rolfe, 2
Ania Lindley = Tainia
Anthogonium Lindley, 1
Ascotainia Ridley = Tainia
Aulostylis Schlechter, 1
Bletia Ruiz & Pavon, 30
Bletilla Reichenbach, 9
Calanthe R. Brown, 150
Calopogon R. Brown, 4
Cephalantheropsis Guillaumin, 7
Eriodes Seidenfaden, 1
Gastrorchis Schlechter, 5
Hancockia Rolfe, 1
Hexalectris Rafinesque, 7
Ipsea Lindley, 2
Jimensia Rafinesque = Bletilla
Mischobulbon Schlechter, 6
Nephelaphyllum Blume, 17
Pachystoma Blume, 6
Phaius Loureiro, 45
Plocoglottis Blume, 40
Spathoglottis Blume, 30
Tainia Blume, 12
Tainiopsis Schlechter = ?
        CHYSINAE SCHLECHTER
Chysis Lindley, 7

Coelogyneae Pfitzer
        THUNIINAE SCHLECHTER
Thunia Reichenbach, 5
        COELOGYNINAE BENTHAM
Acoridium Rolfe = Dendrochilum
Basigyne J. J. Smith, 1
Bracisepalum J. J. Smith, 2

Bulleyia Schlechter, 1
Chelonistele Pfitzer, 11
Coelogyne Lindley, 100
Dendrochilum Blume, 100
Dickasonia L. O. Williams, 1
Entomophobia de Vogel, 1
Forbesina Ridley, 1
Geesinkorchis de Vogel, 1
Gynoglottis J. J. Smith, 1
Ischnogyne Schlechter, 1
Kalimpongia U. C. Pradhan = Dickasonia
Nabaluia Ames, 3
Neogyna Reichenbach, 1
Otochilus Lindley, 4
Panisea Steudel, 7
Pholidota Hooker, 30
Platyclinis Bentham = Dendrochilum
Pleione D. Don, 15
Pseudacoridium Ames, 1
Sigmatogyne Pfitzer, 2
Zetagyne Ridley 1 (= Panisea?)

Epidendreae I (New World)
        SOBRALIINAE SCHLECHTER
Elleanthus Presl, 115
Epilyna Schlechter, 3 (∿ Elleanthus)
Fregea Reichenbach = Sobralia
Sertifera Lindley & Reichenbach, 6
Sobralia Ruiz & Pavon, 95
        ARPOPHYLLINAE DRESSLER (1990c)
Arpophyllum La Llave & Lexarza, 5
        MEIRACYLLIINAE DRESSLER (1960)
Meiracyllium Reichenbach, 2
        COELIINAE DRESSLER (1990c)
Coelia Lindley, 5
Bothriochilus Lemaire = Coelia
        LAELIINAE BENTHAM
Acrorchis Dressler, 1
Alamania La Llave & Lexarza, 1
Amblostoma Scheidweiler = Epidendrum

Artorima Dressler & Pollard, 1
Auliza Salisbury = Epidendrum
Barkeria Knowles & Westcott, 14
Basiphyllaea Schlechter, 3
Brassavola R. Brown, 17
Briegeria Senghas = Jacquiniella
Broughtonia R. Brown, 5
Cattleya Lindley 45
Cattleyopsis Lemaire = Broughtonia
Caularthron Rafinesque, 3
Constantia Barbosa Rodrigues, 4
Costaricaea Schlechter = Scaphyglottis
Dilomilis Rafinesque, 4
Dimerandra Schlechter, 8
Dinema Lindley = Encyclia
Diothonaea Lindley = Epidendrum
Domingoa Schlechter, 3
Encyclia Hooker, 235
Epidanthus L. O. Williams = Epidendrum
Epidendropsis Garay = Epidendrum
Epidendrum Linnaeus, 800
Hagsatera González T., 2
Helleriella Hawkes, 2 (∿ Platyglottis)
Hexadesmia Brongniart = Scaphyglottis
Hexisea Lindley, 2 (∿ Scaphyglottis)
Homalopetalum Rolfe, 4
Hormidium Heynhold = Encyclia
Huebneria Schlechter = Orleanesia
Isabelia Barbosa Rodrigues, 2
Isochilus R. Brown, 10
Jacquiniella Schlechter, 12
Kalopternix Garay = Epidendrum
Laelia Lindley, 69 (∿ Cattleya)
Laeliopsis Lindley = Broughtonia
Lanium Lindley = Epidendrum
Leaoa Schlechter & Porto = Scaphyglottis
Leptotes Lindley, 3
Loefgrenianthus Hoehne, 1
Myrmecophila Rolfe, 6
Nageliella L. O. Williams, 2
Nanodes Lindley = Epidendrum
Neocogniauxia Schlechter, 2
Neolehmannia Kränzlin = Epidendrum
Neowilliamsia Garay = Epidendrum
Nidema Britton & Millspaugh, 2
Octadesmia = Dilomilis
Oerstedella Reichenbach, 32
Orleanesia Barbosa Rodrigues, 8
Pachystele Schlechter = Scaphyglottis
Physinga Lindley = Epidendrum
Pinelia Lindley, 3 (∿ Homalopetalum)
Platyglottis L. O. Williams, 1
Pleuranthium Bentham = Epidendrum
Ponera Lindley, 11

Pseudolaelia Campos-Porto & Brade, 6
Pygmaeorchis Brade, 2
Psychilus Rafinesque, 15
Quisqueya Dod, 4 (∿ Broughtonia)
Reichenbachanthus Barbosa Rodrigues, 5
Renata Ruschi = Pseudolaelia
Rhyncholaelia Schlechter, 2
Scaphyglottis Poeppig & Endlicher, 85
Schomburgkia Lindley, 14 (∿ Laelia, Cattleya)
Sophronitella Schlechter = Isabelia
Sophronitis Lindley, 6
Stenoglossum Kunth = Epidendrum
Tetragamestus Reichenbach = Scaphyglossum
Tetramicra Lindley, 11

PLEUROTHALLIDINAE LINDLEY

Acostaea Schlechter, 4
Andreettaea Luer = Pleurothallis
Apatostelis Garay = Stelis
Barbosella Schlechter, 15
Barbrodria Luer, 1
Brachionidium Lindley, 35
Brenesia Schlechter = Pleurothallis
Chamelophyton Garay, 1
Condylago Luer, 1
Cryptophoranthus Barbosa Rodrigues = Pleurothallis
Dubois-Reymondia Karsten = Myoxanthus
Dracula Luer, 90
Dresslerella Luer, 8
Dryadella Luer, 3
Frondaria Luer, 1
Kraenzliniella Kuntze = Pleurothallis
Lepanthes Swartz, 460
Lepanthopsis Ames, 40
Luerella Braas = Masdevallia
Masdevallia Ruiz & Pavon, 380
Myoxanthus Poeppig & Endlicher, 45
Octomeria R. Brown, 135
Ophidion Luer, 4
Pabstiella Brieger & Senghas = Pleurothallis
Phloeophila Hoehne & Schlechter = Pleurothallis
Physothallis Garay = Pleurothallis
Physosiphon Lindley = Pleurothallis
Platystele Schlechter, 80
Pleurothallis R. Brown, 1120
Pleurothallopsis Porto & Brade = Octomeria
Porroglossum Schlechter, 30
Restrepia Kunth, 30
Restrepiella Garay & Dunsterville, 1

[Epidendroid continued**]**
Restrepiopsis Luer, 15
Rodrigoa Braas = Masdevallia
Salpistele Dressler, 6
Scaphosepalum Pfitzer, 30
Stelis Swartz, 370
Teagueia Luer, 6
Triaristella Luer = Trisetella
Trichosalpinx Luer, 90
Trisetella Luer, 20
Yolanda Hoehne = Brachionidium

Epidendreae II (Old World)
GLOMERINAE SCHLECHTER
Aglossorhyncha Schlechter, 6

Agrostophyllum Blume, 85
Earina Lindley, 7
Giulianettia Rolfe = Glossorhyncha
Glomera Blume, 50
Glossorhyncha Ridley, 80
Ischnocentrum Schlechter, 1
Sepalosiphon Schlechter, 1
ADRORHIZINAE SCHLECHTER
Adrorhizon Hooker f., 1
Sirhookera O. Kuntze, 2
POLYSTACHYINAE PFITZER
Hederorkis Thouars, 2
Imerinaea Schlechter, 1
Neobenthamia Rolfe, 1
Polystachya Hooker, 150

Dendrobioid Subclade

Podochileae Pfitzer
ERIINAE BENTHAM
Ascidieria Seidenfaden, 1
Ceratostylus Blume, 70
Cryptochilus Wallich, 6
Dendrolirium Blume = Eria
Epiblastus Schlechter, 20
Eria Lindley, 500
Mediocalcar J. J. Smith, 20
Porpax Lindley, 11
Sarcostoma Blume, 2
Stolzia Schlechter, 4
Trichotosia Blume, 45 ≃ Eria
Tylostylis Blume = Eria
Xiphosium Lindley = Eria
PODOCHILINAE BENTHAM & HOOKER
Appendicula Blume, 50
Chilopogon Schlechter, 3
Chitonochilus Schlechter, 1
Cyphochilus Schlechter, 7
Poaephyllum Ridley, 9
Podochilus Blume, 60
THELASIINAE SCHLECHTER
Chitonanthera Schlechter, 7
Kerigomnia Van Royen = Chitonanthera
Octarrhena Thwaites, 35
Oxyanthera Brongniart, 6
Phreatia Lindley, 160
Rhynchophreatia Schlechter, 5
Thelasis Blume, 20
Vanroemeria J. J. Smith, 1 ?
RIDLEYELLINAE
Ridleyella Schlechter, 1

Dendrobieae Endlicher
DENDROBIINAE LINDLEY
Cadetia Gaudichaud, 67
Dendrobium Swartz, 900
Diplocaulobium Kränzlin, 94
Ephemerantha Hunt & Summerhayes = Flickingeria
Epigeneium Gagnepain, 12
Flickingeria Hawkes, 70
Katharinea Hawkes = Epigeneium
Pseuderia Schlechter, 4
BULBOPHYLLINAE SCHLECHTER
Bulbophyllum Thouars, 1000
Chaseella Summerhayes, 1
Cirrhopetalum Lindley = Bulbophyllum
Codonosiphon Schlechter, 3 (∿ Bulbophyllum)
Dactylorhynchus Schlechter, 1 (∿ Bulbophyllum)
Drymoda Lindley, 2
Epicranthes Blume = Bulbophyllum
Genyorchis Schlechter, 6
Hapalochilus Senghas, 50 (∿ Bulbophyllum)
Hyalosema Rolfe = Bulbophyllum
Ione Lindley = Sunipia
Jejosephia Rao & Mani, 1 ∿ Trias
Monomeria Lindley, 2
Monosepalum Schlechter, 3 (∿ Bulbophyllum)
Pedilochilus Schlechter, 15
Saccoglossum Schlechter, 2
Sunipia Smith, 18

Tapeinoglossum Schlechter, 2 (∿ Bulbophyllum)
Trias Lindley, 10

Vandeae Lindley
  AERIDINAE PFITZER
Abdominea J. J. Smith, 1
Acampe Lindley, 5
Adenoncos Blume, 15
Aerides Loureiro 20
Amesiella Garay, 1
Arachnis Blume, 11
Armodorum Breda, 4
Ascocentrum Schlechter, 5
Ascochilopsis Carr, 1
Ascochilus Ridley, 5
Ascoglossum Schlechter, 2
Ascolabium Ying, 1
Biermannia King & Pantling, 7
Bogoria J. J. Smith, 4
Brachypeza Garay, 7
Calymmanthera Schlechter, 5
Camarotis Lindley = Micropera
Ceratocentron Senghas 1
Ceratochilus Blume, 1
Chamaeanthus Schlechter, 2
Cheirorchis C. E. Carr = Cordiglottis
Chiloschista Lindley, 18
Chroniochilus J. J. Smith, 6
Cleisocentron Brühl, 3
Cleisomeria E. Don, 2
Cleisostoma Blume, 80
Cordiglottis J. J. Smith, 7
Cottonia Wight, 1
Cryptopylos Garay, 1
Dimorphorchis D. Don, 2
Diplocentrum Lindley, 2
Diploprora Hooker f., 2
Doritis Lindley, 2 (∿ Phalaenopsis)
Dryadorchis Schlechter, 3
Drymoanthus Nicholls, 3
Dyakia E. Christenson, 1
Eparmatostigma Garay, 1
Esmeralda Reichenbach, 2
Euanthe Schlechter = Vanda
Gastrochilus D. Don, 50
Grosourdya Reichenbach, 10
Gunnarella Senghas, 20
Haraella Kudo, 1
Holcoglossum Schlechter, 9
Hygrochilus Pfitzer, 1
Hymenorchis Schlechter, 10
Kingidium P. F. Hunt = Phalaenopsis
Kingiella Rolfe = Phalaenopsis

Lesliea Seidenfaden, 1
Loxoma Garay, 3
Luisia Gaudichaud, 40
Macropodanthus L. O. Williams, 4
Malleola J. J. Smith, 35
Megalotis Garay, 1
Micropera Lindley, 17
Microsaccus Blume, 11
Microtatorchis Schlechter, 45
Mobilabium Rupp, 1
Neofinetia Hu, 1
Nothodoritis Tsi, 1
Omoea Blume, 2
Ornithochilus Lindley, 3
Papilionanthe Schlechter, 10
Papillilabium Dockrill, 1
Paraphalaenopsis Hawkes, 4
Parapteroceras Averyanov, 5
Parasarcochilus Dockrill = Pteroceros
Pelatantheria Ridley, 5
Pennilabium J. J. Smith, 11
Peristeranthus T. E. Hunt, 1
Phalaenopsis Blume, 44
Phragmorchis L. O. Williams, 1
Plectorhiza Dockrill, 3
Pomatocalpa Breda, 35
Porphyrodesme Schlechter, 3
Porrorachis Garay, 2
Proteroceras Joseph & Vayravelu, 1
Pteroceras Hasskarl, 28
Renanthera Loureiro, 15
Renantherella Ridley, 2 ∿ Renanthera
Rhinerrhiza Rupp, 1
Rhynchogyna Seidenfaden & Garay, 2
Rhynchostylis Blume, 3
Robiquetia Gaudichaud, 40
Saccolabiopsis J. J. Smith, 13
Saccolabium Blume, 4
Sarcanthus Lindley = Cleisostoma
Sarcanthopsis Garay, 7
Sarcochilus R. Brown, 14
Sarcoglyphis Garay, 10
Sarcophyton Garay, 2
Schistotylus Dockrill, 1
Schoenorchis Blume, 20
Sedirea Garay & Sweet, 2
Seidenfadenia Garay, 1
Smithsonia Saldanha, 3
Smitinandia Holttum, 3
Staurochilus Pfitzer, 4
Stauropsis Reichenbach = Trichoglottis
Stereochilus Lindley, 6
Taeniophyllum Blume, 170
Thrixspermum Loureiro, 140

[Dendrobioid continued]
Trachoma Garay = Tuberolabium
Trichoglottis Blume, 60
Trudelia Garay, 5
Tuberolabium Yamamoto, 12
Uncifera Lindley, 7
Vanda Jones, 45
Vandopsis Pfitzer, 5
Ventricularia Garay, 1
Xenicophyton Garay, 1
    ANGRAECINAE SUMMERHAYES
Aeranthes Lindley, 30
Ambrella H. Perrier, 1
Angraecum Bory, 200
Bonniera Cordemoy, 2
Calyptrochilum Kränzlin, 10
Campylocentrum Bentham, 55
Cryptopus Lindley, 4
Dendrophylax Reichenbach, 6
Harrisella Fawcett & Rendle, 3
  (~ *Campylocentrum*)
Jumellea Schlechter, 60
Lemurella Schlechter, 4
Lemurorchis Kränzlin, 1
Listrostachys Reichenbach, 1
Neobathiea Schlechter, 5
Oeonia Lindley, 6
Oeoniella Schlechter, 5
Ossiculum Laan & Cribb, 1
Perrierella Schlechter, 1
Polyradicion Garay, 4 (≃ Dendrophylax)
Polyrrhiza Pfitzer = Polyradicion
Sobennikoffia Schlechter, 4
    AERANGIDINAE SUMMERHAYES
Aerangis Reichenbach, 60
Ancistrorhynchus Finet, 14

Angraecopsis Kränzlin, 15
Azadehdelia Braem, 1
Barombia Schlechter = Aerangis
Beclardia A. Richard, 2
Bolusiella Schlechter, 6
Cardiochilus Cribb, 1
Chamaeangis Schlechter, 13
Chauliodon Summerhayes, 1
Cribbia Senghas = Azadehdelia
Cyrtorchis Schlechter, 16
Diaphananthe Schlechter, 20
Dinklageella Mansfeld, 2
Distylodon Summerhayes, 1
Eggelingia Summerhayes, 2
Encheiridion Summerhayes, 3
Eurychone Schlechter, 2
Holmesia Cribb, 1
Listrostachys Reichenbach, 3
Margelliantha Cribb, 5
Microcoelia Lindley, 23
Microterangis Senghas, 6
Mystacidium Lindley, 9
Nephrangis Summerhayes, 1
Plectrelminthus Rafinesque, 1
Podangis Schlechter, 1
Rangaeris Summerhayes, 6
Rhaesteria Summerhayes, 1
Rhipidoglossum Schlechter, 34
Sarcorhynchus Schlechter, 3
Solenangis Schlechter, 5
Sphyrarhynchus Mansfeld, 1
Summerhayesia Cribb, 2
Taeniorhiza Summerhayes, 1
Triceratorhynchus Summerhayes, 1
Tridactyle Schlechter, 38
Ypsilopus Summerhayes, 5

Misfits and Leftovers

    ARUNDINAE DRESSLER 1990c
Arundina Blume, 1
Dilochia Lindley, 6
    COLLABIINAE SCHLECHTER
Chrysoglossum Blume, 6
Collabium Blume, 10
Diglyphosa Blume, 12
Pilophyllum Schlechter = Chrysoglossum
  ?
    CLADERIA HOOKER F., 1

    ERIOPSIS LINDLEY, 4
    POGONIINAE PFITZER
Cleistes L. C. Richard, 55
Duckeella Porto & Brade, 3
Isotria Rafinesque, 2
Pogonia Jussieu, 2
Pogoniopsis Reichenbach, 2
    THAIA SEIDENFADEN, 1
    XERORCHIS SCHLECHTER, 2

Literature Cited

Abe, K. 1972. Contributions to the embryology of the family Orchidaceae. VII. A comparative study of the orchid embryo sac. *Science Reports of the Tohôku University*, Ser. IV (Biol.) 36:179–201.

Ackerman, J. D. 1977. Biosystematics of the genus *Piperia* Rydb. (Orchidaceae). *Botanical Journal of the Linnaean Society* 75:245–270.

Ackerman, J. D. 1981. Pollination biology of *Calypso bulbosa* var. *occidentalis* (Orchidaceae): a food-deception system. *Madroño* 28:101–110.

———. 1983. Euglossine bee pollination of the orchid, *Cochleanthes lipscombiae*: a food source mimic. *American Journal of Botany* 70:830–834.

———. 1986a. Mechanisms and evolution of food-deceptive pollination systems in orchids. *Lindleyana* 1:108–113.

———. 1986b. Coping with the epiphytic existence: pollination strategies. *Selbyana* 9:52–60.

Ackerman, J. D., and M. R. Mesler. 1979. Pollination biology of *Listera cordata* (Orchidaceae). *American Journal of Botany* 66:820–824.

Ackerman, J. D., and A. M. Montalvo. 1990. Short-and long-term limitations to fruit production in a tropical orchid. *Ecology* 71:263–272.

Ackerman, J. D., and N. H. Williams. 1980. Pollen morphology of the tribe Neottieae and its impact on the classification of the Orchidaceae. *Grana* 19:7–18.

———. 1981. Pollen morphology of the Chloraeinae (Orchidaceae: Diurideae) and related subtribes. *American Journal of Botany* 68:1392–1402.

Adams, B. R. 1988. New species and combinations in the genus *Scaphyglottis* (Orchidaceae). *Phytologia* 64:249–258.

Adams, P. B., and S. D. Lawson. 1987. Pollination of *Dendrobium kingianum* Bidw. by the honey bee (*Apis mellifera*). *The Orchadian* 8:250–251.

———. 1988. Multiple bee pollinators of *Dendrobium kingianum* Bidw. in the natural habitat. *The Orchadian* 9:103–107.

Agnew, J. D. 1986. Self compatibility/incompatibility in some orchids of the subfamily Vandoideae. *Plant Breeding* 97:183–186.

Albert, V. A., M. W. Chase, and A. W. Coleman. In press. A molecular phylogenetic evaluation of floral evolution in *Cypripedium reginae* Walter, *Paphiopedilum delenatii* Guillaumin, and *Phragmipedium schlimii* (Lindley & Reichb. f.) Rolfe (Cypripedioideae: Orchidaceae). *Lindleyana*.

Allen, P. H. 1959. *Mormodes lineatum*: a species in transition. *American Orchid Society Bulletin* 28:411–414.

Andersen, T. F., B. Johansen, I. Lund, F. N. Rasmussen, H. Rasmussen, and I. Sørensen. 1988. Vegetative architecture of *Eria*. *Lindleyana* 3:117–132.

Andersson, L. 1990. The driving force: species concepts and ecology. *Taxon* 39:375–382.

Aoyama, M. 1989. Karyomorphological studies in *Cymbidium* and its allied genera, Orchidaceae. *Bulletin of the Hiroshima Botanical Garden* 11:1–121.

Aoyama, M., and K. Karasawa. 1988. Karymorphological studies on *Lycaste*, Orchidaceae. *Bulletin of the Hiroshima Botanical Garden* 10:7–45.

Aoyama, M., and R. Tanaka. 1986. Karyomorphological studies in *Gastrodia elata* and *G. confusa*. *La Kromosomo* II-42:1336–1340.

Aoyama, M., R. Tanaka, and T. Itoh. 1987. Karyomorphological observations on saprophytic *Lecanorchis japonica* Bl. *The Journal of the Orchid Society of India* 1:51–55.

Aoyama, M., R. Tanaka, S. Takaki, and K. Kojima. 1986. Cytogenetic studies on intergeneric hybrids of *Cymbidium* I, F1 hybrid of *Cymbidium floribundum* (2n = 40) × *Eulophella Rolfei* (2n = 52). *La Kromosomo* II-41:1290–1297.

Aoyama, M., and R. Tanaka. 1986. Karyomorphological studies in *Gastrodia elata* and *G. confusa*. *La Kromosomo* II-42:1336–1340.

Arditti, J., and M. H. Fisch. 1977. Anthocyanins of the Orchidaceae: distribution, heredity, functions, synthesis, and localization. In *Orchid Biology Reviews and Perspectives*, Vol. 1. Ed. J. Arditti. Ithaca: Comstock.

Arekal, G. D., and K. A. Karanth. 1981. The embryology of *Epipogium roseum* (Orchidaceae). *Plant Systematics and Evolution* 138:1–7.

Arends, J. C., and F. M. van der Laan. 1986. Cytotaxonomy of the Vandeae. *Lindleyana* 1:33–41.

Atwood, J. T. Jr. 1984. The relationships of the slipper orchids (Subfamily Cypripedioideae). *Selbyana* 7:129–247.

———. 1985. Pollination of *Paphiopedilum rothschildianum*: brood-site deception. *National Geographic Research* 1985(spring):247–254.

———. 1986. The size of the Orchidaceae and the systematic distribution of epiphytic orchids. *Selbyana* 9:171–186.

Balogh, P. 1979. Morfología del polen de la tribu Cranichideae Endlicher subtribu Spiranthinae Bentham (Orchidaceae). *Orquídea* (Méx.) 7:241–260.

———. 1982. Generic redefinition in subtribe Spiranthinae (Orchidaceae). *American Journal of Botany* 69:1119–1132.

Barthlott, W. 1976. Morphologie der Samen von Orchideen im Hinblick auf taxonomische und funktionelle Aspekte. *Proceedings of the 8th World Orchid Conference*, Frankfurt, 444–455.

Barthlott, W., and B. Ziegler. 1981. Mikromorphologie der Samenschalen als systematisches Merkmal bei Orchideen. *Berichte der Deutschen Botanischen Gesellschaft* 94:267–273.

Bates, R. 1977. Pollination of orchids 3. *Native Orchid Society of South Australia Newsletter* 1(5):7.

_____ . 1981. Observation of pollen vectors on a putative hybrid swarm of *Microtis* R. Br. *The Orchadian* 7:14.

_____ . 1984a. Pollination of *Prasophyllum elatum* R. Br. (with notes on associated biology). *The Orchadian* 8:14–17.

_____ . 1984b. The genus *Microtis* R. BR. (Orchidaceae): a taxonomic revision with notes on biology. *Journal of the Adelaide Botanical Garden* 7:45–89.

Beardsell, D. V., and P. Bernhardt. 1983. Pollination biology of Australian terrestrial orchids. In *Pollination '82*. Ed. E. G. Williams, R. B. Knox, J. H. Gilbert, and P. Bernhardt. Parkville, Victoria: University of Melbourne Press.

Beardsell, D. V., M. A. Clements, J. F. Hutchinson, and E. G. Williams. 1986. Pollination of *Diuris maculata* R. Br. (Orchidaceae) by floral mimicry of the native legumes *Daviesia* spp. and *Pultenaea scabra* R. Br. *Australian Journal of Botany* 34:165–173.

Beer, J. G. 1863. *Beiträge zur Morphologie und Biologie der Familie der Orchideen.* Vienna: Carl Gerold's Sohn.

Benzing, D. H. 1986a. The genesis of orchid diversity: emphasis on floral biology leads to misconceptions. *Lindleyana* 1:73–89.

_____ . 1986b. The vegetative basis of vascular epiphytism. *Selbyana* 9:23–43.

Benzing, D. H., W. E. Friedman, G. Peterson, and A. Renfrow. 1983. Shootlessness, velamentous roots, and the pre-eminence of Orchidaceae in the epiphytic biotope. *American Journal of Botany* 70:121–133.

Benzing, D. H., and D. W. Ott. 1981. Vegetative reduction in epiphytic Bromeliaceae and Orchidaceae: its origin and significance. *Biotropica* 13:131–140.

Bernhardt, P., and P. Burns-Balogh. 1983. Pollination and pollinarium of *Dipodium punctatum* (Sm.) R. Br. *Victorian Naturalist* 100:197–199.

_____ . 1986a. Floral mimesis in *Thelymitra nuda* (Orchidaceae). *Plant Systematics and Evolution* 151:187–202.

_____ . 1986b. Observations of the floral biology of *Prasophyllum odoratum* (Orchidaceae, Spiranthoideae). *Plant Systematics and Evolution* 153:65–76.

Berry, P. E., and R. N. Calvo. 1991. Pollinator limitation and position dependent fruit set in the high Andean orchid *Myrosmodes cochleare* (Orchidaceae). *Plant Systematics and Evolution* 174:93–101.

Bino, R. J., A. Dafni, and A. D. J. Meeuse. 1982. The pollination ecology of *Orchis galilaea* (Bornm. et Schulze) Schltr. (Orchidaceae). *New Phytologist* 90:315–319.

Bock, I. 1986. Revision der Gattung *Comparettia* Poepp. & Endl. *Die Orchidee* 37:192–196, 199–206, 255–263.

_____ . 1988. Die Gattung *Rodriguezia* Ruiz & Pavon. *Die Orchidee* 39:145–150.

Bockemühl, L. 1983–1988. Die Gattung *Odontoglossum* HBK. Studien zu einer natürlichen Gliederung. *Die Orchidee* 34–39, in many parts.

Boyden, T. C. 1982. The pollination biology of *Calypso bulbosa* var. *americana* (Orchidaceae): Initial deception of bumblebee visitors. *Oecologia* 55:178–184.

Braem, G. J. 1986a. *Tolumnia*—der neue, aber doch alte, Name für die "Variegaten" Oncidien. *Die Orchidee* 37:55–59.

_____ . 1986b. Cattleya. *The Brazilian Bifoliate Cattleyas.* Hildesheim: Brücke-Verlag Kurt Schmersow.

_____ . 1986c. Cattleya. *II. The Unifoliate Cattleyas.* Hildesheim: Brücke-Verlag Kurt Schmersow.

_____ . 1988. The chromosome numbers of the taxa usually referred to as "Variegata Oncidiums" (Genera *Braasiella, Hispaniella, Olgasis & Tolumnia*) with a review of the chromosome numbers of the Orchidaceae and special reference to the Oncidiinae. *Schlechteriana* 1:23–33.

Brantjes, N. B. M. 1981. Ant, bee and fly pollination in *Epipactis palustris* (L.) Crantz (Orchidaceae). *Acta Botanica Neerlandica* 30:59–68.

Brieger, F. G., and E. Lückel. 1983–1984. Der *Miltonia*-Komplex—Eine Neubeurteilung. *Die Orchidee* 34:128–134, 216–219; 35:41–46.

Burns-Balogh, P. 1984. Classification of the tribe Diurideae (Orchidaceae) I. Subtribe Prasophyllinae Schlechter. *Selbyana* 7:318–327.

_____ . 1986. A synopsis of Mexican Spiranthinae. *Orquídea* (Méx.) 10:76–96.

Burns-Balogh, P., and P. Bernhardt. 1985. Evolutionary trends in the androecium of the Orchidaceae. *Plant Systematics and Evolution* 149:119–134.

_____ . 1988. Floral evolution and phylogeny in the tribe *Thelymitreae* (Orchidaceae: Neottioideae). *Plant Systematics and Evolution* 159:19–47.

Burns-Balogh, P., and V. A. Funk. 1986. A phylogenetic analysis of the Orchidaceae. *Smithsonian Contributions to Botany* 61:1–79.

Burns-Balogh, P., and M. Hesse. 1988. Pollen morphology of the cypripedioid orchids. *Plant Systematics and Evolution* 158:165–182.

Burns-Balogh, P., and H. Robinson. 1983. Evolution and phylogeny of the *Pelexia* alliance (Orchidaceae: Spiranthoideae: Spiranthinae). *Systematic Botany* 8:263–268.

Burns-Balogh, P., D. L. Szlachetko, and A. Dafni. 1987. Evolution, pollination, and systematics of the tribe Neottieae (Orchidaceae). *Plant Systematics and Evolution* 156:91–115.

Buth, D. G. 1984. The application of electrophoretic data in systematic studies. *Annual Review of Ecology and Systematics* 15:501–522.

Butzin, F. 1971. Die Namen der supragenerischen Einheiten der Orchidaceae. *Willdenowia* 6:301–340.

Cady, L. 1965. The flying duck orchids. *Australian Plants* 3:174–177.

Carlquist, S. 1987. Presence of vessels in wood of *Sarcandra* (Chloranthaceae): comments on vessel origins in angiosperms. *American Journal of Botany* 74:1765–1771.

Catling, P. M. 1982. Breeding systems of northeastern North American *Spiranthes* (Orchidaceae). *Canadian Journal of Botany* 60:3017–3039.

_____ . 1983. Pollination of northeastern North American *Spiranthes* (Orchidaceae). *Canadian Journal of Botany* 61:1080–1093.

_____ . 1987. Notes on the breeding systems of *Sacoila lanceolata* (Aublet) Garay (Orchidaceae). *Annals of the Missouri Botanical Garden* 74:58–68.

_____ . 1990. Auto-pollination in the Orchidaceae. In *Orchid Biology, Reviews and Perspectives*, Vol. 5. Ed. J. Arditti. Portland, OR: Timber Press.

Catling, P. M., and V. R. Catling. 1989. Observations on the pollination of *Platanthera huronensis* in southwest Colorado. *Lindleyana* 4:78–84.

Catling, P. M., and G. Knerer. 1980. Pollination of the small white lady's-slipper (*Cypripedium candidum*) in Lambton County, southern Ontario. *The Canadian Field-Naturalist* 94:435–438.

Cauwet-Marc, A.-M., and M. Balayer. 1984a. Les Orchidées du Bassin Méditerranéen. Contribution à l'étude caryologique des espèces des Pyrénées-Orientales (France) et contrées limitrophes—Tribu des *Neottieae* Lindl. *Bulletin de la Societé botanique de France* 131, *Lettres botaniques* (2): 121–137.

_____ . 1984b. Les genres *Orchis* L., *Dactylorhiza* Necker ex Newski, *Neotinea* Reichb. et *Traunsteinera* Reichb.: Caryologie et proposition de phylogénie et d'évolution. *Botanica Helvetica* 94:391–406.

_____ . 1986. Les orchidées du bassin méditerranean. Contribution a l'étude caryologique des espèces des Pyrénées-Orientales (France) et contrées limitrophes. II: Tribu des *Ophrydae* Lindl. *pro parte*. *Bulletin de la Societé botanique de France* 133, *Lettres botaniques* 1986:265–277.

Charanasri, U., and H. Kamemoto. 1977. Self incompatibility in the *Oncidium* alliance. *Hawaii Orchid Journal* 6(3):12–15.

Chase, M. W. 1985. Pollination of *Pleurothallis endotrachys*. *American Orchid Society Bulletin* 54:431–434.

_____ . 1986a. A reappraisal of the oncidioid orchids. *Systematic Botany* 11:477–491.

_____ . 1986b. Pollination ecology of two sympatric, synchronously flowering species of *Leochilus* in Costa Rica. *Lindleyana* 1:141–147.

_____ . 1986c. A monograph of *Leochilus* (Orchidaceae). *Systematic Botany Monographs* 14. 97 pp.

_____ . 1987a. Obligate twig epiphytism in the Oncidiinae and other neotropical orchids. *Selbyana* 10:24–30.

_____ . 1987b. Systematic implications of pollinarium morphology in *Oncidium* Sw., *Odontoglossum* Kunth, and allied genera (Orchidaceae). *Lindleyana* 2:8–28.

_____ . 1987c. Revisions of *Hybochilus* and *Goniochilus* (Orchidaceae). *Contributions from the University of Michigan Herbarium* 16:109–127.

Chase, M. W., M. R. Duvall, H. G. Hills, M. T. Clegg, and V. A. Albert. In press. DNA sequences and phylogenetics of the Orchidaceae and other lilioid monocots. *Selbyana*

Chase, M. W., and H. G. Hills. 1992. Orchid phylogeny, flowers sexuality, and fragrance-seeking. *BioScience* 42:43–49.

Chase, M. W., and R. G. Olmstead. 1988. Isozyme number in subtribe Oncidiinae (Orchidaceae): an evaluation of polyploidy. *American Journal of Botany* 75:1080–1085.

Chase, M. W., and J. D. Palmer. 1988. Chloroplast DNA variation, geographical distribution, and morphological parallelism in subtribe Oncidiinae (Orchidaceae). *American Journal of Botany* 75(6II):163–164 (abstract).

_____ . 1992. Floral morphology and chromosome number in subtribe Oncidiinae (Orchidaceae): evolutionary insights from a phylogenetic analysis of chloroplast DNA restriction site variation. In *Molecular Systematics in Plants*. Ed. D. E. Soltis, P. S. Soltis, and J. J. Doyle. London: Chapman and Hall.

Chase, M. W., and J. S. Pippen. 1988. Seed morphology in the Oncidiinae and related subtribes (Orchidaceae). *Systematic Botany* 13:313–323.

_____ . 1990. Seed morphology and phylogeny in subtribe Catasetinae (Orchidaceae). *Lindleyana* 5:126–133.

Chatterji, A. K. 1986. Chromosomes in orchid phylogeny and classification. In *Biology, Conservation, and Culture of Orchids*. Ed. S. P. Vij. Affiliated East-West Press Private Limited.

Chen, S. C., 1979. Notes on bisexual and unisexual forms of *Satyrium ciliatum* Ldl. *Acta Phytotaxonomica Sinica* 17:54–60.

_____ . 1982. The origin and early differentiation of the Orchidaceae. *Acta*

Phytotaxonomica Sinica 20:1–22.

Chen, S. C., and K. Lang. 1986. *Cypripedium subtropicum*, a new species related to *Selenipedium*. *Acta Phytotaxonomica Sinica* 24:317–322. (partially reprinted in *Orchid Digest* 53:115–118. 1989).

Chen, S. C., and Z. Tsi. 1987. *Eria medongensis*, a probably peloric form of *Eria coronaria*, with a discussion on peloria, in Orchidaceae. *Acta Phytotaxonomica Sinica* 25:329–339.

Cheng, C. S., and C. Z. Tang. 1986. A revision of the genus *Vanda* (Orchidaceae) of China. *Acta Botanica Yunnanica* 8:213–221. (also *Orchid Digest* 52:38–46. 1988)

Christenson, E. A. 1986. Nomenclatural changes in the Orchidaceae subtribe Sarcanthinae. *Selbyana* 9:167–170.

_____. 1987a. An infrageneric classification of *Holcoglossum* Schltr. (Orchidaceae: Sarcanthinae) with a key to the genera of the *Aerides-Vanda* alliance. *Notes of the Royal Botanical Garden Edinburgh* 44:249–256.

_____. 1987b. The taxonomy of *Aerides* and related genera. *Proceedings of the 12th World Orchid Conference*, Tokyo, 35–40.

Clements, M. 1988. Orchid mycorrhizal associations. *Lindleyana* 3:73–86.

_____. 1989. Catalogue of Australian Orchidaceae. *Australian Orchid Research* 1:1–160.

Clements, M., and P. Cribb. 1984. The underground orchids of Australia. *Kew Magazine* 1: 84–91.

Clemesha, S. C. 1968. Australia's flying duck orchids. *American Orchid Society Bulletin* 37:668–670.

Comber, J. B. 1984. An unusual orchid from Java. *Orchid Review* 92:280–281.

Correa, M. N. 1956. Las especies argentinas del género *Gavilea*. *Boletín de la Sociedad Argentina de Botánica* 6:73–86.

Cribb, P. J. 1983a. A revision of *Dendrobium* Sect. *Latouria* (Orchidaceae). *Kew Bulletin* 38:229–306.

_____. 1983b. *Dendrobium* sect. *Ceratobium* (Orchidaceae) in the Pacific Islands. *Kew Bulletin* 37:577–590.

_____. 1986. A revision of *Dendrobium* sect. *Spatulata* (Orchidaceae). *Kew Bulletin* 41:615–691.

_____. 1987. *The Genus* Paphiopedilum. Kew: Royal Botanic Gardens; Portland, OR: Timber Press.

Cribb, P. J., and I. Butterfield. 1988. *The Genus* Pleione. Kew: Royal Botanic Gardens; London: Christopher Helm; Portland, OR: Timber Press.

Cribb, P. J., and C. Z. Tang. 1982. *Spathoglottis* (Orchidaceae) in Australia and the Pacific Islands. *Kew Bulletin* 36:721–729.

Cribb, P. J., C. Z. Tang, and J. Butterfield. 1985–1986. Die Gattung *Pleione*. *Die Orchidee* 36:25–30, 47–50, 129–133, 143–147, 201–202, 234–236; 37:17–22, 59–62, 154–158.

Currah, R. S., S. Hambleton, and A. Smreciu. 1988. Mycorrhizae and mycorrhizal fungi of *Calypso bulbosa*. *American Journal of Botany* 75:739–752.

Dafni, A. 1983. Pollination of *Orchis caspia*—a nectarless plant which deceives the pollinators of nectariferous species from other plant families. *Journal of Ecology* 71:467–474.

_____. 1984. Mimicry and deception in pollination. *Annual Review of Ecology and Systematics* 15:259–278.

_____. 1986. Floral mimicry-mutualism and undirectional exploitation of insects by plants. In *Insects and the Plant Surface*. Ed. B. Juniper and R. Southwood.

London: Edward Arnold.

_____ . 1987. Pollination in *Orchis* and related genera: evolution from reward to deception. In *Orchid Biology, Reviews and Perspectives*, Vol. 4. Ed. J. Arditti. Ithaca: Cornell University Press.

Dafni, A., and D. M. Calder. 1987. Pollination by deceit and floral mimesis in *Thelymitra antennifera* (Orchidaceae). *Plant Systematics and Evolution* 158:11–22.

Dafni, A., and Y. Ivri. 1979. Pollination ecology of, and hybridization between, *Orchis coriophora* L. and *O. collina* Sol. ex Russ. (Orchidaceae) in Israel. *New Phytologist* 83:181–187.

_____ . 1981a. Floral mimicry between *Orchis israelitica* Baumann and Dafni (Orchidaceae) and *Bellevalia flexuosa* Boiss. (Liliaceae). *Oecologia* 49:229–232.

_____ . 1981b. The flower biology of *Cephalanthera longifolia* (Orchidaceae) — pollen imitation and facultative floral mimicry. *Plant Systematics and Evolution* 137:229–240.

Dafni, A., Y. Ivri, and N. B. M. Brantjes. 1981. Pollination of *Serapias vomeracea* Briq. (Orchidaceae) by imitation of holes for sleeping solitary male bees (Hymenoptera). *Acta Botanica Neerlandica* 30:69–73.

Dahlgren, R. M. T., and H. T. Clifford. 1982. *The Monocotyledons: a Comparative Study*. London: Academic Press.

Dahlgren, R. M. T., H. T. Clifford, and P. F. Yeo. 1985. *The Families of the Monocotyledons, Structure, Evolution, and Taxonomy*. Berlin: Springer-Verlag.

Dahlgren, R. M. T, and F. N. Rasmussen. 1983. Monocotyledon evolution: characters and phylogenetic estimation. *Evolutionary Biology* 16:255–395.

Dannenbaum, C., M. Wolter, and R. Schill. 1989. Stigma morphology of the orchids. *Botanisches Jahrbücher für Systematik, Pflanzengeschichte und Pflanzengeographie* 110:441–460.

Darwin, C. 1888. *The Various Contrivances by which Orchids Are Fertilised by Insects*. 2nd ed. London: John Murray.

Del Prete, C. 1984. The genus "*Ophrys*" L. (Orchidaceae): a new taxonomic approach. *Webbia* 38:209–220.

Dieringer, G. 1982. The pollination ecology of *Orchis spectabilis* L. (Orchidaceae). *Ohio Journal of Science* 82:218–225.

Dixon, K. W. 1985. The underground orchids of Australia—an appraisal. *The Orchadian* 8:75–79.

Dodson, C. H. 1975. *Dressleria* and *Clowesia*, a new genus and an old one revived in the Catasetinae (Orchidaceae). *Selbyana* 1:130–137.

Dodson, C. H., and R. L. Dressler, 1972. Two undescribed genera in the Orchidaceae—Oncidiinae. *Phytologia* 24:285–292.

Dodson, C. H., and R. Escobar R. 1987. The Telipogons of Costa Rica. *Orquideología* 17:1–137.

Donoghue, M. J. 1985. A critique of the biological species concept and recommendations for a phylogenetic alternative. *Bryologist* 88:172–181.

Donoghue, M. J., and P. D. Cantino. 1988. Paraphyly, ancestors, and the goals of taxonomy: a botanical defense of cladism. *Botanical Review* 54:107–128.

Donoghue, M. J., J. A. Doyle, J. Gauthier, A. G. Kluge, and T. Rowe. 1989. The importance of fossils in phylogeny reconstruction. *Annual Review of Ecology and Systematics* 20:431–460.

Donoghue, M. J., and M. J. Sanderson. 1992. The suitibility of molecular and morphological evidence in reconstructing plant phylogeny. In *Molecular*

Systematics in Plants. Ed. D. E. Soltis, P. S. Soltis and J. J. Doyle. London: Chapman and Hall.

Doyle, J. J. 1987. Plant systematics at the DNA level: promises and pitfalls. *International Organization of Plant Biosystematics Newsletter* 8:3–7

Dransfield, J., J. B. Comber, and G. Smith. 1986. A synopsis of *Corybas* (Orchidaceae) in West Malaysia and Asia. *Kew Bulletin* 41:575–613.

Dressler, R. L. 1960. The relationships of *Meiracyllium* (Orchidaceae). *Brittonia* 12:222–225.

_____. 1961. The structure of the orchid flower. *Missouri Botanical Garden Bulletin* 49:60–69.

_____. 1968. Observations on orchids and euglossine bees in Panama and Costa Rica. *Revista de Biología Tropical* 15:143–183.

_____. 1971. Nomenclatural notes on the Orchidaceae. V. Phytologia 21: 440–443.

_____. 1974. El género *Hexisea. Orquídea* (Méx.) 4:191–200.

_____. 1979. the subfamilies of the Orchidaceae. Selbyana 5:197–206.

_____. 1980. Orquídeas huérfanas. I. *Wullschlaegelia*: una nueva tribu, Wullschlaegeliae. *Orquídea* (Méx.) 7:277–282.

_____. 1981. *The Orchids: Natural History and Classification.* Cambridge, Mass.: Harvard University Press.

_____. 1983a. *Palmorchis* in Panama mit einer neuen Art, *Palmorchis nitida*, an einem unerwarteten Standort. *Die Orchidee* 34:25–31.

_____. 1983b. Classification of the Orchidaceae and their probable origin. *Telopea* 2:413–424.

_____. 1984. La delimitación de géneros en el complejo *Epidendrum. Orquídea* (Méx.) 9:277–298.

_____. 1986. Recent advances in orchid phylogeny. *Lindleyana* 1:5–20.

_____. 1989a. Rostellum and viscidium: divergent definitions. *Lindleyana* 4:48–49.

_____. 1989b. The vandoid orchids: a polyphyletic grade? *Lindleyana* 4:89–93.

_____. 1989c. Pollinia-presentation in *Dendrobium* Section *Pedilonum. Selbyana* 11:35–38.

_____. 1990a. The Neottieae in orchid classification. *Lindleyana* 5:102–109.

_____. 1990b. The Spiranthoideae, grade or subfamily? *Lindleyana* 5:110–116.

_____. 1990c. The major clades of the Orchidaceae–Epidendroideae. *Lindleyana* 5:117–125.

Dressler, R. L., and S. L. Cook. 1988. Conical silica bodies in *Eria javanica. Lindleyana* 3:224–225.

Dressler, R. L., and C. H. Dodson. 1960. Classification and phylogeny in the Orchidaceae. *Annals of the Missouri Botanical Garden* 47:25–68.

Dressler, R. L., and G. A. Salazar. 1991. Viscarium, a term for the glue-bearing area of the rostellum. *Orchid Research Newsletter* 17:11–12.

Duncan, T. 1980. Cladistics for the practicing taxonomist—an eclectic view. *Systematic Botany* 5:136–148.

_____. 1986. Semantic fencing: a final riposte with a Hennigian crutch. *Taxon* 35:110–122.

Dunsterville, G. C. K., and L. A. Garay. 1966. *Venezuelan Orchids Illustrated.* IV. London: Andre Deutsch Ltd.

Dupuis, C. 1984. Willi Hennig's impact on taxonomic thought. *Annual Review of Ecology and Systematics* 15:1–24.

Du Puy, D. J. 1987. The taxonomic hierarchy and its application in the genus *Cymbidium* Sw. (Orchidaceae). *Proceedings of the 12th World Orchid Conference,* Tokyo, 29–34.

Du Puy, D. J., and P. Cribb. 1988. *The genus* Cymbidium. London: Christopher Helm.

Felsenstein, J. 1983. Parsimony in systematics: biological and statistical issues. *Annual Review of Ecology and Systematics* 14:313–333.

Fink, W. L. 1986. Microcomputers and phylogenetic analysis. *Science* 234:1135–1139.

Firth, J. 1965. Orchids in Tasmania. *Australian Plants* 3:178–179.

Folsom, J. P. 1984. Una reinterpretación del estatus y relaciones de las taxa del complejo de *Platanthera ciliaris*. *Orquídea* (Méx.) 9:321–345.

Freudenstein, J. V. 1991a. A systematic study of endothecial thickenings in the Orchidaceae. *American Journal of Botany* 78:766–781.

_____. 1991b. Variation in stipe morphology among the genera of the Corallorhizinae (Orchidaceae): systematic implications for the Epidendroideae. *American Journal of Botany* 78(6II):187–188 (abstract).

Fritz, A.-L. 1990. Deceit pollination of *Orchis spitzelii* (Orchidaceae) on the Island of Gotland in the Baltic: a suboptimal system. *Nordic Journal of Botany* 9:577–587.

Frölich, D., and W. Barthlott. 1988. Mikromorphologie der epicuticularen Wachse und das System der Monokotylen. *Tropische und subtropische Pflanzenwelt* 63: 135 pp.

Fryxell, P. A. 1957. Mode of reproduction of higher plants. *Botanical Review* 23:135–233.

Garay, L. A. 1972. On the origin of the Orchidaceae, II. *Journal of the Arnold Arboretum* 53:202–215.

_____. 1982. A generic revision of the Spiranthinae. *Botanical Museum Leaflets* 28:277–425.

_____. 1986. Olim Vanillaceae. *Botanical Museum Leaflets* 30:223–237.

Garnet, J. R. 1940. Observations on the pollination of orchids. *Victorian Naturalist* 56:191–197.

George, A. S. 1979. Rediscovery of the underground orchid *Rhizanthella gardneri* R. S. Rogers. *The Orchadian* 6:112.

_____. 1981. *Rhizanthella*: the underground orchid of Western Australia. *Proceedings of the Orchid Symposium held as a satellite function of the 13th International Botanical Congress,* Sydney, Australia, pp.77–78.

Gibbs, R. D. 1974. *Chemotaxonomy of Flowering Plants,* III. Montreal: McGill-Queen's University Press.

Gill, D. E. 1990. Fruiting failure, pollinator inefficiency, and speciation in orchids. In *Speciation and Its Consequences*. Ed. D. Otte and J. A. Endler. Sunderland, Massachusetts: Sinauer Associates.

Gould, S. J. 1977. *Ever Since Darwin, Reflections in Natural History*. New York: W. W. Norton.

Greenwood, E. W. 1982. Tipos de viscidio en Spiranthinae. *Orquídea* (Méx.) 8:283–310.

Gregg, K. B. 1983. Variation in floral fragrances and morphology: incipient speciation in *Cycnoches? Botanical Gazette* 144:566–576.

_____. 1989. Reproductive biology of the orchid *Cleistes divaricata* (L.) Ames var. *bifaria* Fernald growing in a West Virginia meadow. *Castanea* 54:57–78.

_____ . 1991. Reproductive strategy of *Cleistes divaricata* (Orchidaceae). *American Journal of Botany* 78:350–360.

Greilhuber, J., and F. Ehrendorfer. 1975. Chromosome numbers and evolution in *Ophrys* (Orchidaceae). *Plant Systematics and Evolution* 124:125–138.

Grime, J. P. 1979. *Plant Strategies and Vegetative Processes.* New York: John Wiley & Sons

Gumprecht, R. 1975. Orchideen in Chile. *Die Orchidee* 26:127–132.

Halbinger, F. 1982. *Odontoglossum* y géneros afines en México y Centroamérica. *Orquídea* (Méx.) 8:155–282.

Hall, A. V. 1982. A revision of the southern African species of *Satyrium. Contributions of the Bolus Herbarium* 10:1–142.

_____ . 1991. A unifying theory for methods of systematic analysis. *Biological Journal of the Linnaean Society* 42:425–456.

Hallé, N. 1965. Deux Orchidées gabonaises présentées d'après des sujets vivants: *Phaius mannii* Reichb. f. et *Manniella gustavi* Reichb. f. *Adansonia* 5:415–419.

_____ . 1977. Orchidacées. *In Flore de la Nouvelle Calédonie et Dépendances.* Ed. A. Aubréville, J.-F. Leroy. Paris: Muséum National d'histoire Naturelle.

_____ . 1986. Les èlatéres des Sarcanthinae et additions aux Orchidaceae de la Nouvelle-Calédonie. *Bulletin du Muséum national d'Histoire naturelle*, Paris, 4e ser., 8, section B, *Adansonia*, no. 3:215–239.

Hashimoto, K. 1981. Chromosome count in *Dendrobium* I. 87 species. *Bulletin of the Hiroshima Botanical Garden* 4:63–80.

_____ . 1987. Karyomorphological studies of some 80 taxa of *Dendrobium* (Orchidaceae). *Bulletin of the Hiroshima Botanical Garden* 9:1–186.

Hashimoto, K., and R. Tanaka. 1983. Karyomorphological observations on some species of *Eria. Bulletin of the Hiroshima Botanical Garden* 6:1–46.

Hashimoto, T. 1990. A taxonomic review of the Japanese *Lecanorchis* (Orchidaceae). *Annals of the Tsukuba Botanical Garden* 9:1–40.

Hegnauer, R. 1963. *Chemotaxonomie der Pflanzen.* Basel und Stuttgart: Birkhäuser Verlag

Hennig, W. 1968. *Elementos de una Sistemática Filogenética.* Buenos Aires: Editorial Universitaria de Buenos Aires.

Hesse, M., and P. Burns-Balogh. 1984. Pollen and pollinarium morphology of *Habenaria* (Orchidaceae). *Pollen et Spores* 26:385–400.

Hesse, M., P. Burns-Balogh, and M. Wolff. 1989. Pollen morphology of the "primitive" epidendroid orchids. *Grana* 28:261–278.

Heusser, C. 1938. Chromosomenverhältnisse bei schweizerischen basitonen Orchideen. *Berichte der Schweizerischen Botanischen Gesellschaft* 48:562–605.

Hillerman, F. E., and A. W. Holst. 1986. *An Introduction to the Cultivated Angraecoid Orchids of Madagascar.* Portland, OR: Timber Press.

Hirmer, M. 1920. Beitrage zur Organographie der Orchideenblute. *Flora* 13:213–309.

Hogan, K. P. 1983. The pollination biology and breeding system of *Aplectrum hyemale* (Orchidaceae). *Canadian Journal of Botany* 61:1906–1910.

Holttum, R. E. 1955. Growth habits of monocotyledons: variations on a theme. *Phytomorphology* 5:399–423.

_____ . 1959. Evolutionary trends in the Sarcanthine orchids. *American Orchid Society Bulletin* 28:747–754.

Howcroft, N. H. S. 1986. The taxonomy of *Spathoglottis* Blume. *The Orchadian* 8:139–148.

Hu, S. Y. 1977. *The genera of Orchidaceae in Hong Kong.* Hong Kong: Chinese University Press.

Huber, H. 1969. Samenmerkmale und Verwandtschaftsverhältnisse der Liliifloren. *Mitteilungen der Botanischen Staatsammlung München* 8:219–566.

Hurusawa, I., and K. Kakadzu. 1982. Vergleichende morphologische Untersuchung des Gynostemiums bei Orchidazeen der Ryukyu-Inseln, I. Teil. *Bulletin College of Science, University of the Ryukyus,* No. 33:27–46.

Inoue, K. 1983. Systematics of the genus *Platanthera* (Orchidaceae) in Japan and adjacent regions with special reference to pollination. *Journal of the Faculty of Science, University of Tokyo,* Sect. III, 13:285–374.

Ishida, G. 1990. Karyomorphological studies in *Calanthe,* Orchidaceae. *Bulletin of the Hiroshima Botanical Garden* 12:1–69.

Ivri, Y., and A. Dafni. 1977. The pollination ecology of *Epipactis consimilis* Don (Orchidaceae) in Israel. *New Phytologist* 79:173–177.

Izaguirre de Artucio, P. 1973. Las especies uruguayas de *Bipinnula* (Orchidaceae). *Boletín de la Sociedad Argentina de Botánica* 15:261–276.

Jenny, R. 1979–1991. Die Gongorinae, 6. *Gongora. Die Orchidee* 30–41: in many parts.

———. 1985. The genus *Gongora. Orchid Digest* 49:135–146, 175–186, 217–224.

———. 1987–(continuing). Die Gongorinae, 7. *Stanhopea. Die Orchidee* 38-, in parts.

Johansen, B. 1990. Incompatibility in *Dendrobium* (Orchidaceae). *Botanical Journal of the Linnaean Society* 103:165–196.

Johansson, D. R. 1974. Ecology of vascular epiphytes in West African rain forest. *Acta Phytogeographica Suecica* 59:1–129.

Jones, D. L. 1981. The pollination of selected Australian orchids. *Proceedings of the Orchid Symposium Held as a Satellite Function of the 13th International Botanical Congress,* Sydney, Australia, pp. 40–43.

———. 1985a. The pollination of *Gastrodia sesamoides* R. Br. in southern Victoria. *Victorian Naturalist* 102:52–54.

———. 1985b. The pollination of *Bulbophyllum weinthalii* R. Rogers. *Victorian Naturalist* 103:99–101.

Jones, D. L., and M. A. Clements. 1987. Reinstatement of the genus *Cyrtostylis* R. Br. and its relationships with *Acianthus* R. Br. (Orchidaceae). *Lindleyana* 2:156–160.

———. 1989. Reinterpretation of the genus *Genoplesium* R. Br. (Orchidaceae: Prasophyllinae). *Lindleyana* 4:139–145.

Jones, D. L., and B. Gray. 1974. The pollination of *Calochilus holtzei* F. Muell. *American Orchid Society Bulletin* 43:604–606.

Jones, K. 1970. Chromosome changes in plant evolution. *Taxon* 19:172–179.

Jones, K., K. Y. Lim, and P. J. Cribb. 1982. The chromosomes of orchids. VII. *Dendrobium. Kew Bulletin* 37:221–227.

Jonsson, L. 1979. The African member of *Taeniophyllum* (Orchidaceae). *Botaniska Notiser* 132:511–519.

———. 1981. A monograph of the genus *Microcoelia* (Orchidaceae). *Symbolae Botanicae Upsaliensis* 23(4).

Kamemoto, H. 1964. Chromosomes and species relationships in the *Vanda* alliance. *Proceedings 4th World Orchid Conference,* Singapore. 107–117.

Karasawa, K. 1979. Karyomorphological studies in *Paphiopedilum,* Orchidaceae. *Bulletin of the Hiroshima Botanical Garden* 2:1–149.

Karasawa, K., and M. Aoyama. 1986. Karyomorphological studies on *Cypripedium* in Japan and Formosa. *Bulletin of the Hiroshima Botanical Garden* 8:1–22.

Karasawa, K., and K. Saito. 1982. A revision of the genus *Paphiopedilum* (Orchidaceae). *Bulletin of the Hiroshima Botanical Garden* 5:1–69.

Kjellsson, G., and F. N. Rasmussen. 1987. Does the pollination of *Dendrobium unicum* Seidenf. involve "pseudopollen"? *Die Orchidee* 38:183–187.

Kjellsson, G., F. N. Rasmussen, and D. Dupuy. 1985. Pollination of *Dendrobium infundibulum, Cymbidium insigne* (Orchidaceae) and *Rhododendrum lyi* (Ericaceae) by *Bombus eximius* (Apidae) in Thailand: a possible case of floral mimicry. *Journal of Tropical Ecology* 1:289–302.

Kumazawa, M. 1956. Morphology and development of the sinker in *Pecteilis radiata* (Orchidaceae). *Botanical Magazine* (Tokyo) 69:455–461.

Kurzweil, H. 1987a. Developmental studies in orchid flowers. I: Epidendroid and vandoid species. *Nordic Journal of Botany* 7:427–442.

———. 1987b. Developmental studies in orchid flowers II: Orchidoid species. *Nordic Journal of Botany* 7:443–451.

———. 1988. Developmental studies in orchid flowers III: Neottioid species. *Nordic Journal of Botany* 8:271–282.

———. 1989a. Floral morphology and ontogeny in *Huttonaea pulchra*. *Lindleyana* 4:1–5.

———. 1989b. An investigation of the floral morphogenesis of *Bonatea speciosa* (Orchidaceae). *South African Journal of Botany* 55:433–437.

———. 1990. Floral morphology and ontogeny in Orchidaceae subtribe Disinae. *Botanical Journal of the Linnaean Society* 102:61–83.

———. 1991. The unusual structure of the gynostemium in the Orchidaceae–Coryciinae. *Botanisches Jahrbücher für Systematik, Pflanzengeschichte und Pflanzengeographie* 112:273–293.

Kurzweil, H., H. P. Linder, and P. Chesselet. 1991. The phylogeny and evolution of the *Pterygodium-Corycium* complex (Coryciinae, Orchidaceae). *Plant Systematics and Evolution* 175:161–223.

Kurzweil, H., and A. Weber. 1991. Floral morphology of southern African Orchideae. I. Orchidinae. *Nordic Journal of Botany* 11:155–178.

———. 1992. Floral morphology of southern African Orchideae. II. Habenariinae. *Nordic Journal of Botany* 12:39–61.

Larson, R. J., and K. S. Larson. 1987. Observations on the pollination biology of *Spiranthes romanzoffiana*. *Lindleyana* 2:176–179.

Liang, H. 1984. Embryological studies of *Gastrodia elata* Blume. *Acta Botanica Sinica* 26:466–472.

Lidén, M., and B. Oxelman. 1989. Species—pattern or process? *Taxon* 38:228–232.

Lim, K.-Y. 1985a. The chromosomes of orchids at Kew. 1. *Bulbophyllum. American Orchid Society Bulletin* 54:190–191.

———. 1985b. The chromosomes of orchids at Kew. 2. *Dendrobium. American Orchid Society Bulletin* 54: 1122–1123.

Linder, H. P. 1981a. Taxonomic studies on the Disinae. 1. A revision of the genus *Brownleea* Lindl. *Journal of South African Botany* 47:13–48.

———. 1981b. Taxonomic studies on the Disinae. 2. A revision of the genus *Schizodium* Lindl. *Journal of South African Botany* 47:339–371.

———. 1981c. Taxonomic studies on the Disinae. III. A revision of *Disa* Berg. excluding sect. *Micranthae* Lindl. *Contributions of the Bolus Herbarium* 9:1–370.

———. 1981d. Taxonomic studies on the Disinae. IV. A revision of *Disa* sect.

Micranthae Lindl. *Bulletin du Jardin Botanique national de Belgique* 51:255–346.

———. 1981e. Taxonomic studies on the Disinae. V. A revision of the genus *Monadenia* Lindl. *Bothalia* 13:339–363.

———. 1981f. Taxonomic studies on the Disinae. VI. A revision of the genus *Herschelia* Lindl. *Bothalia* 13:365–388.

———. 1986. Notes on the phylogeny of the Orchidoideae, with particular reference to the Diseae. *Lindleyana* 1:51–64.

Linder, H. P., and H. Kurzweil. 1990. Floral morphology and phylogeny of the Disinae (Orchidaceae). *Botanical Journal of the Linnaean Society* 102:287–302.

Linder, H. P., and G. Williamson. 1986. Notes on the orchids of southern Tropical Africa, II. *Oligophyton drummondii*, gen. et sp. nov. *Kew Bulletin* 41:313–317.

Lister, P. 1987. Observations on the pollination of *Dendrobium linguiforme* Sw. *The Orchadian* 8:258–261.

Lownes, A. E. 1920. Notes on *Pogonia trianthophora*. *Rhodora* 22:53–55.

Lückel, E., and G. J. Braem. 1982. *Psychopsis* und *Psychopsiella*: eine alte und eine neue Gattung der *Oncidium*-Verwandtschaft. *Die Orchidee* 33:1–7.

Luer, C. A. 1972. *The Native Orchids of Florida*. New York: New York Botanical Garden.

———. 1986a. Systematics of the Pleurothallidinae (Orchidaceae). *Monographs in Systematic Botany* 15:1–81.

———. 1986b. Systematics of *Masdevallia* (Orchidaceae). *Monographs in Systematic Botany* 16:1–63.

———. 1986c. Systematics of *Pleurothallis* (Orchidaceae). *Monographs in Systematic Botany* 20:1–109.

———. 1987. Systematics of *Acostaea*, *Condylago* and *Porroglossum* (Orchidaceae). *Monographs in Systematic Botany* 24:1–91.

———. 1988. Systematics of *Dresslerella* and *Scaphosepalum* (Orchidaceae). *Monographs in Systematic Botany* 26:1–111.

———. 1990. Systematics of *Platystele* (Orchidaceae). *Monographs in Systematic Botany* 38:1–135.

———. 1991. Systematics of *Lepanthopsis*, *Octomeria* subgenus *Pleurothallopsis*, *Restrepiella*, *Restrepiopsis*, *Salpistele* and *Teagueia*. *Monographs in Systematic Botany* 39:1–161.

Lund, I. D. 1987a. The genus *Panisea* (Orchidaceae), a taxonomic revision. *Nordic Journal of Botany* 7:511–527.

———. 1987b. The genus *Cremastra* (Orchidaceae), a taxonomic revision. *Nordic Journal of Botany* 8:197–203.

Lüning, B. 1974. Alkaloids of the Orchidaceae. In *The Orchids: Scientific Studies*. Ed. C. L. Withner. New York: John Wiley & Sons.

Madison, M. 1981. Vanilla beans and bees. *Bulletin of the Marie Selby Botanical Garden* 8:8.

Malguth, R. 1901. *Biologische Eigenthümlichkeiten der Früchte epiphytischer Orchideen*. Ph.D. Thesis, Breslau.

Mansfeld, R. 1935. Zur Terminologie der Pollinienanhängsel der Orchideen. *Repertorium Specierum Novarum Regni Vegetabilis* 38:199–205.

———. 1937. Über das System der Orchidaceae–Monandrae. *Notizblatt des Botanischen Gartens und Museums zu Berlin-Dahlem* 13:666–676.

———. 1954. Über die Verteilung der Merkmale innerhalb der Orchidaceae–Monandrae. *Flora* 142:65–80.

Martínez, A. 1985. The chromosomes of orchids. VIII Spiranthinae and

Cranichidinae. *Kew Bulletin* 40:139–147.

Mayr, E. 1954. Change of genetic environment and evolution. In *Evolution as a Process*. Ed. J. Huxley, A. C. Hardy, and E. B. Ford. London: George Allen and Unwin Ltd.

———. 1982. *The Growth of Biological Thought Diversity, Evolution and Inheritance*. Cambridge, Massachusetts: Belknap Press of Harvard University Press.

Meacham, C. A., and G. F. Estabrook. 1985. Compatibility methods in systematics. *Annual Review of Ecology and Systematics* 16:431–446.

Medley, M. E. 1979. *Some Aspects of the Life History of* Triphora trianthophora *(Sw.) Rydb. (Three Birds Orchid) with Special Reference to Its Pollination*. Master's Thesis, Andrews University.

Mehra, P. N. 1983. *Cytology of Orchids of Khasi and Jaintia Hills*. New Delhi: Kapur/Rajbandhu.

Mehra, P. N., and S. P. Vij. 1972. Cytological studies in the east Himalayan Orchidaceae. 1. Neottieae. *Caryologia* 25:237–251.

Mehrhoff, L. A. 1983. Pollination in the genus *Isotria* (Orchidaceae). *American Journal of Botany* 70:1444–1453.

Minderhoud, M. E., and E. F. de Vogel. 1986. A taxonomic revision of the genus *Acriopsis* Reinwardt ex Blume (Acriopsidinae, Orchidaceae). *Orchid Monographs* 1:16.

Møller, J. D., and H. Rasmussen. 1984. Stegmata in Orchidales: character state distribution and polarity. *Botanical Journal of the Linnaean Society* 89:53–76.

Morales, G. L. 1986. El género *Pterichis* en Colombia. *Orquideología* 16:53–79.

Nageswara Rao, A. 1987. A note on the sectional delineation in the genus *Epipogium* R. Br. (Orchidaceae). *Indian Orchid Journal* 2:13–18.

Nakata, M., and T. Hashimoto. 1983. Karyomorphological studies on species of *Pleurothallis*, Orchidaceae. *Annals of the Tsukuba Botanical Garden* 2:11–32.

Newton, G. D., and N. H. Williams. 1978. Pollen morphology of the Cypripedioideae and the Apostasioideae (Orchidaceae). *Selbyana* 1:169–182.

Nicholls, W. H. 1969. *Orchids of Australia*. Melbourne: Nelson.

Nilsson, L. A. 1978a. Pollination ecology and adaptation in *Platanthera chlorantha* (Orchidaceae). *Botaniska Notiser* 131:35–51.

———. 1978b. Pollination ecology of *Epipactis palustris* (Orchidaceae). *Botaniska Notiser* 131:355–368.

———. 1979a. Anthecological studies on the lady's Slipper, *Cypripedium calceolus* (Orchidaceae). *Botaniska Notiser* 132:329–347.

———. 1979b. The pollination ecology of *Herminium monorchis* (Orchidaceae). *Botaniska Notiser* 132:537–549.

———. 1980. The pollination ecology of *Dactylorhiza sambucina* (Orchidaceae). *Botaniska Notiser* 133:367–385.

———. 1981. The pollination ecology of *Listera ovata* (Orchidaceae). *Nordic Journal of Botany* 1:461–480.

———. 1983a. Anthecology of *Orchis mascula* (Orchidaceae). *Nordic Journal of Botany* 3:157–179.

———. 1983b. Processes of isolation and introgressive interplay between *Platanthera bifolia* (L.) Rich and *P. chlorantha* (Custer) Reichb. (Orchidaceae). *Botanical Journal of the Linnaean Society* 87:325–350.

———. 1983c. Mimesis of bellflower (*Campanula*) by the red helleborine orchid (*Cephalanthera rubra*). *Nature* 305:799–800.

———. 1984. Anthecology of Orchis morio (Orchidaceae) at its outpost in the north. *Nova Acta Regiae Societatis Scientiarum Upsaliensis*, Serie V:C, 3:167–179.

Nilsson, L. A., and L. Jonsson. 1985. The pollination specialization of *Habenaria decaryana* H. Perr. (Orchidaceae) in Madagascar. *Bulletin du Muséum national d'Histoire naturelle*, Paris, Sect. B, *Adansonia* 2: 161–166.

Nilsson, L. A., L. Jonsson, L. Rason, and E. Randrianjohany. 1985. Monophily and pollination mechanisms in *Angraecum arachnites* (Orchidaceae) in a guild of long-tongued hawk-moths (Sphingidae) in Madagascar. *Biological Journal of the Linnaean Society* 26(1):1–19.

_____. 1986. The pollination of *Cymbidiella flabellata* (Orchidaceae) in Madagascar: A system operated by sphecid wasps. *Nordic Journal of Botany* 6:411–422.

Nilsson, L. A., and E. Rabakonandrianina. 1988. Hawk-moth scale analysis and pollination specialization in the epilithic Malagasy endemic *Aerangis ellisii* (Reichenb. fil.) Schltr. (Orchidaceae). *Botanical Journal of the Linnaean Society* 97:49–61.

Nishimura, G. 1991. Comparative morphology of cotyledonous orchid seedlings. *Lindleyana* 6:140–146.

Nordal, I. 1987. Cladistics and character weighting: a contribution to the compatibility versus parsimony discussion. *Taxon* 36:59–60.

Ogura, T. 1953. Anatomy and morphology of the subterranean organs in some Orchidaceae. *Journal of the Faculty of Science of the University of Tokyo: Botany* 6:135–157.

Okada, H. 1988. Karyomorphological observations of *Apostasia nuda* and *Neuwiedia veratrifolia* (Apostasioideae) Orchidaceae. *Japanese Journal of Botany* 63:344–350.

Palmer, J. D. 1987. Chloroplast DNA evolution and biosystematic uses of chloroplast DNA variation. *The American Naturalist* 130(Suppl.):6–29.

Palmer, J. D., R. K. Jansen, H. J. Michaels, M. W. Chase, and J. R. Manhart. 1988. Chloroplast DNA variation and plant phylogeny. *Annals of the Missouri Botanical Garden* 75:1180–1206.

Patt, J. M., M. W. Merchant, D. R. E. Williams, and B. J. D. Meeuse. 1989. Pollination biology of *Platanthera stricta* (Orchidaceae) in Olympic National Park, Washington. *American Journal of Botany* 76:1097–1106.

Peakall, R. 1984. Observations on the pollination of *Leporella fimbriata* (Lindl.) A. S. George. *The Orchadian* 8:44–45.

Peakall, R., and A. J. Beattie. 1989. Pollination of the orchid *Microtis parviflora* R. Br. by flightless worker ants. *Functional Ecology* 3:515–522.

Peakall, R., A. J. Beattie, and S. H. James. 1987. Pseudocopulation of an orchid by male ants: a test of two hypotheses accounting for the rarity of ant pollination. *Oecologia* 73: 522–524.

Peakall, R., and S. H. James. 1989. Chromosome numbers of some Australian terrestrial orchids. *Lindleyana* 4:85–88.

Pettersson, B. 1989. Pollination in the African species of *Nervilia* (Orchidaceae). *Lindleyana* 4:33–41.

_____. 1990. *Nervilia* (Orchidaceae) in Africa. *Acta Universitatis Upsaliensis* 281.

Pijl, L. van der, and C. H. Dodson. 1966. *Orchid Flowers Their Pollination and Evolution*. Coral Gables: University of Miami Press.

Pinto, A. V., and M. A. S. Costa. 1988. Cianogênese: o ciclo circanual em Orchidaceae. *Bradea* 5:73–85.

Platnick, N. I. 1988. Programs for quicker relationships. *Nature* 335:310.

Poggio, L., C. A. Naranjo, and K. Jones. 1986. The chromosomes of orchids IX. *Eulophia. Kew Bulletin* 41:45–49.

Porembski, S., and W. Barthlott. 1988. Velamen radicum micromorphology and classification of Orchidaceae. *Nordic Journal of Botany* 8:117–137.

Pradhan, G. M. 1983. *Vanda cristata. American Orchid Society Bulletin* 52:464–468.

Pridgeon, A. M. 1978. Una revisión de los géneros *Coelia* y *Bothriochilus. Orquídea* (Méx.) 7:57–94.

———. 1982a. Diagnostic anatomical characters in the Pleurothallidinae (Orchidaceae). *American Journal of Botany* 69:921–938

———. 1982b. Numerical analyses in the classification of the Pleurothallidinae (Orchidaceae). *Botanical Journal of the Linnaean Society* 85:103–131.

———. 1987. The velamen and exodermis of orchid roots. In *Orchid Biology, Reviews and Perspectives,* Vol. 4. Ed. J. Arditti. Ithaca: Cornell University Press.

Pridgeon, A. M., W. L. Stern, and D. H. Benzing. 1983. Tilosomes in roots of Orchidaceae. I. Morphology and systematic occurrence. *American Journal of Botany* 70:1365–1377.

de Queiroz, K., and M. J. Donoghue. 1988. Phylogenetic systematics and the species problem. *Cladistics* 4:317–338.

Rao, P. R. M., and S. Sood. 1979. Life history of *Satyrium nepalense* (Orchidaceae). *Norwegian Journal of Botany* 26:285–294.

Rasmussen, F. N. 1977. The genus *Corymborkis* Thou. (Orchidaceae) a taxonomic revision. *Botaniska Tidsskrift* 71:161–192.

———. 1982. The gynostemium of the neottioid orchids. *Opera Botanica* 65:1–96.

———. 1985. The gynostemium of *Bulbophyllum ecornutum* (J. J. Smith) J. J. Smith (Orchidaceae). *Botanical Journal of the Linnaean Society* 91:447–456.

———. 1986a. On the various contrivances by which pollinia are attached to viscidia. *Lindleyana* 1:21–32.

———. 1986b. Ontogeny and phylogeny in Orchidaceae. *Lindleyana* 1:114–124.

Rasmussen, F. N., and H. Rasmussen. 1979. Notes on the morphology and taxonomy of *Diceratostele gabonensis* (Orchidaceae). *Bulletin du Jardin Botanique national de Belgique* 49:139–148.

Rasmussen, H. 1981. The diversity of stomatal development in Orchidaceae subfamily Orchidoideae. *Botanical Journal of the Linnaean Society* 82:381–393.

———. 1982. Branching pattern and inflorescence bud displacement in *Flickingeria* (Orchidaceae). *Nordic Journal of Botany* 2:235–248.

———. 1985. 'Ramicaul'—an improvement within monocotyledon terminology? *Taxon* 34:654–658.

———. 1987. Orchid stomata—structure, differentiation, function, and phylogeny. In *Orchid Biology, Reviews and Perspectives,* Vol. 4. Ed. J. Arditti. Ithaca: Cornell University Press.

Rauh, W., W. Barthlott, and N. Ehler. 1975. Morphologie und Funktion der Testa staubförmiger Flugsamen. *Botanisches Jahrbücher für Systematik, Pflanzengeschichte und Pflanzengeographie* 96:353–374.

Reeves, L. M., and T. Reeves. 1984. Life history and reproduction of *Malaxis paludosa* in Minnesota. *American Orchid Society Bulletin* 53:1280–1291.

Renz, J. 1980. Probleme der Orchideengattung *Habenaria. Die Orchidee* 31:64–72, 93–98

Rico-Gray, V., and L. B. Thien. 1987. Some aspects of the reproductive biology of *Schomburgkia tibicinis* Batem. (Orchidaceae) in Yucatan, Mexico. *Brenesia* 28:13–24.

Rivett, R. A. 1979. Chromosomal and morphological investigation of two *Bulbophyllum* species. *The Orchadian* 6:52–55

Rodríguez C., R. L. 1986. *Orquídeas de Costa Rica*. San José: Editorial Universidad de Costa Rica.

Rolfe, R. A. 1909–1912. The evolution of the Orchidaceae. *Orchid Review* 17–20: in many parts.

Romero, G. A. 1990. Phylogenetic relationships in subtribe Catasetinae (Orchidaceae). *Lindleyana* 5:160–181.

Romero, G. A., and C. E. Nelson. 1986. Sexual dimorphism in *Catasetum* orchids: forcible pollen emplacement and male flower competition. *Science* 232:1538–1540.

Rübsamen, T. 1986. Morphologische, embryologische und systematische Untersuchungen an Burmanniaceae und Corsiaceae (Mit Ausblick auf die Orchidaceae–Apostasioideae). *Dissertationes Botanicae* 92:1–310.

Saether, O. A. 1986. The myth of objectivity—post-Hennigian deviations. *Cladistics* 2:1–13.

Sauleda, R. P. 1988. A revision of the genus *Psychilus* Rafinesque (Orchidaceae). *Phytologia* 65:1–33.

_____. 1989. The genus *Psychilus* Rafinesque (Orchidaceae). *Orchid Digest* 53:163–174.

Sauleda, R. P., and R. M. Adams. 1989. A revision of the West Indian orchid genera *Broughtonia* Robert Brown, *Cattleyopsis* Lemaire and *Laeliopsis* Lindley. *Orchid Digest* 53:39–42.

Schick, B. 1988. Zur Anatomie und Biotechnik des Bestäubungsapparates der Orchideen. I. *Dactylorhiza majalis* (Rchb.) Hunt & Summerh., *Disa uniflora* Bergius und *Oncidium hastatum* Lindl. *Botanisches Jahrbücher für Systematik, Pflanzengeschichte und Pflanzengeographie* 110:215–262.

_____. 1989. Zur Anatomie und Biotechnik des Bestäubungsapparates der Orchideen. II. *Epipactis palustris* (L.) Crantz und *Listera ovata* (L.) R. Br. *Botanisches Jahrbücher für Systematik, Pflanzengeschichte und Pflanzengeographie* 110:289–323.

Schick, B., G. Kunze, and J. Bond. 1987. Rostelldifferenzierung und Pollinarien-bildung europäischer Orchideen. II. Über das Propollinarium von *Malaxis monophyllos* (L.) Sw. *Die Orchidee* 38: 194–197.

Schick, B., K.-H. Seack, and J. Kaeding. 1987. Rostelldifferenzierung und Pollinarienbildung europäischer Orchideen. IV. Kinematographische Doku-mentation der Extrusion des Rostellklebstoffs von *Listera ovata* (L.) R. Br. *Die Orchidee* 38:251–255.

Schill, R. 1978. Palynologische Untersuchungen zur systematischen Stellung der Apostasiaceae. *Botanische Jahrbücher für Systematik, Pflanzengeschichte und Pflanzengeographie* 99:353–362.

Schill, R., and W. Pfeiffer. 1977. Untersuchungen an Orchideenpollinien unter besonderer Berücksichtigung ihrer Feinskulpturen. *Pollen et Spores* 19:5–118.

Schill, R., and M. Wolter. 1986. On the presence of elastoviscin in all subfamilies of the Orchidaceae and the homology to Pollenkitt. *Nordic Journal of Botany* 6:321–324.

Schlegel, M., G. Steinbrück, K. Hahn, and B. Röttger. 1989. Interspecific relation-ship of ten European orchid species as revealed by enzyme electrophoresis. *Plant Systematics and Evolution* 163:107–119.

Schlechter, R. 1926. Das System der Orchidaceen. *Notizblatt des Botanischen Gartens und Museums zu Berlin-Dahlem* 9:563–591.

_____. 1970–continuing. *Die Orchideen.* 3rd ed. Ed. F. G. Brieger, R. Maatsch, and K. Senghas. Berlin: Paul Parey.

Schmid, R., and M. J. Schmid. 1977. Fossil history of the Orchidaceae. In *Orchid Biology: Reviews and Perspectives,* vol. 1. Ed. J. Arditti. Ithaca: Comstock Publishing Associates.

Sehgal, R. N., and S. Sehgal. 1989. Cytology of orchids from khasi and jaintia hills: Genera *Liparis* Rich. and *Oberonia* Lindl. *Journal of the Orchid Society of India* 3:55–59.

Seidenfaden, G. 1978. Orchid genera in Thailand. VI. Neottioideae Lindl. *Dansk Botanisk Arkiv* 32(2):1–195.

_____. 1979. Orchid genera in Thailand. VIII. *Bulbophyllum* Thou. *Dansk Botanisk Arkiv* 33(3):1–228.

_____. 1982. Orchid genera in Thailand. X. *Trichotosia* Bl. and *Eria* Lindl. *Opera Botanica* 62:1–157.

_____. 1983. Orchid genera in Thailand. XI. Cymbidieae Pfitz. *Opera Botanica* 72:1–124

_____. 1985. Orchid genera in Thailand. XII. *Dendrobium* Sw. *Opera Botanica* 83:1–295.

_____. 1986. Orchid genera in Thailand. XIII. Thirty-three epidendroid genera. *Opera Botanica* 89:1–216.

_____. 1988. Orchid genera in Thailand. XIV. fifty-nine vandoid genera. *Opera Botanica* 95:1–398.

Senghas, K. 1965. Die Gattung *Rangaeris. Die Orchidee* 16:176–183.

_____. 1984. *Adrorhizon* und *Sirhookera,* zwei kaum bekannte Orchideengattungen. *Die Orchidee* 35:133–141.

_____. 1987. Die Gattung *Scelochilus,* mit einer neuen Art, *Scelochilus rubriflora,* aus Peru. *Die Orchidee* 38:114–123.

_____. 1988. Eine neue Gliederung der Subtribus Aeridinae (= Sarcanthinae). *Die Orchidee* 39:219–223.

Senghas, K., and H. Sundermann. 1970. Probleme der Orchideengattung *Epipactis. Die Orchidee, Sonderheft.*

_____. 1980. Probleme der Evolution bei europäischen und mediterranen Orchideen. *Die Orchidee, Sonderheft.* Hildescheim: Brücke-Verlag Kurt Schmersow.

_____. 1986. Probleme der Taxonomie, Verbreitung und Vermehrumg europäischer und mediterraner Orchideen, II. *Die Orchidee, Sonderheft.*

Sera, T. 1990. Karyomorphological studies on *Goodyera* and its allied genera in Orchidaceae. *Bulletin of the Hiroshima Botanical Garden* 12:71–144.

Sheviak, C. J., and M. L. Bowles. 1986. The prairie fringed orchids: a pollinator-isolated species pair. *Rhodora* 88:267–290.

Shindo, K., and H. Kamemoto. 1962. Genome relationships of *Neofinetia* Hu and some allied genera of Orchidaceae. *Cytologia* 27:402–409.

Siegerist, E. S. 1986. The genus *Dimerandra. Botanical Museum Leaflets* 30:199–222.

Simpson, G. G. 1953. *The Major Features of Evolution.* New York: Columbia Univ. Press.

Sinoto, Y. 1962. Chromosome numbers in *Oncidium* alliance. *Cytologia* 27:307–313.

Slater, A. T., and D. M. Calder. 1988. The pollination biology of *Dendrobium speciosum* Smith: a case of false advertising? *Australian Journal of Botany* 36:145–158.

Slaytor, M. B. 1977. The distribution and chemistry of alkaloids in the Orchidaceae. In *Orchid Biology Reviews and Perspectives*, Vol. 1. Ed. J. Arditti. Ithaca: Comstock Publishing Associates.

Sood, S. K., and P. R. Mohana Rao. 1988. Studies in the embryology of the diandrous orchid *Cypripedium cordigerum* (Cypripedieae, Orchidaceae). *Plant Systematics and Evolution* 160:159–168.

Soto Arenas, M. A., and E. W. Greenwood. 1989. Undesirable technical terminology—a current example. *Orchid Research Newsletter* 13:8–9.

Southwick, E. E. 1984. Photosynthate allocation to floral nectar: a neglected energy investment. *Ecology* 65:1775–1779.

Stearns, S. C. 1977. The evolution of life history traits: A critique of the theory and a review of the data. *Annual Review of Ecology and Systematics* 8:145–171.

Stebbins, G. L. 1969. Comments on the search for a 'perfect system.' *Taxon* 18:357–484.

Steiner, K. E. 1987. Oil-producing orchids and oil-collecting bees in southern Africa. Abstract, *Annals Congress South African Association of Botany*, Durban, p. 81.

———. 1989. The pollination of *Disperis* (Orchidaceae) by oil-collecting bees in southern Africa. *Lindleyana* 4:164–183.

Stergianou, K. K. 1989. Habit differentiation and chromosome evolution in *Pleione* (Orchidaceae). *Plant Systematics and Evolution* 166:253–264.

Stern, W. L., and A. M. Pridgeon. 1984. Ramicaul, a better term for the pleurothallid "secondary stem." *American Orchid Society Bulletin* 53:397–401.

———. 1985. Stem structure and its bearing on the systematics of Pleurothallidinae (Orchidaceae). *Botanical Journal of the Linnaean Society* 91:457–471.

Stewart, J. 1976. The Vandaceous group in Africa and Madagascar. *Proceedings of the Eighth World Orchid Conference*, Frankfurt 239–248.

Stewart, J. 1986. Stars of the islands—a new look at the genus *Aerangis* in Madagascar and the Comoro Islands. *American Orchid Society Bulletin* 55:792–802, 903–909, 1008–1015, 1117–1125.

———. 1989. The genus *Stenoglottis*. *Kew Magazine* 6:9–22.

Stojanow, N. 1916. Über die vegetative Fortpflanzung der Ophrydineen. *Flora* 109:1–39.

Stoutamire, W. 1967. Flower biology of the Lady's-slippers (Orchidaceae: *Cypripedium*). *Michigan Botanist* 6:159–175.

———. 1975. Pseudocopulation in Australian terrestrial orchids. *American Orchid Society Bulletin* 44:226–233.

———. 1982. Australian *Pterostylis*—the greenhoods. *American Orchid Society Bulletin* 51:796–803.

———. 1983. Wasp-pollinated species of *Caladenia* (Orchidaceae) in southwestern Australia. *Australian Journal of Botany* 31:383–394.

———. 1985. *Spiculaea ciliata* Lindl., the Australian elbow orchid. *Canadian Orchid Journal* 3(3):19–23.

Summerhayes, V. S. 1957. The genus *Eulophidium* Pfitz. *Bulletin du Jardin Botanique d'l'Etat* (Brussels) 27:391–403.

Swamy, B. G. L. 1948. Vascular anatomy of orchid flowers. *Botanical Museum Leaflets* 13:61–95.

Sweet, H. R. 1970. Orquídeas andinas poco conocidas III—*Xerorchis*, Schltr. *Orquideología* 5:89–95.

Szlachetko, D. L. 1991. Thelymitroideae, a new subfamily within Orchidaceae. *Fragmenta Floristica et Geobotanica* 36:33–49.

———. 1980. *The Genus* Phalaenopsis. Pomona: Orchid Digest, Inc.

Tan, K. W. 1969. The systematic status of the genus *Bletilla* (Orchidaceae). *Brittonia* 21:202–214.

Tanaka, R. 1971. Types of resting nuclei in Orchidaceae. *Botanical Magazine,* Tokyo 84:118–122.

Tanaka, R., M. Aoyama, and S. Takaki. 1987. Documented intergeneric hybrids involving *Cymbidium. American Orchid Society Bulletin* 56:130–134.

Tanaka, R., K. Karasawa, and G. Ishida. 1981. Karyomorphological observations on *Calanthe* of Japan. *Bulletin of the Hiroshima Botanical Garden* 4:9–62.

Tanaka, R., and M. Yokota. 1982. Karyotype of *Cephalanthera subaphylla,* Orchidaceae. *Chromosome Information Service* 33:34–36.

Tang, T., F. Wang, and K. Lang. 1982. Materia ad floram Orchidacearum Sinensium—*Amitostigma* Schltr. *Acta Phytotaxonomica Sinica* 20:78–86.

Tohda, H. 1983. Seed morphology in Orchidaceae. I. *Dactylorchis, Orchis, Ponerorchis, Chondradenia* and *Galeorchis. Science Reports, Tohoku University* IV 38:253–268.

———. 1985. Seed morphology in Orchidaceae. II. Tribe Cranichideae. *Science Reports, Tohoku University* IV 39(1):21–43.

———. 1986. Seed morphology in Orchidaceae. III. Tribe Neottieae. *Science Reports, Tohoku University* IV 39:103–119.

Van Royen, P. 1983. The genus *Corybas* (Orchidaceae) in its eastern area. *Phanerogamarum Monographiae* 16. Vaduz: J. Cramer

Verdcourt, B. 1968. African Orchids. Pt. 31: New taxa of *Disperis* from East and Central Africa. *Kew Bulletin* 22:93–99.

Verdcourt, B. 1986. A key to the East African species of *Disperis* (Orchidaceae) with two new species. *Kew Bulletin* 41:51–57.

Vermeulen, P. 1959. The different structure of the rostellum in Ophrydeae and Neottieae. *Acta Botanica Neerlandica* 8:338–355.

———. 1965. The place of *Epipogium* in the system of Orchidales. *Acta Botanica Neerlandica* 14:230–241.

———. 1966. The system of the orchidales. *Acta Botanica Neerlandica* 15:224–253.

Vermeulen, J. J. 1987. A taxonomic revision of the continental African Bulbophyllinae. *Orchid Monographs* 2:1–300.

Veyret, Y. 1980. Précisions botaniques sur l'*Uleiorchis ulei* (Cogn.) Handro (Orchidaceae). *Adansonia* II 20:141–143.

———. 1981. Quelques aspects du pistil et de son devenir chez quelques Sobraliinae (Orchidaceae) de Guyane. *Bulletin du Múseum national d'Histoire naturelle,* Paris, 4e sér., 3, sect. B, *Adansonia* 1:75–83.

Vij, S. P. 1989. Chromosomes and speciation in Indian orchids. *The Journal of the Orchid Society of India* 3:11–24.

Vij, S. P., and N. Shekhar. 1987. Cytological Investigations in Indian cymbidiums. *The Journal of the Orchid Society of India* 1:29–43.

de Vogel, E. F. 1969. Monograph of the tribe Apostasieae (Orchidaceae). *Blumea* 17:313–350.

———. 1986. Revisions in Coelogyninae (Orchidaceae). II. The genera *Bracisepalum, Chelonistele, Entomophobia, Geesinkorchis* and *Nabaluia. Orchid Monographs* 1:17–86.

————. 1988. Revisions in Coelogyninae (Orchidaceae). III. The genus *Pholidota*. *Orchid Monographs* 8:1–118.

Vogel, S. 1959. Organographie der Blüten Kapländischer Ophrydeen. *Abhandlungen der Akademie der Wissenschaften und der Literatur, Mathematisch-Naturwissenschaftliche Klasse* (Mainz) 1959:268–532.

————. 1962. Duftdrüsen im Dienste der Bestäubung. *Abhandlungen der Akademie des Wissenschaften und der Literatur, Mathematisch-Naturwissenschaftliche Klasse* (Mainz) 1963:602–763.

————. 1974. Ölblumen und ölsammelnde Bienen. *Tropische und subtropische Pflanzenwelt* 7:1–267.

————. 1981. Bestäubungskonzepte der Monokotylen und ihr Ausdruck im System. *Berichte der Deutschen Botanischen Gesellschaft* 94:667–675.

Vogt, C. A. 1990. Pollination in *Cypripedium reginae* (Orchidaceae). *Lindleyana* 5:145–150.

Vöth, W. 1984. *Echinomyia magnicoris* Zett. Bestäuber von *Orchis ustulata* L. *Die Orchidee* 35:189 192.

————. 1987. Bestäubungsbiologische Beobachtungen an *Orchis militaris* L. *Die Orchidee* 38:77–84.

Wagner, W. H., Jr. 1980. Origin and philosophy of the groundplan-divergence method of cladistics. *Systematic Botany* 5:173–193.

Wallace, B. J. 1978. On *Cryptostylis* pollination and pseudocopulation. *The Orchadian* 5:168–169.

————. 1980. Cantharophily and the pollination of *Peristeranthus hillii*. *The Orchadian* 6:214–215.

Warcup, J. H. 1981. The mycorrhizal relationships of Australian orchids. *New Phytologist* 87:371–381.

————. 1985. *Rhizanthella gardneri* (Orchidaceae), its Rhizoctonia endophyte and close association with *Melaleuca uncinata* (Myrtaceae) in western Australia. *New Phytologist* 99:273–280.

Wheeler, Q. D., and K. C. Nixon. 1990. Another way of looking at the species problem: a reply to de Queiroz and Donoghue. *Cladistics* 6:77–81.

Whitehead, A. N. 1957. *The Concept of Nature*. Ann Arbor: University of Michigan Press.

Wiehler, H. 1983. A synopsis of the neotropical Gesneriaceae. *Selbyana* 6:1–1219.

Williams, C. A. 1979. The leaf flavonoids of the Orchidaceae. *Phytochemistry* 18:803–813.

Williams, N. H. 1975. Stomatal development in *Ludisia discolor* (Orchidaceae): mesoperigenous subsidiary cells in the monocotyledons. *Taxon* 24:281–288.

————. 1979. Subsidiary cells in the Orchidaceae: their general distribution with special reference to development in the Oncidieae. *Botanical Journal of the Linnaean Society* 78:41–66.

Williams, N. H., and C. R. Broome. 1976. Scanning electron microscope studies of orchid pollen. *American Orchid Society Bulletin* 45:699–707.

Williams, N. H., and W. M. Whitten. 1983. Orchid floral fragrances and male euglossine bees: methods and advances in the last sesquidecade. *Biological Bulletin* 164:355–395.

Wimber, D. E. 1987. A glimpse of *Disa* chromosome numbers. *Orchid Review* 95:162–165.

Wirth, M. 1964. *Supraspecific Variation and Classification in the Oncidiinae*

(Orchidaceae). Ph.D. Thesis. Washington University, St. Louis.

Withner, C. 1988. *The Cattleyas and Their Relatives,* vol. I, *The Cattleyas.* Portland, OR: Timber Press.

———. 1990. *The Cattleyas and Their Relatives,* vol. II, *The Laelias.* Portland, OR: Timber Press.

Wolter, M., and R. Schill. 1985. On acetolysis resistant structures in the Orchidaceae—why fossil record of orchid pollen is so rare. *Grana* 24:139–143.

———. 1986. Ontogenie von Pollen, Massulae und Pollinien bei den Orchideen. *Tropische und subtropische Pflanzenwelt* 56:1–93.

Xu, S., Y. Xiao, and Z. Yang. 1987. Studies on the ontogeny of the pollinium of *Goodyera procera. Acta Botanica Sinica* 29:573–579.

Yokota, M. 1987. Karyotypes and phylogeny in Orchidinae and allied subtribes. *Proceedings of the 12th World Orchid Conference,* Tokyo. 70–79.

Zavada, M. S. 1990. A contribution to the study of pollen wall ultrastructure of orchid pollinia. *Annals of the Missouri Botanical Garden* 77:785–801.

Zimmerman, J. K., and T. M. Aide. 1989. Patterns of fruit production in a neotropical orchid: pollinator vs. resource limitation. *American Journal of Botany* 76:67–73.

Index

301